Non-Thermal Plasma Technology for Polymeric Materials

Non-Thermal Plasma Technology for Polymeric Materials

Applications in Composites, Nanostructured Materials, and Biomedical Fields

Edited by

SABU THOMAS

MIRAN MOZETIČ

UROŠ CVELBAR

PETR ŠPATENKA

PRAVEEN K. M

ELSEVIER

Elsevier
Radarweg 29, PO Box 211, 1000 AE Amsterdam, Netherlands
The Boulevard, Langford Lane, Kidlington, Oxford OX5 1GB, United Kingdom
50 Hampshire Street, 5th Floor, Cambridge, MA 02139, United States

Notices
Knowledge and best practice in this field are constantly changing. As new research and
experience broaden our understanding, changes in research methods, professional practices,
or medical treatment may become necessary.

Practitioners and researchers must always rely on their own experience and knowledge in
evaluating and using any information, methods, compounds, or experiments described herein.
In using such information or methods they should be mindful of their own safety and the safety
of others, including parties for whom they have a professional responsibility.

To the fullest extent of the law, neither the Publisher nor the authors, contributors, or editors,
assume any liability for any injury and/or damage to persons or property as a matter of products
liability, negligence or otherwise, or from any use or operation of any methods, products,
instructions, or ideas contained in the material herein.

British Library Cataloguing-in-Publication Data
A catalogue record for this book is available from the British Library

Library of Congress Cataloging-in-Publication Data
A catalog record for this book is available from the Library of Congress

ISBN: 978-0-12-813152-7

For Information on all Elsevier publications
visit our website at https://www.elsevier.com/books-and-journals

Publisher: Matthew Deans
Acquisition Editor: Edward Payne
Editorial Project Manager: Thomas Van Der Ploeg
Production Project Manager: Sruthi Satheesh
Cover Designer: Greg Harris

Typeset by MPS Limited, Chennai, India

CONTENTS

7. Plasma Treatment of Powders and Fibers   193

Taťana Vacková, Petr Špatenka and Syam Balakrishna

8. Plasma Treatment of Polymeric Membranes   211

Ahmed Al-Jumaili, Surjith Alancherry, Daniel Grant, Avishek Kumar,
Kateryna Bazaka and Mohan V. Jacob

9. Selective Plasma Etching of Polymers and Polymer Matrix Composites   241

Harinarayanan Puliyalil, Gregor Filipič and Uroš Cvelbar

LIST OF CONTRIBUTORS

Surjith Alancherry
Electronics Materials Lab, College of Science and Engineering, James Cook University, Townsville, QLD, Australia

Ahmed Al-Jumaili
Electronics Materials Lab, College of Science and Engineering, James Cook University, Townsville, QLD, Australia

Venu Anand
Plasma Processing Laboratory, Department of Instrumentation and Applied Physics, Indian Institute of Science, Bengaluru, Karnataka, India; Department of Electronics and Communication Engineering, National Institute of Technology Calicut, Kozhikode, Kerala, India

Syam Balakrishna
Department of Material Science, Faculty of Mechanical Engineering, Czech Technical University of Prague, Prague, Czech Republic

Kateryna Bazaka
Electronics Materials Lab, College of Science and Engineering, James Cook University, Townsville, QLD, Australia; School of Chemistry, Physics, Mechanical Engineering, Queensland University of Technology, Brisbane, QLD, Australia

Uroš Cvelbar
Jozef Stefan Institute, Ljubljana, Slovenia

Ladislav Cvrček
Department of Materials Engineering, Faculty of Mechanical Engineering, Czech Technical University in Prague, Prague, Czech Republic

Gregor Filipič
Jozef Stefan Institute, Ljubljana, Slovenia

Marija Gorjanc
Faculty of Natural Sciences and Engineering, University of Ljubljana, Ljubljana, Slovenia

Mohan Rao Gowravaram
Plasma Processing Laboratory, Department of Instrumentation and Applied Physics, Indian Institute of Science, Bengaluru, Karnataka, India

Daniel Grant
Electronics Materials Lab, College of Science and Engineering, James Cook University, Townsville, QLD, Australia

Yves Grohens
FRE CNRS 3744, IRDL, University of Southern Brittany, Lorient, France;
University of South Brittany, Laboratory IRDL PTR1, Research Center
"Christiaan Huygens," Lorient, France

Marta Horáková
Department of Materials Engineering, Faculty of Mechanical Engineering,
Czech Technical University in Prague, Prague, Czech Republic

Joanna Izdebska-Podsiadły
Department of Printing Technology, Institute of Mechanics and Printing,
Faculty of Production Engineering, Warsaw University of Technology,
Warszawa, Poland

Mohan V. Jacob
Electronics Materials Lab, College of Science and Engineering, James Cook
University, Townsville, QLD, Australia

Saravanakumar Jagannathan
JN-UTM Cardiovascular Engineering Centre, Faculty of Biosciences and
Medical Engineering, Universiti Teknologi Malaysia, Johor Bahru, Malaysia

Jemy James
School of Pure and Applied Physics, Mahatma Gandhi University, Kottayam,
Kerala, India; FRE CNRS 3744, IRDL, University of Southern Brittany,
Lorient, France; International and Inter University Centre for Nanoscience and
Nanotechnology (IIUCNN), Mahatma Gandhi University, Kottayam, Kerala,
India

Blessy Joseph
International and Inter University Centre for Nanoscience and Nanotechnology
(IIUCNN), Mahatma Gandhi University, Kottayam, Kerala, India; FRE CNRS
3744, IRDL, University of Southern Brittany, Lorient, France

K.S. Joshy
International and Inter University Centre for Nanoscience and Nanotechnology
(IIUCNN), Mahatma Gandhi University, Kottayam, Kerala, India

Lavanya Jothi
Department of Chemistry, Indian Institute of Space Science and Technology,
Thiruvananthapuram, Kerala, India

Jomon Joy
School of Chemical Sciences, Mahatma Gandhi University, Kottayam, Kerala,
India

Nandakumar Kalarikkal
School of Pure and Applied Physics, Mahatma Gandhi University, Kottayam,
Kerala, India; School of Chemical Sciences, Mahatma Gandhi University,
Kottayam, Kerala, India; International and Inter University Centre for
Nanoscience and Nanotechnology (IIUCNN), Mahatma Gandhi University,
Kottayam, Kerala, India

Shrikaant Kulkarni
Department of Chemical Engineering, Vishwakarma Institute of Technology, Pune, Maharashtra, India

Avishek Kumar
Electronics Materials Lab, College of Science and Engineering, James Cook University, Townsville, QLD, Australia

Jorge López-García
Faculty of Technology, Tomas Bata University, Zlín, Czech Republic

Miran Mozetič
Jozef Stefan Institute, Ljubljana, Slovenia

Gomathi Nageswaran
Department of Chemistry, Indian Institute of Space Science and Technology, Thiruvananthapuram, Kerala, India

Parvathy Nancy
School of Pure and Applied Physics, Mahatma Gandhi University, Kottayam, Kerala, India

C.V. Pious
School of Chemical Sciences, Mahatma Gandhi University, Kottayam, Kerala, India

Praveen K. M
International and Inter University Centre for Nanoscience and Nanotechnology (IIUCNN), Mahatma Gandhi University, Kottayam, Kerala, India; University of South Brittany, Laboratory IRDL PTR1, Research Center "Christiaan Huygens," Lorient, France; Department of Mechanical Engineering, SAINTGITS College of Engineering, Kottayam, Kerala, India

Gregor Primc
Jozef Stefan Institute, Ljubljana, Slovenia

Harinarayanan Puliyalil
National Institute of Chemistry (D-13), Ljubljana, Slovenia; IMEC, Leuven, Belgium

Ashin Shaji
Institute of Physics, Slovak Academy of Sciences, Slovak, Slovak Republic

S. Snigdha
International and Inter University Centre for Nanoscience and Nanotechnology (IIUCNN), Mahatma Gandhi University, Kottayam, Kerala, India

Petr Špatenka
Department of Material Science, Faculty of Mechanical Engineering, Czech Technical University of Prague, Prague, Czech Republic

Rajesh Thomas
Plasma Processing Laboratory, Department of Instrumentation and Applied Physics, Indian Institute of Science, Bengaluru, Karnataka, India; International Iberian Nanotechnology Laboratory, Braga, Portugal

Sabu Thomas
School of Chemical Sciences, Mahatma Gandhi University, Kottayam, Kerala, India; International and Inter University Centre for Nanoscience and Nanotechnology (IIUCNN), Mahatma Gandhi University, Kottayam, Kerala, India

K.H. Thulasi Raman
Plasma Processing Laboratory, Department of Instrumentation and Applied Physics, Indian Institute of Science, Bengaluru, Karnataka, India

Tat'ana Vacková
Department of Material Science, Faculty of Mechanical Engineering, Czech Technical University of Prague, Prague, Czech Republic

Alenka Vesel
Jozef Stefan Institute, Ljubljana, Slovenia

Guillaume Vignaud
FRE CNRS 3744, IRDL, University of Southern Brittany, Lorient, France

Rok Zaplotnik
Jozef Stefan Institute, Ljubljana, Slovenia

CHAPTER 1

Relevance of Plasma Processing on Polymeric Materials and Interfaces

Praveen K. M[1,2,3], C.V. Pious[4], Sabu Thomas[1] and Yves Grohens[2]
[1]International and Inter University Centre for Nanoscience and Nanotechnology (IIUCNN), Mahatma Gandhi University, Kottayam, Kerala, India
[2]University of South Brittany, Laboratory IRDL PTR1, Research Center "Christiaan Huygens," Lorient, France
[3]Department of Mechanical Engineering, SAINTGITS College of Engineering, Kottayam, Kerala, India
[4]School of Chemical Sciences, Mahatma Gandhi University, Kottayam, Kerala, India

1.1 INTRODUCTION

With the growing global energy crisis and ecological risks, polymers are playing a vital role in five major areas where the needs of society face huge technological challenges. These are: energy, sustainability, health care, security and informatics, and defence and protection [1]. In order to fulfil the aforesaid diverse societal needs, advanced polymeric materials are developed either by synthesizing new polymers or through the modification of existing polymeric materials. In most of the cases, researchers opt for the modification of existing polymers rather than synthesizing new polymers due to the considerable cost aspects. Polymers are a major class of materials and possess a wide range of mechanical, physical, chemical, and optical properties. Polymers have increasingly replaced metallic components in various applications. These substitutions reflect the advantages of polymers in terms of corrosion resistance, low electrical and thermal conductivity, low density, high strength-weight ratio (particularly when reinforced), noise reduction, wide choice of colors and transparencies, ease of manufacturing and complexity of design possibilities, and relatively low cost.

The interface is defined as the boundary between two layers of materials with different chemistries and/or microstructures. The interphase is described as the volume of material affected by the interaction at the interface. The term interphase is considered as a three-dimensional zone which is different from a two-dimensional interface and it is now widely used in the adhesion community to indicate the presence of a

Non-Thermal Plasma Technology for Polymeric Materials
DOI: https://doi.org/10.1016/B978-0-12-813152-7.00001-9

1

chemically or mechanically altered zone between adjacent phases. In the interphase zone, one can observe the gradation of properties from one phase to another, rather than the abrupt change necessitated by the acceptance of a two-dimensional interface [2].

Adhesions of polymeric materials have a direct relationship with surface/interfacial properties. The adhesion at the interface/interphases has prime importance when advanced polymeric materials are developed via surface modification techniques. Adhesion problems exist with the bonding of polymers and the adhesion of coatings to polymer surfaces, while interface problems between different phases of melt-blended polymers and polymer composites are recurring and challenging areas in polymer-based industries. With a thorough understanding of the surface and interface phenomenon in polymer multiphase systems, one can control the interfacial properties by engineering the interface/interphase region leading to the development of advanced materials with multifunctional properties.

The first section of this chapter provides a concise introduction to polymeric materials, followed by a description of their structure, properties, and applications. Then it addresses the salient features of polymer surfaces and interfacial problems with a brief overview of surface phenomena and adhesion phenomena in the bonding of polymers. Finally, the major part of this chapter covers the relevance of plasma processing on polymeric materials and interfaces. This part starts with a basic introduction, continues with discussions on the mechanism of plasma surface modification, effects of plasma treatment on specific polymers, various characterization techniques, and applications of plasma in polymeric materials.

1.2 STRUCTURE, PROPERTIES, AND APPLICATIONS OF POLYMERS

Polymers are defined as macromolecules, which are formed by the joining of repeatable structural units. The repeatable structural units are derived from monomers and are linked to each other by covalent bonds. The properties of a polymer depend on the chemical structures of monomeric units, molecular weight, molecular weight distribution, and its architecture. The processes of the formation of polymers from respective monomers is known as polymerization. Two major classifications of polymerization processes are condensation (step-growth or step-reaction polymerization) and addition polymerization (chain-growth or chain-reaction polymerization). In the former, the bonding takes place by condensation reaction resulting in the loss of small molecules which are often water. In the later, bonding takes place by the reaction of unsaturated monomers without reaction by-products. The final properties of the product made from polymers depends highly on the inherent properties of the polymer/s and the additives used for modification The influence of molecular weight, structure (linear, branched, cross-linked, or network), degrees of polymerization and crystal-linity, glass transition temperature, and additives are briefed below.

The molecular weight and molecular weight distribution of polymers strongly influence the final properties of the end products. The mechanical properties such as the tensile and the impact strength, the resistance to cracking, and rheological measurements such as viscosity (in the molten state) of the polymer etc., increases with increase in molecular weight. Summing up of individual molecular weights of the monomeric units in a given chain gives the molecular weight of the polymer. It is important to keep in mind that the average chain length is proportional to the molecular weight of the given polymer. That is, higher the molecular weight of a given polymer, the greater the average chain length. The size of a polymer chain can be expressed in terms of the degree of polymerization which is defined as the ratio of the molecular weight of the polymer to the molecular weight of the repeating unit. The covalent bonds which link the monomeric units together are called primary bonds and the polymer chains are held together by secondary bonding. In a given polymeric material, the increase in strength and viscosity of the polymer with molecular weight is attributed to the fact that, as the length of the polymer chain increases, the greater is the energy needed to overcome the combined strength of the secondary bonding with primary bonding (covalent). Depending upon the type of chains, the polymers are classified into linear polymers, branched polymers, cross-linked polymers, and network polymers. If the repeating units in a polymer chain are all of the same type, the molecule is called a homopolymer. Copolymers contain two types of polymers and terpolymers contain three types.

Crystallinity plays a critical role in delivering enhanced property to polymeric materials and their products. By controlling the rate of solidifi-cation during cooling and the chain structure, it is possible to achieve desired degrees of crystallinity to polymers. The improved hardness, stiffness, and low ductility values of polymeric materials are outcomes of increased crystallinity. The optical properties of polymers are also influenced by the degree of crystallinity. At low temperatures, the amorphous polymers are hard, rigid, brittle, and glassy while at high temperatures they are rubbery or leathery. The temperature at which a transition occurs is called the glass-transition temperature (T_g). The information on T_g can predict the nature of the polymer at its service temperature that is, whether it rigid and glassy, or flexible and rubbery.

The two major classes of polymers are thermoplastics and thermosets. When the temperature of a certain polymer is raised above the T_g, or melting point, T_m, they can be easily molded into desired shapes. The increase in temperature weakens the secondary bonding via the thermal vibration of the long molecules which, in turn, enables more free movement of adjacent chains particularly when subjected to external shaping forces. These types of polymers are referred as thermoplastics. If the temperature of a thermoplastic polymer is raised above its T_g, it becomes leathery and on increasing the temperature further, it turns rubbery. At higher temperatures, say above T_m for crystalline thermoplastics, it becomes a molten fluid and its viscosity decreases with increasing temperature. At this stage, it can be molded into different shapes. When thermoplastics are stretched, the long–chain molecules tend to align in the general direction of the elongation. This process is known as orientation. Due to the viscoelastic behavior, thermoplastics are particularly susceptible to creep and stress relaxation. When subjected to tensile or bending stresses, some thermoplastics may develop localized wedge-shaped and/or narrow regions of highly deformed material. This process is known as crazing. It typically contains about 50% voids. With increasing tensile load on the specimen, these voids coalesce to form a crack, which eventually can lead to a fracture of the polymer. The environmental factors such as the presence of solvents, lubricants, or water vapor, residual stresses in the material, radiation, etc., can increase the crazing behavior in certain polymers. An important feature of certain thermoplastics is their ability to absorb water (hygroscopic nature). With increasing moisture absorption, the T_g, the yield stress, and the elastic modulus of the polymer are lowered severely. Compared to metals, plastics generally are characterized by low thermal and electrical conductivity, low specific gravity, and a high coefficient of thermal expansion. Certain polymers can be stretched or compressed to very large strains, and then, when subjected to heat, light, or a chemical environment, they recover their shape. These are known as shape memory polymers [1].

When the long-chain molecules in a polymer are cross-linked in a three-dimensional manner, the structure in effect becomes one giant molecule with strong covalent bonding. These polymers are called thermosetting polymers or thermosets, because the complete network is formed during polymerization and the shape of the part is permanently set. This curing (cross-linking) reaction is irreversible. Thermosetting polymers do not have a sharply defined T_g. Because of the strong covalent bonding, the strength and hardness of a thermoset are not affected by temperature or by rate of deformation. Thermosets generally possess better mechanical, thermal, and chemical properties, as well as electrical resistance and dimensional stability, than thermoplastics.

In order to impart certain specific properties, polymers usually are compounded with additives such as plasticizers (to impart flexibility and softness), fillers (to improve the strength, hardness, toughness, abrasion resistance, dimensional stability, or stiffness of plastics), colorants, flame retardants, and lubricants, etc.

Another important classification of polymers is elastomers which covers a large family of amorphous polymers having a low T_g. They have a characteristic ability to undergo large elastic deformations without rupture. Its soft nature and low elastic modulus make it useful for applications in tires, seals, footwear, hoses, belts, and shock absorbers. Nowadays the two important considerations in polymers are recyclability and bio-degradability. Numerous attempts continue to be made to produce fully biodegradable plastics by utilizing various agricultural waste (agro wastes), plant carbohydrates, plant proteins, and vegetable oils. The long-term performance of biodegradable plastics during their life span has not been assessed fully. There is also an increasing concern that the emphasis on biodegradability will divert attention from the issue of the recyclability of plastics and the efforts for conservation of materials and energy [1]. Another significant category of polymers are composites. A composite material is a combination of two or more chemically and physically different phases separated by a distinct interface, in such a manner that its properties and structural performance are superior to those of the constituents acting independently [3]. Composite materials having at least one polymeric phase can be termed polymer composites. Polymer composites with nanoparticles are referred to as polymer nanocomposites. Polymer nanocomposites find suitability in many applications such as biomedical [4,5], optical [5], electronic [6], sensor [7,8], aerospace [9], and nuclear applications [10].

1.3 POLYMER SURFACES AND INTERFACIAL PROBLEMS

To meet specific applications in par with growing demands of polymer industries, the nature of the polymeric material formulations are too complex to allow for the simultaneous optimization of bulk and surface properties. Due to the complexity in polymer formulations, it is also difficult to obtain both bulk and surface properties through a single polymer processing operation. Pre/posttreatment of polymeric surfaces has, therefore, become a common practice in the fabrication of many polymeric materials and its end products. The ultimate aim of polymer surface modification is, either to increase the potential for surface interactions (e.g., adhesion promotion or to obtain a hydrophilic surface) or to decrease the degree to which the surface interacts with a given material (e.g., release or antifouling applications or to obtain a hydrophobic surface). Hydrophilicity, hardness, adhesion, antifouling, antifogging, lubrication, biocompatibility, and conductivity, among others, are considered to be the main properties in demand.

The surfaces and interfaces of polymers are known to have important roles in the final properties and end-use applications of polymeric products. In the case of inhomogeneous polymer blends and composites, the interfaces between different phases are the decisive factor for the final mechanical properties [11]. Not only the strength properties, even the esthetic properties of the polymer product—which might critically influence the decision for the purchase of the product—and its corrosion or scratch resistance, texture, etc., are determined by surface composition and structure.

Hydrophilicity of a solid surface is an important property because controlling the surface wettability is crucial in many practical applications [12]. Surfaces with water contact angles larger than 150 degrees are termed as super hydrophobic surfaces. These type of surfaces have enormous applications in areas such as antisticking, anticontamination [13], and self-cleaning [14], etc. Based on the observations inspired from nature, it can be inferred that, the special functionalities of polymeric surfaces are usually not governed by the intrinsic property of materials, but are more likely related to the unique microstructures or nanostructures. The latest research trend in this domain is to focus on the responsive wettability of surfaces. By combining special topographic structures onto functional surfaces with responsive wettability helps in designing surfaces with responsive switching between superhydrophilicity and super hydrophobicity [15]. A thorough knowledge of surface phenomenon in polymers as well as the adhesion phenomena in the bonding of polymers is essential to design the surface modification strategies suitable for respective material surfaces with specific properties.

1.3.1 Surface Phenomena in Polymers

The past few years have witnessed dedicated research on surface modification of polymeric materials because of its enormous applications in applied surface science and engineering. This includes applications in biomaterials, coatings, and thin films, demanding smooth, colored, or scratch-resistant surfaces, appearance, durability, and stability which are directly or indirectly linked with external interfaces. Also, the gross behavior (bulk properties) of the material requires properties such as toughness or impact resistance of polymers, which are connected with defect, void, or crack formation which, in turn, involves internal interfaces. The polymer surfaces are different in nature when compared with its bulk, partly due to the difference in molecular weight, addition of functional modifiers such as surfactants, compatabilizers, coupling agents, antioxidants, fire retardants, etc., segregated lower energy components such as lubricants, and migration of additives across the interfaces while contacting with another surfaces. The most accepted method to read the surface characteristics and to measure the surface energy state is the determination of surface tension. This is done by examining the outermost layers of atoms which are typically below 0.2 nm in depth [11]. The work which is required to increase the size of the surface of a phase is referred to as the surface tension. If the phase is a solid, the equivalent term surface free energy is commonly used. If the adjacent phase is a liquid or a solid, reference is made to interfacial tension. Compared with the bulk phase, a molecule at the surface of a material meets fewer molecules with which it can form interactions. Presence at the surface is, therefore, less beneficial from an energy point of view. However, surface tension reflects the strength of bonding within the bulk material. The energy necessary to break these bonds is called the surface energy. In thermodynamics perspective, the surface tension can be defined as the partial differential of the Gibbs free energy (G) of the system with the respect to the area (A), at the constant temperature (T) and pressure (P) [16].

$$\Upsilon = \left[\frac{\partial G}{\partial A}\right] T, P \qquad (1.1)$$

In the thermodynamic consideration of surface energy, the surface free energy (G^S) can be defined as the excess of free energy associated with the surface (A), between the total free energy (G) of the system (G) and the one in the bulk (G^B) [16].

$$G^S = \frac{G - G^B}{A} \tag{1.2}$$

Hard solids are referred as "high energy surfaces" (e.g., metals and their oxides and ceramics, having surface tension $\sim 500-5000 \ mJ/m^2$) due to its very different binding forces (covalent, ionic, metallic). Weak molecular solids and liquids are referred as low energy surfaces (surface tension $< 100 \ mJ/m^2$) due to van der Waals forces and hydrogen bonding. Most of the polymers are considered as low energy surfaces where the interactions between the chains are dominated by van der Waals forces and hydrogen bonds. Surface and interfacial tensions of polymers significantly affect the wetting and coating processes in the melt blending of polymers and its composites, the behavior of colloidal dispersions, adhesion, friction, and biocompatibility and also in corrosion and adsorption processes. The surface tension of polymers, on the one hand, is determined from the contact angle analysis by positioning a liquid on the sample surface as shown in Fig. 1.1. On the other hand the interfacial tension between different polymers and between a polymer and a substrate is much more difficult to obtain than surface tension. The reason for difficulty in measurement of interfacial tension is due to high viscosities of materials, long time scales for achievement of equilibrium and sample decomposition, etc. However, the interface width and compatibility of polymer materials can be determined on the basis of model assumptions using mean field theory [11].

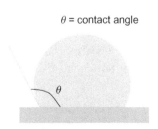

θ = contact angle

Figure 1.1 Schematic illustration of contact angle measurement.

1.3.2 Adhesion Phenomena in Bonding of Polymers

According to ASTM International, "adhesion" can be defined as a state in which two surfaces are held together by interfacial forces, which may consist of valence forces, interlocking forces, or both [16]. Though adhesion is an interfacial phenomenon, the experimentally measured adhesion depends on many factors in addition to events at the adherend — adherate interface. The material which adheres to the adherent is known as adherate. Thin films, thick films, coatings, and paints, etc., are common examples of adherates. An adhesive is a special type of adherate in which it adheres to two adherends [17]. The materials which are commonly used as adhesives fall under the category of polymeric materials. Only high molecular weight materials having long-chain molecules are expected to possess pronounced adhesive properties [18]. It is important to mention that, rather than relating adhesion to wetting, where it is incorrectly approached the bonding as a thermodynamically reversible phenomenon, the testing of bond strength by analyzing type of failure though determination of the type of separation is a more practical approach for quantifying adhesive bond strength. Adhesion reflects not only fundamental adhesion (which refers to the forces between atoms at the interface) but also the mechanical response of the adhesive, substrate, and interfacial region. (Practical adhesion refers to the results of the various tests of adhesion of the adhesive joint, coating, or composites, such as shear, peel, tensile strength obtained from the related tests). The prime requirement for a successful bond between two phases is to establish contact at a molecular or atomic level. It can be seen that there are no unique mechanisms of adhesion applicable to all adhesion cases and each particular case of adhesive bonding needs to be studied and explained. In order to derive some useful predictions associated with adhesion phenomena at polymer—polymer interfaces, mechanical theory, adsorption theory, chemical theory, diffusion theory, electrostatic theory, rheological theory, and weak boundary layer theory (WBL), etc., are studied by various researchers. The mechanical theory of adhesion relates the interlocking (on a macroscale a well as on a microscale) of the solidified adhesive material with the roughness and irregularities of the surface. However, the attainment of good adhesion with smooth adherands (such as glass, where surface roughness is comparatively low) restricts the general validity of mechanical theory. The adsorption theory proposes that the formation of a strong bond between adhesive and adherent is attained via intermolecular forces. On the basis of the magnitude of bond energy, the types of bonds are classified into primary bonds (ionic, covalent, metallic) and secondary bonds (van de Waals, etc.). Chemical theory considers molecular bonding as the most widely accepted mechanism for explaining adhesion between two surfaces in close contact. This theory explains that intermolecular forces between adhesive and substrate are due to dipole—dipole interactions, van der Waals forces, and chemical interactions (i.e., ionic, covalent, and metallic bonding). According to diffusion theory, diffusion of polymeric chains across the interface determines the adhesion strength. Appreciable diffusion does not take place in systems involving one or more hard solids (such as metals, glass, oxides, etc.,) under normal conditions of temperature and time scale. According to electrostatic theory, when two dissimilar polymeric materials are brought into contact, a charge transfer takes place which results in the formation of an electrical double layer. According to rheological theory, adhesion occurs due to the interpenetration of substrates across an interface. WBL explains premature adhesive failure due to the existence of WBL's at the surface [19]. The right blend of knowledge of the chemical composition of the polymer surfaces along with adhesion strength tests will uncover the correlations between the chemical bonding mechanism and adhesion. By modifying the surface chemical structures, the adhesion strength can be tailored for specific applications.

1.4 SURFACE MODIFICATION STRATEGIES IN POLYMER MULTIPHASE SYSTEMS

One of the principal focus areas of research in polymer surface science and interfacial engineering has been on improving adhesion at the polymer—polymer interface in order to achieve useful polymer blends and tailored composites. Multiphase polymer blends or composites often have poor physical properties because of low adhesion at the polymer—polymer interface. It is known that, for compatibilization, the favorable enthalpic interactions between the segments of a block copolymer and a homopolymer substrate will enhance adhesion by virtue of interpenetration of the copolymer segments into that phase. As the availability of such preformed copolymers is limited, a potentially viable alternative is to form these copolymers in situ by reactive processing during blending, coextrusion, or thermal lamination using functionalized polymers. That is, the functionally reactive groups can be attached directly to the two immiscible polymers of interest or attached to other polymers which form miscible mixtures with these phases. The chemical reactions at the polymer—polymer interface are also responsible for promoting adhesion. These principles were widely adopted for developing polymer blends/composites which shows toughness characteristics [20]. The surface modification of polymeric materials has been of great research interest in the past few years because of its importance in applications such as biomaterials and coatings [21]. Among the various known techniques, three major surface modification principles which allow the optimization of functional polymer architectures are surface segregation, surface structure, and surface reorganization. These functional polymer architectures can be used for producing adhesive and release surfaces, the suppression of dewetting by the use of functional additives to lower the surface tension of a coating, and the creation of smart polymer surfaces with selective adhesion properties [22]. If the goal of surface modification is to create a hydrophobic surface, then the functional group usually has a lower surface free energy than the polymer backbone, and if adhesion promotion is the objective of surface modification, then the functional group will have higher surface energy than the polymer backbone. In order to design functionalized polymers fundamental knowledge of different facets are required. It includes the appropriate selection of chemistry to introduce specific functional groups, different techniques to incorporate functional groups, processing, rheology, interfacial adhesion, mechanical properties, morphological analysis of blends, and fracture characteristics, etc.

1.5 PLASMA PROCESSING OF POLYMERIC MATERIALS

Plasma-based technologies are emerging to be a proven technology which provides an efficient, economic, and versatile solution to adhesion problems. Plasma technologies find increasing applications in the domain of polymer chemistry in terms of both materials processing and synthesis. Plasma-based technologies can be used to synthesize a new class of polymeric materials with a unique set of properties from an organic or mixed organic — inorganic precursor, for which the structure—property relationships can be varied considerably depending on the final application. Using plasma as a surface modification technique, polymer surfaces can be activated with specific functional groups and thereby design the surfaces for various applications of interest (i.e., for the treatment of polymer substrates by a reactive or inert gas aiming at a specific surface functionalization).

The plasma is defined as a partially ionized gas containing ions, electrons, various neutral species, and photons at many different levels of excitation. Khehta et al. proposed a more accurate definition of plasma based on the works of Grill and Bittencourt as "a quasi-neutral gas of charged and neutral particles characterized by a collective behaviour" [23]. When various plasma species interact with a polymer surface, a number of physical and chemical changes occur at the plasma—surface interface. If the objective is to modify the polymer surface using plasma, upon plasma treatment, the free electrons gain energy from an imposed electric field, colliding with neutral gas molecules and transferring energy. These collisions and transfer of energy form free radicals, atoms, and ions, which then interact with the solid surfaces placed in the plasma—that is, the energy and momentum of the plasma-activated species—will be transferred to the lattice atoms at the solid surface. This leads to drastic modifications of the molecular structure at the surface, providing the desired surface properties [24] (Fig. 1.2).

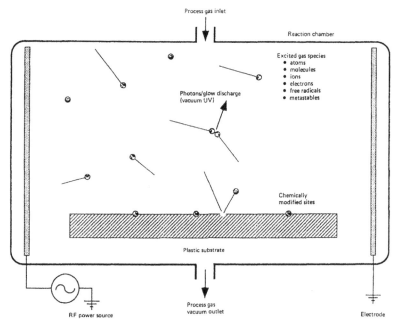

Figure 1.2 Schematic of the surface modification of plastic in a gas plasma reactor. *Reproduced with permission from Elsevier.*

The plasma process causes changes only to the few molecular layers from the surface of the material subjected to plasma treatment. The plasma treatment also changes the molecular weight of the surface layer by scissioning (reducing the length of molecules), branching, and crosslinking. The type of gas used and the composition of the material surface have a major role in deciding the intended change in the surface of the material. The selection of the process gas determines how the plasma treatment will alter the surface of polymeric material.

Oxygen plasma treatment is regarded as the most commonly used plasma process to enhance adhesion. The oxygen plasma is aggressive and forms numerous components, such as O^+, O^-, O_2^+, O_2^-, O, and O_3, etc. During plasma processing these components recombine and results in release of energy and photons. The photons in the ultra violet region have enough energy to break the polymer's carbon—carbon and carbon—hydrogen bonds. If the energy of the activated species is not sufficient to break the bonds at the surface, it will transfer a part of its energy onto the surface, resulting in localized heating. In addition to this, the incoming high energy ions also result in ion implantation by impinging into the material. When neutral atoms, having very low kinetic energy, are interacting with the material surface, the chemical changes are induced mostly by the recombination of activated species on the surface. The fluorinated plasmas are known to reduce the surface energy and increase chemical inertness, whereas noble gas plasmas induce surface crosslinking. One of the great advantages of plasma treatment in comparison with the chemical method is that "plasma by-products" do not require special handling and are readily removed by a vacuum system. To understand the complex reaction between the plasma and the surface of the polymeric material, the understanding of the plasma parameters and their effect on the surface modification is essential. It is important to consider the effect of all the plasma properties including microscopic plasma parameters like ion energy, ion flux, and degree of ionization, etc., in addition to macroscopic properties, such as excitation source power and time of exposure, to analyze the surface modification of polymers. Optical emission spectroscopy and Langmuir probe analysis are commonly used techniques to characterize plasma and diagnose plasma parameters [25].

The three competing mechanisms which alter polymers upon plasma treatment are ablation, crosslinking and functionalization. The above three mechanisms occur simultaneously to initiate the required change.

The extent of each mechanism depends on the chemistry and the process variables.

1.5.1 Ablation

Ablation results in the evaporation of surface material. Upon treatment, the plasma breaks the covalent bonds of the polymer backbone by bombardment with high-energy particles such as free radicals, electrons, and ions. As a result, long molecules become shorter, their volatile oligomers and monomers boil off (ablate) and are swept away with the exhaust. The bond breaking occurs on the polymer surface usually with the oxygen plasma [24].

1.5.2 Crosslinking

Surface crosslinking is usually carried out with an inert process gas (oxygen-free argon or helium) and/or noble gases. As there are no oxygen by-products in the plasma, the molecule can form a bond with a nearby free radical on a different chain (crosslink) [24].

1.5.3 Functionalization (Activation)

In the activation process, the functional groups present on the surface of polymeric material are replaced with atoms or chemical groups from the plasma. In this process, upon plasma treatment the plasma breaks the polymer's backbone or groups from the backbone, creating free radicals on the surface. The new groups formed on the polymer surface can alter its adhesion characteristics. That is, the type of atoms or groups substituted during plasma treatment determines how the surface will behave [24].

From the engineering point of view, the modified polymer surfaces are very important as they serve as interfaces of blends/composites which are used for high-performance applications [25].

1.6 PLASMA TREATMENT ON SPECIFIC POLYMERS

1.6.1 Engineering Polymers

Engineering polymers belong to a group of polymeric materials referred to as technical thermoplastics which can be used permanently at temperatures between 100°C and 150°C. It possesses good mechanical and thermal characteristics, high dimensional stability, good chemical resistance, and resistance to wear. Plasma treatment provides an efficient and economical solution to obtain quality adhesive bonds with engineering plastics such as polyamides [26], polyesters [27], polycarbonates [28], polyacetals, aromatic esters, imides [29], and liquid crystal polymers. Upon plasma treatment, it is reported that the mode of failure often changes from adhesive (interfacial) to cohesive in the adherend.

1.6.2 Commodity Polymers

Nowadays plasma-based technologies are used widely for the surface treatment of commodity resins. Hansen et al. reported that plasma treatment of polyethylene and Nylon 6 surfaces greatly improves their bondability, with bond failures occurring in the polymer rather than at the adhesive/polymer interface [30]. Large numbers of studies have been carried out to check the viability of plasma treatment of polyethylene [31−34] and polypropylene [35−39].

1.6.3 Biopolymers/Biodegradable Plastics

Numerous attempts are underway to develop fully biodegradable plastics by the use of various agricultural waste (agrowastes), plant carbohydrates, plant proteins, and vegetable oils. It includes biopolymers made from cereal substitutes, such as rice grains or wheat flour, made almost entirely from starch extracted from potatoes, wheat, rice, or corn, made from coffee beans and rice hulls that are dehydrated and molded under high pressure and temperature, made from potato starch, limestone, and cellulose, etc. Surface modification techniques to improve adhesion and biocompatibility of these polymer surfaces are of paramount importance in the medical field. Selective modification of the near-surface micro-structure and chemical state of biopolymers can be achieved with the radio frequency plasma-enhanced surface modification method provides an effective means for altering the biochemical properties of polymer surfaces without affecting the bulk properties [40−43].

1.6.4 Engineered Composites

Due to their attractive properties, especially high strength-to-weight and stiffness-to-weight ratios, engineered composites are widely used in structural applications where weight savings are a major consideration. A critical factor in engineered composites is the strength of the bond between the reinforcements and matrix. Due to poor chemical compatibility, strong adhesion at their interface is needed for the effective transfer of stress and bond distribution throughout the interface. Reinforcement fibers derived from high performance polymeric materials provide potential applications in engineered composites. But its structure has limited property development in structural composites. Plasma helps to re-engineer the surface chemistry of any polymer [44–47] to maximize its adhesive qualities which, in turn, makes it possible to use the engineered composites for structural applications.

1.6.5 Fluoropolymers

Fluoropolymers are a group of polymers in which part or all of the hydrogen atom/s has been replaced by fluorine atom/s. Fluoropolymers usually possess outstanding temperature and chemical resistance properties which, in turn, make it difficult to modify for bonding or for surface decoration. Due to handling and disposal problems, the previous option based on metallic sodium etchant has limited usage and paves the way for alternate treatment techniques. The plasma treatment is found to be a reliable option for the treatment of fluoro polymer surfaces. The surface altered by plasma becomes completely wettable. It enables the surface to be apt for printing on, and also bonding by, structural adhesives. The peel strength of adhesive bonded fluoropolymers by plasma treatment is found to be greater than by using a chemical treatment technique [23,48–50].

1.7 OVERVIEW OF CHARACTERIZATION TECHNIQUES FOR THE DIAGNOSIS OF PLASMA PROCESSED SURFACES

The enhanced surface reactivity is characterized in the laboratory by different techniques. It includes wettability analysis, water absorption studies, spectroscopic analysis such as Fourier-transform infrared spectroscopy (FTIR) and X-ray photoelectron spectroscopy (XPS), etc., microscopic examination using scanning electron microscopy (SEM) and atomic force microscopy (AFM), etc.

1.7.1 Wettability Analysis

Wettability is the ability of a liquid to spread over a surface. It can be measured by the contact angle between the liquid and the surface. There is a direct relationship between contact angle and surface energy—that is, the contact angle decreases with surface energy. Chapter 10, Wettability Analysis and Water Absorption Studies of Plasma Activated Polymeric

Materials, gives more detailed information about the wettability analysis of plasma activated polymeric materials.

1.7.2 Water Absorption Studies

The water absorption capacity of plasma-treated polymer surfaces can be studied using the water absorbency time test. The water absorptive capacity method provides a measure of the amount of liquid held within a polymer sample after specified times of immersion and drainage [25]. Chapter 10, Wettability Analysis and Water Absorption Studies of Plasma Activated Polymeric Materials, gives more detailed information about the water absorption studies of plasma activated polymeric materials.

1.7.3 Fourier-Transform Infrared Spectroscopy

Plasma surface modifications are confined to a few nanometers below the surface. FTIR spectroscopy in ATR mode (attenuated total reflection) is one of the suitable methods used to bring out the finer surface information of plasma activated surfaces [24]. Chapter 12, Spectroscopic and Mass Spectrometry Analysis of Plasma Activated Polymeric Materials, gives more detailed information about the FTIR analysis of plasma activated polymeric materials.

1.7.4 X-Ray Photoelectron Spectroscopy

XPS analysis is a commonly adopted technique to determine elemental composition of the plasma activated surfaces and O/C ratio, etc. FTIR can only give qualitative information whereas XPS provides quantitative chemical analysis of the plasma activated surfaces. Chapter 12, Spectroscopic and Mass Spectrometry Analysis of Plasma Activated Polymeric Materials, gives more detailed information about the XPS analysis of plasma activated polymeric materials.

1.7.5 Secondary Ion Mass Spectrometry

The secondary ion mass spectrometry (SIMS) analysis is destructive technique (sacrifices the surface particles that are analyzed by the mass spectrometer) for analyzing the mass/charge ratio of ionized particles produced from the sample upon bombardment of an energetic (25 keV) primary ion beam (Bi^+, Cs^+, $C60^+$). It is a useful technique to evaluate the crosslinking degree of a plasma polymer by means of time-of-flight secondary ion mass spectrometry (ToF-SIMS) analysis. Chapter 12,

Spectroscopic and Mass Spectrometry Analysis of Plasma Activated Polymeric Materials, gives more detailed information about the SIMS analysis of plasma activated polymeric materials.

The morphological features of the plasma activated surfaces can be examined using microscopic techniques. It helps to obtain information about the surface topography and composition in a micrometer scale. Chapter 11, Microscopic Analysis of Plasma Activated Polymeric Materials, deals with the details regarding the sample preparation, analysis, and interpretation of the data from the microscopic analysis of the plasma activated/modified polymeric materials. It includes initial observations of the surface using optical microscope followed by detailed evaluation using a scanning electron microscope, atomic force microscope and transmission electron microscope.

1.8 CONCLUSIONS

Modified polymeric surfaces are gaining wide acceptance especially in biomedical industries. Concurrently the plasma based technologies are also emerging as polymer surface modification technique which allows the surface of a polymer to be modified without affecting its bulk properties. Plasma based technologies can also be used to synthesize a new class of polymeric materials, with a unique set of properties from an organic or mixed organic − inorganic precursor, for which the structure−property relationships can be varied considerably depending on the final application. By controlling the plasma parameters and by understanding the plasma chemistry, it is possible to effectively modify surfaces of polymeric materials.

REFERENCES

[1] S. Kalpakjian, S. Schmid, Manufacturing, Engineering and Technology SI 6th Edition-Serope Kalpakjian and Stephen Schmid: Manufacturing, Engineering and Technology. Digital Designs, 2006.

[2] D.A. Jesson, J.F. Watts, The interface and interphase in polymer matrix composites: effect on mechanical properties and methods for identification, Polym. Rev. 52 (3) (2012) 321−354.

[3] C.V. Pious, S. Thomas, Polymeric materials—structure, properties, and applications, in: J. Izdebska, S. Thomas, (Eds.), Printing on Polymers: Fundamentals and Applications, first ed., 2015, pp. 21 − 40; G. Reina, J.M. González-Domínguez, A. Criado, E. Vázquez, A. Bianco, M. Prato, Promises, facts and challenges for graphene in biomedical applications. Chem. Soc. Rev. 46 (15) (2017), 4400−4416.

[4] Y. Zare, I. Shabani, Polymer/metal nanocomposites for biomedical applications, Mater. Sci. Eng. C 60 (2016) 195−203.

[5] S.W. Song, Y. Jeong, S. Kwon, Photocurable polymer nanocomposites for magnetic, optical, and biological applications, IEEE J. Sel. Top. Quant. Elect. 21 (4) (2015) 324−335.

[6] X. Qi, C. Tan, J. Wei, H. Zhang, Synthesis of graphene−conjugated polymer nano-composites for electronic device applications, Nanoscale 5 (4) (2013) 1440−1451.

[7] J. Abraham, P. Xavier, S. Bose, S.C. George, N. Kalarikkal, S. Thomas, Investigation into dielectric behaviour and electromagnetic interference shielding effectiveness of conducting styrene butadiene rubber composites containing ionic liquid modified MWCNT, Polymer (Guildf) 112 (2017) 102−115.

[8] R. Ahmad, N. Griffete, A. Lamouri, N. Felidj, M.M. Chehimi, C. Mangeney, Nanocomposites of gold nanoparticles molecularly imprinted polymers: chemistry, processing, and applications in sensors, Chem. Mater. 27 (16) (2015) 5464−5478.

[9] A.C. de Leon, Q. Chen, N.B. Palaganas, J.O. Palaganas, J. Manapat, R.C. Advincula, High performance polymer nanocomposites for additive manufacturing applications, React. Funct. Polym. 103 (2016) 141−155.

[10] J. Kim, B.C. Lee, Y.R. Uhm, W.H. Miller, Enhancement of thermal neutron atten-uation of nano-B 4 C,-BN dispersed neutron shielding polymer nanocomposites, J. Nucl. Mater. 453 (1) (2014) 48−53.

[11] M. Stamm, Polymer surface and interface characterization techniques., Polym. Surf. Interf. (2008) 1−16.

[12] T. Sun, L. Feng, X. Gao, L. Jiang, Bioinspired surfaces with special wettability, Acc. Chem. Res. 38 (8) (2005) 644−652.

[13] R. Blossey, Self-cleaning surfaces--virtual realities, Nat. Mater. 2 (5) (2003) 301.

[14] J. Vasiljević, M. Gorjanc, B. Tomšič, B. Orel, I. Jerman, M. Mozetič, et al., The sur-face modification of cellulose fibres to create super-hydrophobic, oleophobic and self-cleaning properties, Cellulose 20 (1) (2013) 277−289.

[15] T. Sun, G. Wang, L. Feng, B. Liu, Y. Ma, L. Jiang, et al., Reversible switching between superhydrophilicity and superhydrophobicity, Angew. Chem. 116 (3) (2004) 361−364.

[16] V. Kovačević, Surface and interface phenomenon in polymers, 2008.

[17] K.L. Mittal, The role of the interface in adhesion phenomena., Polym. Eng. Sci. 17.7 (1977) 467−473.

[18] S.S. Voyutskii, V.L. Vakula, The role of diffusion phenomena in polymer-to-polymer adhesion., J. Appl. Polym. Sci. 7.2 (1963) 475−491.

[19] F. Awaja, M. Gilbert, G. Kelly, B. Fox, P.J. Pigram, Adhesion of polymers, Progress Polym. Sci. 34 (9) (2009) 948−968.

[20] D.R. Paul, J.W. Barlow, Polymer-polymer interfaces in blends and composites, The University of Texas at Austin, Department of Chemical Engineering and Center for Polymer Research.

[21] J. Chen, H. Zhuang, J. Zhao, J.A. Gardella, Solvent effects on polymer surface struc-ture, Surf. Interf. Analysis 31 (8) (2001) 713−720.

[22] J.T. Koberstein, Molecular design of functional polymer surfaces, J. Polym. Sci. Part B: Polym. Phys. 42 (16) (2004) 2942−2956.

[23] F. Khelifa, S. Ershov, Y. Habibi, R. Snyders, P. Dubois, Free-radical-induced grafting from plasma polymer surfaces, Chem. Rev. 116 (6) (2016) 3975−4005.

[24] S.L. Kaplan, P.W. Rose, Plasma surface treatment of plastics to enhance adhesion, Int. J. Adhesion Adhesives 11 (2) (1991) 109−113.

[25] S. Guruvenket, G.M. Rao, M. Komath, A.M. Raichur, Plasma surface modification of polystyrene and polyethylene, Appl. Surf. Sci. 236 (1) (2004) 278−284.

[26] K.M. Praveen, S. Thomas, Y. Grohens, M. Mozetič, I. Junkar, G. Primc, et al., Investigations of plasma induced effects on the surface properties of lignocellulosic natural coir fibres, Appl. Surf. Sci. 368 (2016) 146−156.

[27] M.R. Wertheimer, H.P. Schreiber, Surface property modification of aromatic polyamides by microwave plasmas, J. Appl. Polym. Sci. 26 (6) (1981) 2087−2096.

[28] A.M. Wrobel, M. Kryszewski, W. Rakowski, M. Okoniewski, Z. Kubacki, Effect of plasma treatment on surface structure and properties of polyester fabric, Polymer (Guildf) 19 (8) (1978) 908−912.

[29] A. Hofrichter, P. Bulkin, B. Drévillon, Plasma treatment of polycarbonate for improved adhesion, J. Vac. Sci. Technol. A 20 (1) (2002) 245−250.

[30] N. Inagaki, S. Tasaka, K. Hibi, Improved adhesion between plasma-treated polyimide film and evaporated copper, J. Adhesion Sci. Technol. 8 (4) (1994) 395−410.

[31] G.P. Hansen, R. Rushing, Achieving optimum bond strength with plasma treatment, Soc. Manuf. Eng., Trans. (1990) 17.

[32] S.Y. Jin, J. Manuel, X. Zhao, W.H. Park, J.H. Ahn, Surface-modified polyethylene separator via oxygen plasma treatment for lithium ion battery, J. Ind. Eng. Chem. 45 (2017) 15−21.

[33] P.B. Jacquot, D. Perrin, R. Léger, B. Gallard, P. Ienny, Impact on peel strength, tensile strength and shear viscosity of the addition of functionalized low density polyethylene to a thermoplastic polyurethane sheet calendered on a polyester fabric, J. Ind. Text. (2017). 1528083716686853.

[34] Y. Liu, Y. Tao, X. Lv, Y. Zhang, M. Di, Study on the surface properties of wood/polyethylene composites treated under plasma, Appl. Surf. Sci. 257 (3) (2010) 1112−1118.

[35] S. Guruvenket, G.M. Rao, M. Komath, A.M. Raichur, Plasma surface modification of polystyrene and polyethylene, Appl. Surf. Sci. 236 (1) (2004) 278−284.

[36] R.M. France, R.D. Short, Plasma treatment of polymers: the effects of energy transfer from an argon plasma on the surface chemistry of polystyrene, and polypropylene. A high-energy resolution X-ray photoelectron spectroscopy study, Langmuir 14 (17) (1998) 4827−4835.

[37] N.Y. Cui, N.M. Brown, Modification of the surface properties of a polypropylene (PP) film using an air dielectric barrier discharge plasma, Appl. Surf. Sci. 189 (1) (2002) 31−38.

[38] X. Yuan, K. Jayaraman, D. Bhattacharyya, Effects of plasma treatment in enhancing the performance of woodfibre-polypropylene composites, Compos. Part A: Appl. Sci. Manuf. 35 (12) (2004) 1363−1374.

[39] B. Bae, B.H. Chun, D. Kim, Surface characterization of microporous polypropylene membranes modified by plasma treatment, Polymer (Guildf) 42 (18) (2001) 7879−7885.

[40] M. Lafuente, E. Martínez, I. Pellejero, M. Carmen Artal, M. Pilar Pina, Wettability control on microstructured polypropylene surfaces by means of O_2 plasma, Plasma Process. Polym. 14 (8) (2017).

[41] K. Komvopoulos, Plasma-enhanced surface modification of biopolymers, World Tribology Congress III, American Society of Mechanical Engineers, 2005, January, pp. 701−702.

[42] K. Bazaka, M.V. Jacob, R.J. Crawford, E.P. Ivanova, Plasma-assisted surface modification of organic biopolymers to prevent bacterial attachment, Acta Biomater. 7 (5) (2011) 2015−2028.

[43] K.N. Pandiyaraj, M.R. Kumar, A.A. Kumar, P.V.A. Padmanabhan, R.R. Deshmukh, M. Bah, et al., Tailoring the surface properties of polypropylene films through cold atmospheric pressure plasma (CAPP) assisted polymerization and immobilization of biomolecules for enhancement of anti-coagulation activity, Appl. Surf. Sci. 370 (2016) 545−556.

[44] J. Hasan, R.J. Crawford, E.P. Ivanova, Antibacterial surfaces: the quest for a new generation of biomaterials, Trends Biotechnol. 31 (5) (2013) 295−304.

[45] G.M. Wu, Oxygen plasma treatment of high performance fibers for composites, Mater. Chem. Phys. 85 (1) (2004) 81−87.

[46] J. Jang, H. Yang, The effect of surface treatment on the performance improvement of carbon fiber/polybenzoxazine composites, J. Mater. Sci. 35 (9) (2000) 2297−2303.

[47] X. Yuan, K. Jayaraman, D. Bhattacharyya, Effects of plasma treatment in enhancing the performance of woodfibre-polypropylene composites, Compos. Part A: Appl. Sci. Manuf. 35 (12) (2004) 1363−1374.

[48] M. Sharma, S. Gao, E. Mäder, H. Sharma, L.Y. Wei, J. Bijwe, Carbon fiber surfaces and composite interphases, Compos. Sci. Technol. 102 (2014) 35−50.

[49] N. Inagaki, S. Tasaka, H. Kawai, Improved adhesion of poly (tetrafluoroethylene) by NH3-plasma treatment, J. Adhesion Sci. Technol. 3 (1) (1989) 637−649.

[50] S.R. Kim, Surface modification of poly (tetrafluoroethylene) film by chemical etching, plasma, and ion beam treatments, J. Appl. Polym. Sci. 77 (9) (2000) 1913−1920.

FURTHER READING

L.M. Siperko, R.R. Thomas, Chemical and physical modification of fluoropolymer surfaces for adhesion enhancement: a review, J. Adhesion Sci. Technol. 3 (1) (1989) 157−173.

CHAPTER 2

Introduction to Plasma and Plasma Diagnostics

Miran Mozetič, Alenka Vesel, Gregor Primc and Rok Zaplotnik
Jozef Stefan Institute, Ljubljana, Slovenia

2.1 NONEQUILIBRIUM STATE OF GAS

Ordinary gas is found in the state of thermal equilibrium. This means that all gaseous particles assume the Boltzmann distribution over kinetic and potential energies. The distribution of gaseous particles over the kinetic energy is calculated as

$$f_E(E) = 2\sqrt{\frac{E}{\pi}}\left(\frac{1}{k_B T}\right)^{3/2}\exp\left(\frac{-E}{k_B T}\right), \tag{2.1}$$

where k_B is the Botlzmann constant ($k_B = 1.38 \times 10^{-23}$ J/K) and T is the so-called "kinetic temperature" of gaseous particles. Other terms also appear in literature such as "neutral gas kinetic temperature" or "translational temperature." The distribution depends only on the gas kinetic temperature and not on the mass of the gaseous particle. The average kinetic energy of a gaseous particle at room temperature is approximately 0.04 eV. The kinetic energy is defined as $E = \frac{1}{2}m\langle v^2\rangle$, thus the particles of higher mass will move slower than light particles. A typical random velocity of molecules such as nitrogen and oxygen is close to 500 m/s at room temperature, whereas the velocity of free electrons (if present in equilibrium gas) is much higher, obviously by the factor of $\sqrt{m/m_e}$, where m and m_e are masses of a molecule and an electron, respectively. The dissociation fraction (number of atoms, molecules, and ions) is proportional to different energies:

1. the number of dissociated atoms in a thermal equilibrium is proportional to the dissociation energy E_D;
2. the ionization fraction is proportional to ionization energy E_i;
3. the number of molecules excited into a given vibrational state (characterized by vibrational quantum number v) is proportional to the vibrational energy E_v;

Non-Thermal Plasma Technology for Polymeric Materials
DOI: https://doi.org/10.1016/B978-0-12-813152-7.00002-0
23

4. the number of molecules excited into separate rotational states of the same vibrational level in a given electronic state to rotational energy E_r. Corresponding equations are as follows:

$$N_A \propto e^{-\frac{E_D}{k_B T_D}}, \tag{2.2a}$$

$$N_i \propto e^{-\frac{E_i}{k_B T_i}}, \tag{2.2b}$$

$$N_v \propto e^{-\frac{E_v}{k_B T_v}}, \tag{2.2c}$$

$$N_r \propto e^{-\frac{E_r}{k_B T_r}} \tag{2.2d}$$

The indexes in Eq. (2.2a) correspond to the excited state of molecules. N_A is the number of atoms created by dissociation of the parent molecule, N_i is the number of ions, N_v is the number of molecules in the v-th vibrational state and N_r is the number of molecules in r-th rotational state. The terms T_D, T_i, T_v, and T_r are the dissociation, ionization, vibrational, and rotational temperatures, respectively. The proportional factors are often complex because of the stepwise processes. For a gas in thermal equilibrium all the temperatures in Eqs. (2.1) and (2.2b) are the same, thus, one can calculate the dissociation and ionization fractions as well as distribution of molecules over vibrational and rotational states if the gas temperature is known.

The nonequilibrium states of gases are much more complex. While the relation (1) is often satisfied (at least in a fair approximation) for given types of gaseous particles (neutral molecules and atoms as well as electrons and ions), the electron temperature is usually much larger than the temperature of other gaseous particles. The relations (2) are rarely satisfied. First, there are large differences between T_D, T_i, T_v, and T_r, and eventually the distribution over vibrational states does not assume the Boltzmann relation. In a typical nonequilibrium gas, the rotational temperature will be close to the kinetic temperature of neutral molecules, while the dissociation, ionization, and vibrational temperatures assume almost arbitrary values depending on particular conditions. Therefore, the term "temperature" should not be used in the description of gas in the nonequilibrium state. A more accurate description of the nonequilibrium gas contains the kinetic temperatures, the ionization and dissociation fractions as well as the vibrational temperature of the gaseous molecules if distribution follows the Boltzmann law (2). Typically, only the first few vibrational states are taken into account when determining the vibrational temperature and they usually follow the Boltzman law. Population of highly excited vibrational states may differ significantly, though.

One can obtain a nonequilibrium state of a gas by heating only specific gaseous particles. In current praxis the heating of charged particles is chosen. Gas is subjected to an electrical field where charged particles gain kinetic energy while neutral particles remain unaffected. The charged particles then transfer a part of their kinetic energy to the neutral particles causing excitation of ro-vibrational and electronic states, dissociation or/and ionization.

Every nonequilibrium state approaches equilibrium spontaneously. This process is often called "thermalization" and is explained as the equalization of the various temperatures in Eqs. (2.1) and (2.2c). The gaseous particles in various states thermalize either by emitting light quanta (radiation) or by transferring their energy in the excited state to colder particles. The transfer could be through super-elastic collisions (transfer of the internal energy to the kinetic energy of colliding gaseous particles) or by accommodation on surfaces of the gas container. The ratio between gas-phase and surface thermalization depends on the mean free path of gaseous particles which in turn depends on the pressure for a given container. The collision frequency is proportional to the pressure and so is the gas-phase thermalization. Some super-elastic collisions require three-body collisions whose frequency increases with the square of the pressure. A good example is a gas-phase association of two atoms to the parent molecule. Such association requires three-body collisions to satisfy conservation of energy and momentum. Because the collision frequencies increase with increasing pressure the life-time of the nonequilibrium gas state decreases with increasing pressure. In contrast, at pressures where the mean free paths are comparable to the dimension of the gas container the surface accommodation prevails and the life-time increases with increasing pressure. Obviously, there is an optimal pressure for sustaining the nonequilibrium state of gas; in a typical laboratory scale container it is of the order of 100 Pa.

2.2 GASEOUS PLASMA AND AFTERGLOW

A typical example of a nonequilibrium state of a gas is gaseous plasma. Gas in the container is exposed to an electric field. Charged particles (there are always some free electrons in gas) are accelerated in the electric field, they gain energy and transfer it to neutral molecules causing excitation to ro-vibrational and electronic states as well as dissociation or/and ionization. The new electrons formed at ionization collisions are further accelerated, cause ionization of neutral molecules, and consequently, they multiply. Soon after applying an electrical field of sufficient strength, the breakdown of the gas is observed—originally nonconductive gas becomes conductive and is transformed to the state of plasma. As the density of charged particles increases, the loss probability increases too; therefore, a steady concentration of charged particles appears. At such conditions, the production of charged particles is equal to the loss in the gas phase and/ or on surfaces. The steady concentration of charged particles appears soon after applying the electrical field, typically in a microsecond or faster. A longer time is sometimes needed to establish a steady concentration of gaseous particles in various excited states, but typically the steady-state of plasma is obtained in a time scale below a millisecond after applying the electrical field. An exception is dusty plasma where changes are observed over a time scale above one second.

Both negatively and positively charged particles are accelerated in the electric field. Positively charged particles are molecular and atomic ions. In a static electric field, they gain energy equal to the voltage difference between positive and negative electrodes ($e_0(V_+ - V_-) = e_0U$), where e_0 is the elemental charge. Eventually an ion collides with a neutral molecule. The ion and molecular masses are almost identical, thus an elastic collision between a fast ion and a slow molecule causes sharing of the initial ion kinetic energy. A substantial fraction of the ion energy is, therefore, transferred to the kinetic energy of a neutral molecule. Consequently, such collisions contribute to an increase of the kinetic temperature as in Eq. (2.1). Such sharing of the kinetic energy is not possible for collisions between electrons and neutral molecules. The electrons also gain the energy of eU when passing the potential difference of U, but do not lose it upon elastic collision with a molecule because of the difference in masses. At the elastic collision, both the momentum and energy are conserved and only a very small fraction of kinetic energy is exchanged between a fast particle of a low mass (electron) and slow heavy particle (charged or neutral atom or molecule). This is the key reason for discrepancy in kinetic temperature between electrons and any other gaseous particles in the nonequilibrium gaseous plasma: The electrons are heated by acceleration in the electrical field, but do not lose their kinetic energy for heating other particles. Instead, they use their energy for ionization, dissociation, and excitation events. Obviously, the electron temperature is higher than any of the temperatures in Eqs. (2.1) and (2.2d) in the nonequilibrium gaseous plasma.

The electrons are much faster than the molecular or atomic ions because of their low mass (even if their temperature is not higher than the temperature of the other gaseous particles). In the gas phase, electrons diffuse (move randomly) until they eventually reach the surface of the plasma container. They leave positively charged particles in the gas phase; therefore, plasma becomes positive against the walls (and the walls become negative against plasma). Soon after ignition of the electrical discharge, the plasma becomes positive enough to retard all but the fastest electrons from the surface. The accumulation of the positive charge lasts until the fluxes of ions and electrons onto the surface equalize, which happens in a microsecond if not sooner. A potential difference between plasma and the surface is established and the requirement for a steady state (equal fluxes of positively and negatively charged particles onto the wall) is satisfied. The potential difference occurs in a linear dimension that is of the order of a Debye length, which has been defined as

$$\lambda_{\mathrm{D}} = \sqrt{\frac{\varepsilon_0 k_{\mathrm{B}} T_{\mathrm{e}}}{e_0^2 n_{\mathrm{e}}}} \tag{2.3}$$

where ε_0 is dielectric permeability ($\varepsilon_0 = 8.85 \times 10^{-12}$ F/m), e_0 is the elementary charge (1.6×10^{-19} As), T_{e} is the electron temperature, and n_{e} is the electron density. The sheath appears next to the walls of the plasma container as well as next to any object or sample immersed into plasma. The sheath is useful because it prevents draining of electrons from the gaseous plasma. Obviously, the sheath thickness should be smaller than the smallest linear dimension of the plasma container. The Debye length is inversely proportional to the electron density and proportional to the electron temperature. Fig. 2.1 shows the Debye length versus the electron density for a range of densities between 10^{14} and 10^{20} m^{-3} and electron temperatures between 1 and 10 eV.

Apart from electronegative gases that contain a substantial concentration of negatively charged molecules or atoms, the major negatively charged particles are free electrons. As stressed above, they lose their kinetic energy in the sheath next to any object placed into a plasma reactor, thus, their influence on the materials' properties is marginal. The negatively charged ions (if present at all) cannot reach the surface of any

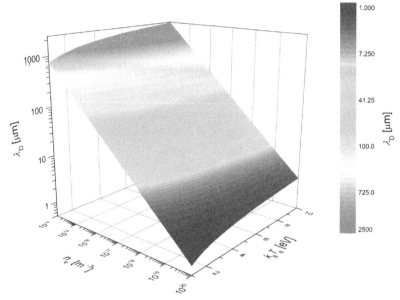

Figure 2.1 The Debye length in gaseous plasma.

material; therefore, their role in modification of surface properties is easily neglected, because the kinetic temperature of massive negative ions is not sufficient to provide negatively charged ions capable of traversing the sheath. In contrast, the positive ions are accelerated across the sheath; therefore, their influence on the materials' modification should always be taken into account. The neutral excited gaseous particles are not affected by the electric field inside the sheath and they reach the surface thermally with a negligible kinetic energy, but with the internal energy corresponding to the energy level of the excited state.

Users of plasma technologies are often concerned about the neutralization of charged particles in the gas phase. Positively and negatively charged particles (ions and electrons) are attracted by the electrostatic force and the collisions between them are frequent. The probability for neutralization, however, is low because of the quantum character and conservation of energy and momentum. The available energy before the collision is a sum of the potential and kinetic energies of colliding particles. The potential energy is equal to the ionization energy if the positively charged ion is in the ground state. The available energy before the collision is, therefore, larger than the ionization energy of the parent atom/molecule. The excessive energy cannot appear as the kinetic energy of a neutral atom/molecule after collision between a positive ion and electron because of the momentum conservation. Therefore, the neutralization should involve an additional particle. One possibility is a radiative neutralization by emission of a light quantum; however, the probability of such an event is very low because of the rules of quantum mechanics. The other possibility is a three-body collision—usually an electron, ion, and a neutral molecule or atom is involved. The three-body collision frequency increases with a square of the pressure and is regarded marginal at pressures below, say, 100 Pa. The effect is, however, the dominant loss mechanism of charged particles at atmospheric pressure, where the frequency is of the order of MHz. That is the reason why the nonequilibrium plasma at atmospheric pressure can be sustained only in strong electrical fields. Such a field is present at the ionization front of a streamer—a short-duration discharge which is limited to a small volume. Three-dimensional, cold nonequilibrium gaseous plasma is, therefore, unlikely to be sustained in large discharge chambers.

A variety of discharges are suitable for sustaining the nonequilibrium gaseous plasma. A simple DC glow discharge is created by placing two electrodes into a plasma chamber. One electrode can simply be the grounded housing of the chamber. The electrodes could be of the same dimensions (typically in the case of a dielectric reactor made from glass) or they could be asymmetric. The latter case is typically applied in metallic chambers. In such cases, the larger electrode (usually the grounded chamber) should be biased negatively and the smaller electrode positively. Such a coupling prevents too high ion fluxes on the smaller electrode and, thus, problems may arise because of arcing (transformation of a rather high-impedance glow discharge to a low-impedance electric arc). In the case of DC discharge, the sheath next to the cathode is biased few at 100 V below plasma potential; therefore, the positive ions bombard the cathode with substantial kinetic energy causing heating of the cathode and sputtering of the cathode material.

More popular are high frequency discharges. Both radiofrequency (RF) and microwave (MW) discharges are widespread sources of the nonequilibrium plasma. Plasma created with a MW discharge is usually concentrated to the region of a high electrical field, therefore, an asymmetry is often observed. Such an effect is less pronounced in the RF-driven plasma because the penetration depth of the RF field into conductive plasma is larger. Large plasma reactors suitable for treatment of polymer materials are usually driven by RF discharges. Electrodeless RF discharges are particularly useful. In such discharges, one can take advantage of a resonant electron heating. Such conditions appear where the electron collision frequency is similar to the frequency of the electric field. At the commonly used industrial frequency of 13.56 MHz, such a resonant heating occurs at the pressure of the order of 100 Pa.

All nonequilibrium systems tend to approach equilibrium when selected heating of specific gaseous particles is omitted. Then the electrons quickly cool down and are lost at neutralization either in the gas phase or on the surfaces of the plasma chamber. The typical time is of the order of a microsecond. Other reactive gaseous particles persist in the plasma reactor long after the charged particles have been neutralized and the state of plasma has been lost. A good example are metastables and atoms of molecular gases. They persist in the chamber over a time span of milliseconds, seconds or even minutes (providing they are not simply pumped away from the gas container) as long as the gas pressure is rather low. At atmospheric pressure, all reactive gaseous species thermalize quickly because of the high probability for three-body collisions. The gas that is left after neutralization of charged particles is not plasma but is still in the nonequilibrium state and is called "afterglow" or "postglow." The afterglow is particularly suitable for modification of the surface properties of polymers. The nonequilibrium gas in the afterglow can be sustained by two mechanisms: either pulsing the discharge or using continuous plasma in a small volume while continuously pumping the gas into a separate chamber—a common expression is "flowing afterglow." Pulsed systems typically operate at plasma on-time of the order of microseconds (to establish a desired nonequilibrium state of the gas), while the off-time is of a longer magnitude. A recommended configuration of the flowing afterglow is presented in Fig. 2.2. Plasma is limited to a rather small volume inside the discharge chamber. Molecular gas in a thermal equilibrium at room temperature passes the plasma region and enters the afterglow chamber. The orifice between the discharge and the afterglow chambers is preferably small; the conduction for gas flow should be much smaller than the pumping speed of the vacuum pump. A strong pressure gradient appears between the discharge and the afterglow chamber, resulting in drifting gas at subsonic velocity from the discharge to the afterglow chamber, therefore, the loss of reactive particles on the surface of the orifice is minimal. One of the best discharges for sustaining a highly nonequilibrium gaseous plasma in a rather small volume at elevated pressures (around 1000 Pa) is a MW discharge in the surfatron mode.

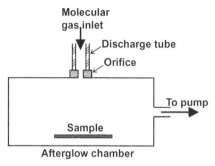

Figure 2.2 A useful configuration for treatment of polymers with neutral reactive particles in the flowing afterglow.

2.3 REACTIVE GASEOUS SPECIES

Nonequilibrium gas, either plasma or its afterglow, is rich in gaseous particles of substantial excitation energy suitable for tailoring surface properties of polymer materials. The particles of substantial affinity for chemical interaction with polymer materials are positively charged ions, neutral atoms in the ground as well as in electronically excited states (if metastable), some types of metastable molecules as well as light quanta that are emitted from gaseous plasma (providing the photon energy is high enough). The light quanta cause bond scission of the surface film of a thickness equal to the penetration depth. For UV radiation the penetration depth in polymers is a fraction of a micrometer. Other particles cause modification of a very thin surface film of the order of nanometer. Surface activation as well as etching is observed as a result of treatment with nonequilibrium gas.

2.3.1 Noble Gases

Pure noble gases are rarely used for tailoring surface properties of polymers. The reason for this is the lack of chemical interaction. Noble gases are characterized by high ionization energies. The electronically excited states are sometimes metastable, which means that the spontaneous relaxation by electrical dipole radiation is forbidden by the rules of quantum mechanics. The radiation life-time is, therefore, long enough that the relaxation frequently appears due to quenching—that is, transfer of internal energy to other particles, usually at super-elastic collisions. Table 2.1 represents the ionization energies for helium, argon, neon krypton, and xenon as well as the energy of the first metastable state and its radiative life-time.

Table 2.1 Ionization and excitation energies for metastable states of the lowest energy and the life-time of the metastable states

Type of particle	Symbol	Excitation energy (eV)	Radiative life-time	References
Ar ground	$^1S_0$	0	Stable[a]	[1]
Ar metastable $1s_5$	$^3P_2$	11.54	56 s	[1,2]
Ar metastable $1s_3$	$^3P_0$	11.72	45 s	[1,2]
Ar ion	$3p^5\ ^2P^\circ_{3/2}$	15.76	Stable[a]	[1]
He ground	$1\ ^1S_0$	0	Stable[a]	[1]
He metastable	$2\ ^3S_1$	19.82	~7700 s	[1]
He ion	$1s\ ^2S_{1/2}$	24.59	Stable[a]	[1]

[a]Nonradiative state.

The particles likely to be found in nonequilibrium plasma created in noble gases are positively charged ions and electrons as well as metastables. The electron temperature in such a plasma is rather high and increases with the increasing ionization energy; it is the highest for helium and the lowest for xenon. The ions can be accelerated in a sheath next to the powered electrode and inflict radiation damage of the surface film. The surface film's thickness depends on the kinetic energy of the positive ions. The radiation damage often causes structural changes as well as rich morphology. Originally smooth polymer surfaces become rough on the submicrometer scale on treatment with energetic ions. In the case where the polymer is not backed by an electrode with high negative potential powered by a high-frequency generator (samples kept at floating potential) the interaction is marginal. The same applies for metastables.

Gas always contains impurities. The common impurity in a plasma reactor is water vapour. The dissociation energy of water molecules as well as OH radicals is below the energy of metastables, therefore, water is likely to dissociate fully in the plasma of noble gases, especially in helium and argon. The ionization energy of O and H radicals is lower than the energy of some metastables, so the impurities are highly ionized. The plasma of noble gases (in particularly helium and argon) is, therefore, used predominantly as the source of metastables that possess high internal energy.

2.3.2 Oxygen

The nonequilibrium state of oxygen is traditionally used for the activation of polymer surfaces. The activation stands for the formation of oxygen-rich functional groups on the polymer surface [3,4]. The groups are polar and, therefore, contribute to the polar component of the polymer surface energy and increases its wettability [5]. Oxygen molecules have two metastable electronic states that are worth stressing $O_2(a)$ and $O_2(b)$: the first is at the excitation energy of 0.978 eV and the second one at 1.627 eV. The excitation energies for oxygen molecules are given in Table 2.2.

The dissociation energy of the oxygen molecule is 5.12 eV. The dissociation could appear at the electron impact with a molecule in the ground state, but in plasma of a rather low electron temperature step-wise dissociation prevails: molecules are first excited to metastable states and these are then dissociated. The dissociation by an electron impact rarely leads to the formation of thermal atoms—those with the same kinetic temperature as neutral molecules. A part of the electron energy is spent for kinetic energies of both atoms that are formed upon dissociative collision of an electron with an oxygen molecule. The atoms then effectively thermalize at the elastic collisions with molecules, especially at elevated pressure. The atoms are stable in a vacuum and the major loss mechanism is by three body collisions (at high pressure) or on the walls of the plasma or afterglow chamber. Table 2.3 shows the recombination coefficients for oxygen atoms on selected surfaces. Obviously, the atom density in a reactor will be large if the reactor is made from materials that exhibit low probability for surface recombination.

Table 2.2 Excitation energies of oxygen molecules and atoms

Type of particle	Symbol	Excitation energy (eV)	Radiative life-time	References
O_2 ground state	$X^3\Sigma^-_g$	0	Stable[a]	[6]
O_2 (a)	$a^1\Delta_g$	0.978	~ 55 min	[6,7]
O_2 (b)	$b^1\Sigma^+_g$	1.627	~ 10 s	[6,8]
O_2^+	$X^2\Sigma_g$	12.059	Stable[a]	[6,9]
O_3	X^1A_1	0	Stable[a]	[6]
O ground	^3P	0	Stable[a]	[1]
O first metastable	^1D	1.967	100 s	[1]
O second metastable	^1S	4.189	1 s	[1]
O^+	$^4S°_{3/2}$	13.618	Stable[a]	[1]

[a]Nonradiative state.

Table 2.3 Recombination coefficients for oxygen atoms on selected surfaces at room temperature

Material	Recombination probability	References
Al_2O_3	0. 0020	[10]
CoO	0.020	[11]
Cu_2O	0.025	[12]
NiO	0.27	[13]
Fe	0.01	[14]
Pt	0.0027	[10]
TiO_2	0.013	[11]
WO_3	0.012	[11]
Pyrex glass	10^{-4}	[15]
Quartz	10^{-4}	[16]
Stainless steel	0.07	[17]
Teflon	$7.5 \ 10^{-5}$	[10]

One peculiarity of oxygen plasma is a high probability for super-elastic collisions between molecules in vibrational states and atoms in the ground state: $O_2(v) + O \rightarrow O_2(v-1) + O$. The excessive energy (difference between the v-th and $(v-1)$-th state is shared between the molecule and atom as increased kinetic energy. Such collisions heat the gas rather effectively and lowers the vibrational temperature. Taking into account just two or three lowest vibrational states, the vibrational temperature (Eq. 2.2a) is just above the kinetic temperature (Eq. 2.1) in highly non-equilibrium oxygen plasma. The population of highly excited states may differ substantially from the Boltzmann law.

Oxygen flowing afterglow (Fig. 2.2) is rather cool and its kinetic temperature (which is equal to the rotational or vibrational temperatures) rarely exceeds 40°C as long as the walls of the afterglow chamber are kept at room temperature. The temperature of a polymer placed in the afterglow chamber is somewhat higher because of exothermic surface reactions: heterogeneous surface recombination of atoms and oxidation of polymer. Little knowledge is available on the relaxation of O_2 molecules in both metastable states on polymer materials. The relaxation should cause heating of the solid material, of course, but the probability for such events is unknown. The density of O–atoms in the ground state is always high in the afterglow chamber providing the reactors are made of materials with a low recombination coefficient (Table 2.3) and the plasma system (Fig. 2.2) is properly designed. The dissociation fraction easily exceeds 50% in glowing plasma and 10% in the flowing afterglow [18].

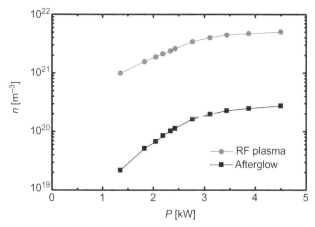

Figure 2.3 The density of O-atoms in RF plasma and its flowing afterglow 2 m away from the plasma source.

Fig. 2.3 represents the O-atom densities in RF plasma and its flowing afterglow 2 m away from the plasma source [19]. The oxygen plasma afterglow is very suitable for functionalization of many polymers because the concentration of polar functional groups is often higher than through the treatment in glowing plasma. Also, in the afterglow the atoms cause negligible etching [20,21].

2.3.3 Hydrogen

Hydrogen afterglow contains only molecules and atoms in the ground electronic state. These gaseous particles are unlikely to interact chemically with most polymers, therefore, the hydrogen afterglow is obsolete and one should rather use glowing plasma for tailoring the surface properties of polymers. Hydrogen molecules or atoms do not have metastable states, therefore, any electronic state of neutral particles is quickly relaxed by radiation. Table 2.4 represents excitation energies of hydrogen molecules and atoms. The dissociation energy of the hydrogen molecule is only 4.5 eV, thus, a dominant mechanism for the formation of neutral H-atoms in the ground state is by electron impact. A peculiarity of hydrogen molecules is the $b^3 \Sigma_u^+$ state which is unstable and dissociates in a short time after formation. Another peculiarity is the existence of the $a^3 \Sigma_g^+$ state with high potential energy. This state relaxes by a photon emission to the dissociative state $b^3 \Sigma_u^+$. The emission appears in the UV range with the maximal photon energy of about 10 eV. Hydrogen plasma is, therefore, an excellent

Table 2.4 Excitation energies of hydrogen molecules and atoms

Type of particle	Symbol	Excitation energy (eV)	Radiative life-time	References
H_2 ground state	$X^1 \Sigma_g^+$	0	Stable[a]	[6]
H_2 metastable	$c^3 \Pi_u$	11.87	1 ms	[22,23]
H ground state	$n = 1$	0	Stable[a]	[1]
H 1st excited state	$n = 2$	10.19	~ 2 ns	[1]
H 2nd excited state	$n = 3$	12.08	~ 30 ns	[1]
H 3rd excited state	$n = 4$	12.74	~ 100 ns	[1]

[a]Nonradiative state.

Table 2.5 Recombination coefficients for hydrogen atoms on selected surfaces at room temperature

Material	Recombination probability	References
Al	$1.8 \ 10^{-3}$	[24]
Co	0.18	[25]
Cu	0.19	[25]
Ni	0.18	[24]
Fe	0.17	[25]
Pt	0.25	[25]
Ti	0.1	[25]
W	0.06	[25]
Pyrex glass	10^{-3}	[26]
Quartz	10^{-3}	[26]
Stainless steel	0.1	[27]
Teflon	$4.5 \ 10^{-4}$	[28]

source of a continuum radiation in the UV range providing the electron temperature is large enough to excite the $a^3 \Sigma_g^+$ state and the dissociation fraction is not too high. These two requirements are contradictory, therefore, a strong continuum radiation does not appear in any hydrogen plasma. One method of stimulating such radiation is the application of a mixture of hydrogen and argon. The excitation energy of the first argon metastable state is just above the excitation energy of H_2 ($b^3 \Sigma_u^+$) state and the pumping of energy from argon metastables is efficient. Another method for increasing the continuum radiation from the hydrogen plasma is the construction of a discharge chamber from a material that acts as a catalytic surface for the heterogeneous recombination of H atoms to parent molecules. Table 2.5 shows recombination coefficients for hydrogen atoms on selected surfaces. If the goal is strong UV radiation, one should use metals such as titanium. On the other hand, if a high dissociation fraction is needed, the best material is quartz glass.

Hydrogen atoms do not have metastable states, therefore, radiation can occur from any excited state directly to the ground state. The radiation is discrete and in the vacuum ultraviolet (VUV) range. In fact, radiation from the first excited state of H-atom to the ground state at 121 nm is defined as the long wavelength boundary of VUV radiation. Both continuum and discrete radiation arising from excited molecules and atoms respectively are suitable for breaking bonds in the surface layer of polymers. The penetration depth of VUV radiation is smaller than for the UV continuum. Synergy between (V)UV radiation and hydrogen atoms leads to the quick removal of oxygen from the surface of polymers, therefore, the treatment is suitable for hydrophobization of moderately hydrophilic polymers rich in oxygen.

Another peculiarity of hydrogen atoms is that the probability for heterogeneous surface recombination of H-atoms to parent molecules on glass surfaces increases significantly with increasing temperatures. Such effects cause the time dependence of the H-atom density in a plasma reactor which is not cooled properly. For example, the H-atom density drops by 30% as the temperature of the plasma reactor made from borosilicate glass increases to 100°C.

2.3.4 Nitrogen and Ammonia

Nitrogen molecules have numerous excited states and some are metastable [29,30]. Table 2.6 represents the excitation energies of selected excited states of nitrogen molecules and atoms. The dissociation energy is much

Table 2.6 Excitation energies of nitrogen molecules and atoms

Type of particle	Symbol	Excitation energy (eV)	Radiative life-time	References
N_2 ground state	$X^1\Sigma_g^+$	0	Stable[a]	[6]
N_2 first metastable	$A^3\Sigma_u^+$	6.17	0.01 s	[6,23]
N_2 second metastable	$a^1\Pi_g$	8.55	0.0001 s	[6,23]
N_2^+	$X^2\Sigma_g^+$	15.6	Stable[a]	[6]
N ground	$^4S^\circ$	0	Stable[a]	[1]
N first metastable	$^2D^\circ$	2.38	~ 10 h	[1]
N second metastable	$^2P^\circ$	3.57	~ 10 s	[1]

[a]Nonradiative state.

higher than for oxygen and hydrogen, thus a step-wise dissociation often prevails. One distinctiveness of nitrogen is its very low probability for super-elastic collisions between vibrationally excited molecules and neutral atoms in the ground state, reaction $N_2(v) + N \rightarrow N_2(v-1) + N$. The consequence is a high vibrational temperature (taking into account the Boltzmann distribution) in nitrogen plasma. In fact, the vibrational temperature is usually several 1000 K although the kinetic (or rotational) temperature is not much above room temperature. Another peculiarity is the existence of the metastable states of high excitation energy (see Table 2.6). The excitation energy of the A-state is over half of the dissociation energy, therefore, two colliding molecules may produce a positively charged molecular ion and a free electron. Not only does it help ionizing molecules in the glowing plasma, but the charged particles are also formed in the flowing afterglow. Furthermore, unlike afterglows of hydrogen and oxygen, which are dark (except for the weak radiation from metastable oxygen molecules in the "b" state), the nitrogen plasma afterglow emits radiation of pink colour.

Another distinctiveness of nitrogen plasma is its poor emission from excited nitrogen atoms. Unlike oxygen or hydrogen where radiation from atomic species often prevails, atomic lines in nitrogen plasma are usually hardly visible. See Section 4.2 for details.

The richness of various excited states makes the interaction between nitrogen plasma and polymers difficult to understand. Nitrogen plasma definitely causes the formation of nitrogen functional groups on the polymer surface, however, they often appear together with the oxygen functional groups although no oxygen is added intentionally to the plasma chamber. One explanation of this effect which has been observed by most researchers in this field is the excellent dissociation of oxygen (or water vapor) which is present in plasma chambers in small quantities [31]. There are numerous nitrogen species of higher excitation energy than the dissociation energy of oxygen or water molecules.

Treatment of polymers with nitrogen plasma usually leads to the preferential formation of amides rather than amino groups [32]. If amino groups are preferred, it is better to use weakly ionized ammonia plasma. Particularly, such plasma contains NH_x radicals that readily interact with the polymer surface [33]. Some authors reported a substantial concentration of amino groups on polymer surfaces using glowing plasma, but it is usually better to use the afterglow of ammonia plasma if functionalization with amino groups is the goal.

2.3.5 Fluorocarbon Gases

With the increasing number of atoms in a single gaseous molecule, the plasma becomes more complex and parameters are difficult to predict. This is especially true for carbon-containing gaseous molecules of more than five atoms. Molecules are dissociated upon plasma conditions, but the recombination of atoms often does not lead to parent molecules. Atomic clusters may form, especially at elevated pressure, because the rules that prevent simple recombination (as in the case of the recombination of O-atoms) do not prevent recombination of an atom with a cluster. Eventually, the clusters grow in size and "dusty plasma" is formed [34]. Dust particles become negatively charged, therefore, there is an efficient drain of the free electrons and plasma may become unstable.

A carbon-containing gas of particular importance for tailoring the surface properties of polymers is tetraflouromethane (CF_4). It is chemically perfectly inert but becomes reactive upon the nonequilibrium state, especially in the state of gaseous plasma. Some data about the CF_4 excited states are presented in Table 2.7. A peculiarity of this gas is that positive CF_4^+ ions are rarely observed. The ionization by electron impact rather leads to simultaneous dissociation, therefore, the CF_3^+ ions are formed together with a free fluorine atom. Such atoms are chemically reactive and interact chemically with polymers. The result is the hydrophobization of the polymer surface. The CF_x radicals may or may not deposit on the surface of a material facing plasma, depending on the F/C ratio [36,37]. In the case when hydrocarbon is added into the gas mixture, one usually observes a thin film of hydrogenated carbon with a high fluorine concentration growing on the surface of a material facing plasma. The hydrogen and fluorine atoms recombine to HF in the gas phase (at high pressure) or on surfaces (predominantly at low pressure) and the remaining radicals condense on the surface forming a hydrophobic coating. The fluorine atoms also interact with the walls of the discharge chamber (in particular when the chamber is made from a glass) and cause etching, therefore, the chamber walls should be covered with a foil made from tetrafluoroethylene (or a similar material) to suppress this unwanted effect.

Table 2.7 Excitation energies of CF_4 molecules and F atoms

Type of particle	Symbol	Excitation energy (eV)	Radiative life-time	References
CF_4 ground	X^1T_1	0	Stable[a]	[35]
F ground	$^2P°$	0	Stable[a]	[1]

[a]Nonradiative state.

An addition of oxygen or water vapor to a gas mixture causes extensive etching of a polymer exposed to plasma sustained in CF_4 [38,39]. This effect is beneficial when a nanostructured surface finish of a polymer is desirable but detrimental in the other cases. Because the excitation energy of many excited states of tetrafluoromethane is high, the added oxygen or water molecules are rapidly dissociated. O and OH formed during dissociation interact with the polymer together with fluorine atoms and form volatile molecules composed of carbon, hydrogen, fluorine, and oxygen, therefore, the etching is usually more efficient than using only oxygen plasma.

2.4 BASIC DIAGNOSTIC TOOLS

The surface finish of polymers exposed to gaseous plasma depends on plasma parameters, in particular the fluxes of reactive gaseous species onto the surface of the material facing plasma. Good control of the processing parameters, therefore, requires knowledge of the density and temperatures of such gaseous species. There is no technique for measuring the densities and temperatures of all species, so complementary techniques need be used for a proper characterization. These techniques include electrical probes [40,41], calorimetry, catalytic probes (CP) [42,43], chemical titration [44], mass spectrometry (MS) [45], optical emission spectroscopy (OES) [46,47], actinometry [48], optical absorption spectroscopy including cavity ring down spectroscopy [49], two-photon absorption laser induced fluorescence (TALIF) [50], VUV absorption spectroscopy [51,52], and Fourier-Transform infrared absorption spectroscopy [53]. This section describes the basics of some of the frequently used techniques which are particularly suitable for characterization of plasma used for tailoring the surface properties of polymers.

2.4.1 Electrical Probes

One of the oldest techniques for plasma characterization is using an electrical probe. In the literature these are often called "Langmuir probes." Both commercial and self-made probes are popular. An electrical probe will give basic information about the electron density and temperature. More sophisticated versions sometimes also allow for the determination of electron energy distribution function.

An electrical probe is a small electrode placed inside the plasma reactor and connected to a DC power supply of an adjustable voltage between approximately -30 V and $+30$ V. The other connection of the power supply is connected to a large electrode also in contact with the gaseous plasma. The larger electrode is often called the "reference electrode." The surface of the reference electrode should be at least a few orders of magnitude larger than the probe to assure that the voltage on the probe is the same as that of the power supply. The current flowing from plasma to the probe is measured as shown in Fig. 2.4A. The probe characteristics (current versus voltage) should be as in Fig. 2.4B. When the electron energy distribution function corresponds to the Boltzmann law (Eq. 2.1) the central part of the curve in Fig. 2.4B is exponential and allows for the determination of the electron temperature using the equation [54]:

$$I_e = I_{e_0} e^{\left(\frac{eV}{kT_e}\right)} \qquad (2.4)$$

In practice, the curve often deviates from exponential. One possible reason is the deviation of the distribution function from Boltzmann law, but more commonly it is because of the plasma's instabilities or stray effects caused by the high frequency electric field. There should be a knee on the curve when the probe assumes plasma potential (see Fig. 2.4)

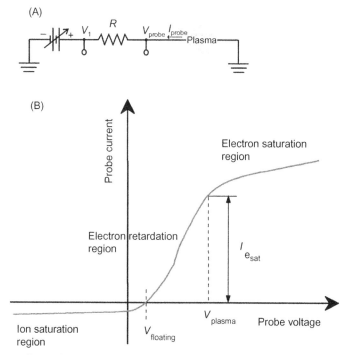

Figure 2.4 Electrical circuit suitable for electrical probes (A) and ideal characteristics (B).

but it is usually not sharp at all. When the probe is above the plasma potential, the electron saturation current (I_{esat}) is observed. In practice, however, the current keeps increasing with the increasing voltage because the thickness of the sheath next to the probe depends on the probe voltage. The electron density is calculated from the saturated current using Eq. (2.5) [55]:

$$n_e = 3.73 \times 10^{13} \frac{I_{esat}}{A_{probe}\sqrt{kT_e}}, \tag{2.5}$$

where A_{probe} is the surface of the probe tip. The ion saturation current is observed at large negative biasing of the probe. As in the case of electron saturation, the current keeps increasing with the increasing negative bias; therefore, the density of positively charged ions is difficult to determine. One possibility is extrapolating of the current to the plasma potential but this method is not accurate because the exact behavior of the ion flow versus the probe voltage is unknown. Still, a simple electrical probe allows for the estimation of the ion density in plasma, which is an important parameter. In the case of rather powerful discharges created by RF or MW generators such single probes often fail because of stray effects.

Simple electrical probes perform rather well in metallic chambers, but are impractical in glass chambers powered by electrodeless discharges. In such a case a double probe is a better choice. A probe comprising of two electrodes is immersed in plasma and connected to a DC power supply enabling adjustable voltage between approximately -30 V and $+30$ V as shown in Fig. 2.5A. In this case, both electrodes are floating on the plasma potential, therefore, they are always negatively biased against the plasma. The characteristic of such a double probe is shown in Fig. 2.5B. The ion saturated current (I_{sat}) is observed at both negative and positive biasing and the curve should have the form of hyperbolical tangens [56]:

$$I = I_{sat}\tanh\left(\frac{eV}{2kT_e}\right) \tag{2.6}$$

The ion density is estimated from the ion saturation currents and the electron temperature from the maximum slope of the characteristics as shown in Fig. 2.5B using the equations [57]:

$$N = \frac{2I_{sat}}{e_o A \sqrt{\frac{k_B T_e}{m_+}}} \tag{2.7}$$

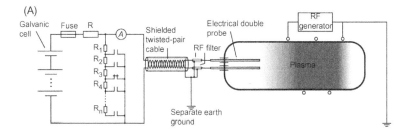

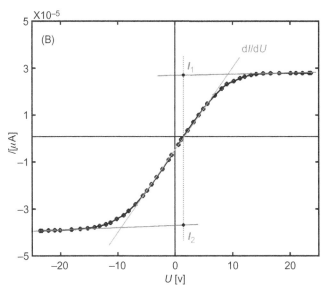

Figure 2.5 Electrical circuit suitable for double electrical probes (A) and ideal characteristics (B).

and

$$kT_e = \frac{e_0 I_1 I_2}{\frac{dI}{dU}(I_1 + I_2)} \tag{2.8}$$

where N is the density of charged particles, A is the probe area, m_+ is the positive ion mass, k is the Boltzmann constant, T_e the electron temperature, e_0 the elementary charge, I_1 the saturated ion current on the first electrode extrapolated to zero net current, I_2 the saturated ion current on the second electrode extrapolated to zero net current, and dI/dU the first derivative of the curve at the inflection point. If electrodes are the same $I_1 = I_2 = I_{sat}$.

In practice the ion saturation current increases with the voltage, therefore, it is difficult to state the exact ion density. One method of suppressing this effect is the application of electrodes of a rather large diameter, definitely at least of an order of magnitude larger than the Debye length (Eq. 2.3). Double probes are often less sensitive to oscillations or other stray effects than single probes. Electrical probes are only suitable for characterization of nonequilibrium plasma at low pressure. At atmospheric pressure, the probe would disturb plasma too much and any reading may lead to a wrong interpretation.

2.4.2 Optical Emission Techniques

Light emitting volume such as gaseous plasma emits light because of excited atoms or excited molecules. An atom or an ion will emit radiation when radiative transition between two states occur. Below the ionization limit are the discrete energy levels, whereas above this limit there is a continuum of levels. Because of this, three types of radiation can occur:

- Line radiation occurs because of electron transitions between discrete bound levels thus leading to line spectra. An outer shell electron falls from an upper bound level to a lower bound level, when the basic selection rules of spectroscopy are satisfied. This is the most common type of radiation.
- Recombination or free-bound radiation occurs when a free electron recombines with the ion. Because the upper level is continuous, the radiation is also continuous, but with some structure because of the discrete nature of the lower levels. This is a very rare phenomenon in the type of plasmas described here.
- Free-free radiation refers to bremsstrahlung emitted from plasma, that is, the transition between two free energy levels. It occurs when a free electron is decelerated and deflected by another charged particle or when a free-electron collides with a larger particle. In a partially ionized gas, there are competing interactions between electrons and neutral particles, and between electrons and ions. Which type of bremsstrahlung process dominates depends on plasma parameters, especially the degree of ionization. The bremsstrahlung spectrum is continuous, because the initial and final states are continuous.

The schematic presentation of three types of radiation are presented in an energy level diagram in Fig. 2.6A.

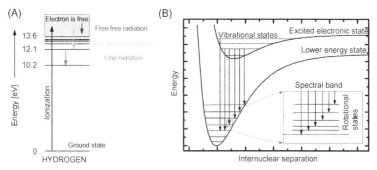

Figure 2.6 Energy level diagram for hydrogen atom with three types of radiation (A). Electronic, vibrational, and rotational states of a molecule with some of the transitions between them (B).

The emission from excited molecules are more complex, because the molecules can occupy not only different electronic states but also vibrational and rotational states (Fig. 2.6B). The ground energy level of a molecule is never null. Each electronic energy level has multiple vibrational levels, and each vibrational energy level consists of multiple rotational levels. All these levels are quantized and, therefore, the molecular spectrum is a line spectrum even though sometimes it may appear as continuous because of the poor resolution of the spectrometer. A group of emission lines that are closely spaced and arranged in a regular sequence are called spectral bands and correspond to a transition between two vibrational levels. Transitions between two electronic states, which consist of multiple bands, is called a system. The schematic presentation of these transitions is presented in Fig. 2.6B.

2.4.2.1 Optical Emission Spectroscopy

Spectra of the light emitted by plasma can be measured with an optical emission spectrometer. The example of the optical path emitted by plasma, collected by collimator with optical fiber and crossing the optical arrangement of a spectrometer is shown in Fig. 2.7.

Thera are a few things that needs to be done before the actual measurements take place. The OES spectrometer needs to be calibrated in two ways. First, wavelengths must be calibrated so there is no shift and second, the OES spectrometer needs to be calibrated for its spectral response. The latter is not necessary, but is recommended. In order to do this, two different light sources are needed: one for a wavelength calibration, such as mercury argon lamp and the other for spectral response

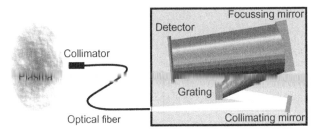

Figure 2.7 The schematics of an optical emission spectrometer and the optical path of the light emitted by plasma.

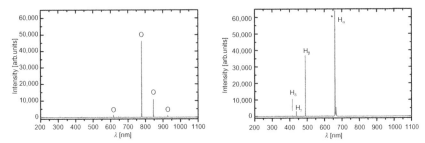

Figure 2.8 OES spectra of an oxygen (left) and hydrogen (right) ICP plasma in H-mode.

calibration, usually a deuterium–halogen lamp. Once the spectrometer is calibrated and positioned, the background needs to be measured at the same integration time as will be used in the actual measurements. This background is then subtracted from the measured spectrum.

Examples of atomic line spectra are presented in Fig. 2.8. It can be seen that the oxygen and hydrogen molecules are totally dissociated in ICP H-mode plasma because only excited atoms emit light.

Examples of an excited diatomic molecule spectra are shown in Fig. 2.9. The OES spectra of nitrogen ICP plasma in E-mode and H-mode are presented. Here four different nitrogen spectral features can be observed. The most prominent are the two N_2 spectral systems: first $(B\ ^3\Pi_g - A\ ^3\Sigma^+_u)$ and second $(C\ ^3\Pi_u - B\ ^3\Pi_g)$ positive systems, which consist of multiple bands (transitions between vibrational states) each marked with a dash. It can be seen, especially in the first positive system, that in E-mode these bands are clearly separated, whereas in H-mode plasma the bands are more fused together. This is because of the fact that the rotational temperature in H-mode is higher than in E-mode plasma and, therefore, the rotational transitions are more pronounced, which

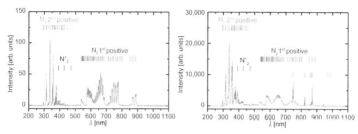

Figure 2.9 OES spectra of nitrogen ICP plasma in E-mode (left) and in H-mode (right).

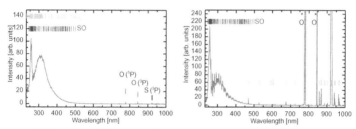

Figure 2.10 OES spectra of SO_2 ICP plasma in E-mode (left) and in H-mode (right).

makes the bands wider. In both E- and H-mode plasma a N_2^+ first negative system (B $^2\Sigma^+_u$–X $^2\Sigma^+_g$) is visible, whereas nitrogen atom lines are only visible in H-mode plasma where more nitrogen molecules are dissociated. Here it should be mentioned that the names negative and positive system simply refers to the occurrence of these bands in the negative glow or the positive column, respectively, of a glow discharge.

The OES spectra of a polyatomic molecules are presented in Figs. 2.10 and 2.11 An SO_2 ICP plasma in E-mode emits light almost only in UV (ultraviolet). Its spectrum is a continuum from approximately 200 −400 nm, because of SO and SO_2 bands, and really poor emission atom lines of O and S in the IR (infrared) part of the spectrum. The same gas at higher powers (H-mode) emits light mostly in IR, that is, the S and O atom lines, where the latter are in saturation in Fig. 2.10. The other spectral feature under these conditions is SO system in UV. The lack of SO_2 system suggests that the dissociation of plasma in H-mode is much higher than in E-mode.

One would assume that a molecule with even more atoms, such as ammonia, would emit only a continuum. However, the spectra in Fig. 2.11 shows that this is not the case. The OES spectra of NH_3 in

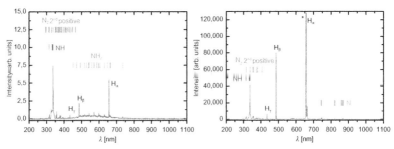

Figure 2.11 OES spectra of NH_3 ICP plasma in E-mode (left) and in H-mode (right).

E-mode clearly shows the NH_2 bands, NH bands, N_2 second positive system, and hydrogen Balmer atom lines. Ammonia dissociation energies are much lower than those of SO_2, that is, why the ammonia plasma dissociation fraction is higher at the same RF power. The presence of nitrogen molecule emission bands proves the presence of nitrogen atoms, because some nitrogen atoms may recombine on the plasma reactor surface, thus, forming N_2 molecules. In H-mode the dissociation fraction is even higher, therefore, the NH_2 bands are not visible anymore. Furthermore, N atom lines also appear. Here it should be stressed that the integration times in the H-mode plasma were approximately 1000-times lower than in the E-mode, whereas the light emitted by H-mode plasma was much more intense.

In Figs. 2.8−2.11 it can be seen that from OES spectrum one can determine which excited species are present in the plasma, but unfortunately one cannot directly determine its density, because the OES technique is qualitative and not quantitative. However, with some additions and assumptions one can also determine the densities of certain species. This method is called actinometry.

Actinometry was first introduced by J.W. Coburn and M. Chen in 1980 as a method for determining the density of fluorine atoms [58]. This method for determining the density of free atoms in ground state uses OES and an actinometer. The latter is a gas that is leaked into a reactive plasma in small concentrations. Usually it is a noble gas with a radiative transition near the transition of a particle whose density needs to be measured. For example, when neutral oxygen atom densities are measured, the most commonly used actinometer is argon.

The density of neutral oxygen atoms can be determined from the OES line intensity ratios, because the density of an argon atom in a gas mixture is known. The emission lines usually used for this determination are: oxygen line at 844.6 nm and argon line at 750 nm. The ratio between emission intensities is [59]:

$$\frac{I_O}{I_{Ar}} \propto \frac{n_O \int_{\varepsilon_O}^{\infty} \sigma_O(\varepsilon) f(\varepsilon) \sqrt{\varepsilon} d\varepsilon}{n_{Ar} \int_{\varepsilon_{Ar}}^{\infty} \sigma_{Ar}(\varepsilon) f(\varepsilon) \sqrt{\varepsilon} d\varepsilon},$$ (2.9)

where I_O and I_{Ar} are peak intensities of emission lines, n_O and n_{Ar} are the number densities of atoms in ground state, σ_O ans σ_{Ar} are the energy-dependent electron excitation cross section for excitations of atoms from ground state, $f(\varepsilon)$ is electron energy distribution function and ε is electron energy. This expression does not depend on electron density, but depends on electron energy distribution function. If the integral terms are constant this can be written:

$$\frac{I_O}{I_{Ar}} = \frac{k_O \, n_O}{k_{Ar} \, n_{Ar}} = k'_O \frac{n_O}{n_{Ar}},$$ (2.10)

where k are constants if, σ_O and σ_{Ar} are identical or if $f(\varepsilon)$ remains constant. This term is the basis of actinometry technique. In real systems the cross sections are relatively unimportant while $f(\varepsilon)$ decreases rapidly with increasing energy and, therefore, the density of electrons which exceed the thresholds is what really counts. If the energy thresholds ε_O and ε_{Ar} are similar, then it can be assumed that k'_O is constant.

The biggest flaw of this technique are its two assumptions. Actinometry is based on assumptions that the excited atom is excited only with collision of an electron and an atom in the ground state and that this excited atom loses its energy only with de-excitation by photon emission. The excitation and de-excitation oxygen examples are, $O(2p^4 \ ^3P) + e^- \rightarrow O^*(3p \ ^3P) + e^-$ and $O^*(3p \ ^3P) \rightarrow O(3 \ s \ ^3S) + h\nu$ (844.6 nm), respectively.

However, excited atoms used for actinometry can be obtained not only with excitation from the ground states, but also with multiple stage excitation or with de-excitation of higher energy states or, in some cases, with dissociative oxidation [60]. Besides that, the excited atoms do not lose energy only with de-excitation, but also with atom collisions. These corrections would make the actinometry technique complicated because in determining the densities of atoms in ground states one would need a rather complex kinetic model [61].

Nevertheless, in some cases actinometry gives good results. Fig. 2.12 shows a comparison between catalytic probe measurements and actinometry measurements.

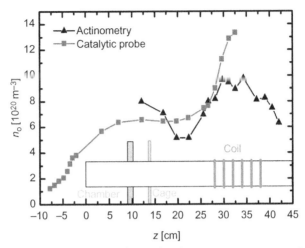

Figure 2.12 O-atom concentration obtained with actinometry compared to the ones obtained with catalytic probe.

NO titration is another way how one can determine the density of atoms with the use of OES. Titration is a quantitative chemical method for determining the unknown concentration of an identified chemical species. In plasma systems the most commonly used titrant is NO and is used to determine the density of O and N atoms. The method is based on reactions between oxygen or nitrogen atoms, generated in plasma, and nitrogen monoxide leaked into flowing afterglow. Because of security reasons usually a mixture of NO and argon is used.

The type of reactions between NO and N atoms depends on the volume flow of the NO. At low flow rates the reactions are [62]:

$$N + NO \rightarrow N_2O,$$
$$N + O + Ar \rightarrow NO^*Ar,$$
$$NO^* \rightarrow NO + h\nu.$$

At higher NO flow rates, reactions are a little different:

$$N + NO \rightarrow N_2 + O,$$
$$O + NO + Ar \rightarrow NO_2^* + Ar,$$
$$NO_2^* \rightarrow NO_2 + h\nu.$$

At low volume flow of NO mostly NO^* is produced, which emits blue light, and at higher flow NO_2^* is produced, which emits green light. The transition between these two regimes is called the extinction point, where the densities of NO and N are the same, and there is no emission of light. From the extinction point nitrogen density is determined. At higher NO flows the intensity of green light linearly increases with the flow. This can be written as:

$$I\left(NO_2^*\right) = K(\lambda) \cdot n_N \cdot n_{NO}$$

where the slope of the line is $K(\lambda) \cdot n_N$.

In an afterglow of an oxygen plasma the following reactions are taking place:

$$O + NO + Ar \rightarrow NO_2^* + Ar,$$
$$NO_2^* \rightarrow NO_2 + h\nu.$$

In the first reaction a metastable molecule NO_2^* is produced, which at the transition back to the ground state emits green light with a peak intensity around $\lambda = 575$ nm. Intensity of this radiation is measured with OES.

The number of the produced metastable molecules NO_2^* depends on the density of the neutral oxygen atoms from the plasma and on the volume flow of NO. Therefore, the intensity of the emitted light $I(NO_2^*)$ is proportional to the density of neutral oxygen atoms n_O and the density of the nitrogen monoxide n_{NO}. This can be written as:

$$I\left(NO_2^*\right) = K(\lambda) \cdot n_{NO} \cdot n_O, \tag{2.11}$$

where $K(\lambda)$ is a constant, which depends on the spectral sensitivity of the spectrometer, photon energy, the emission probability, and the rate coefficient of the reactions.

Constant $K(\lambda)$ is similar to the one determined with nitrogen atoms. When the constant is determined, the density of oxygen atoms can be calculated from the line slope where NO flow rates are varied.

This method is not used often because of the toxicity of NO and NO_2. Also, the method can only be used in the flowing afterglow, whereas in the discharge region the NO would be dissociated because of the electron collisions. The density of the neutral atoms inside the plasma can only be determined with a suitable model and with measurements in the flowing afterglow.

2.4.2.2 Stark Broadening

Besides the wavelength of the emission lines (determining the excited species), and their intensities (in some cases for determining the densities), one can also use the shape of the emission lines in order to determine some of the plasma parameters. The shape of the emission line is a mixture of instrumental broadening, natural broadening, Doppler broadening, self-absorption broadening, and pressure broadening, which is a combination of van der Waals broadening, resonance broadening, and Stark broadening. Analysis of the Stark broadening of spectral lines emitted by the plasma is the most commonly used method to measure electron density. From a technical point of view, the method is much more affordable than other techniques such as Thomson scattering or IR heterodyne interferometry, and can provide information on electron density with high accuracy. In order to determine the electron density, one must estimate the Stark broadening by fitting the experimental profile of the line with a Voigt function and using deconvolution of Doppler, van der Waals, resonance, and self-absorption broadening.

The optical spectrum can also be used for the determination of electronic, vibrational, and rotational temperature with the use of Boltzmann plot and computer simulation comparison. For example, a vibrational temperature can be determined from the Boltzmann plot, $\ln(I_{v'v''}\varepsilon(\lambda)\lambda_{v'v''}/A_{v'v''})$ versus $G(v')$, where $I_{v'v''}\varepsilon(\lambda)$, $\lambda_{v'v''}$ and $G(v')$ are the integrated band intensities corrected for spectral sensitivity, band origin wavelength and the vibrational energy term, respectively. For a simulation of OES spectra some software, such as Specair [63] and Lifbase [64], etc., are commercially available. A comparison of a simulated spectra of nitrogen plasma and a measured one is presented in Fig. 2.13. Here best fit was achieved with $T_{\rm rot} = 400K$ and $T_{\rm vib} = 8800K$.

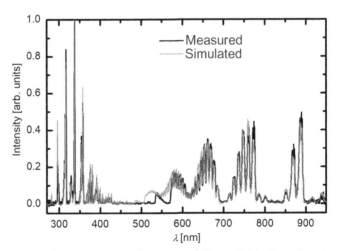

Figure 2.13 Simulated spectra with commercially available Specair software and measured spectra of a nitrogen ICP plasma in E-mode.

2.4.3 Optical Absorption Techniques

Contrary to OES, absorption techniques do not observe emitted light by the plasma, but light absorbed by either the plasma or plasma afterglow. In some cases, these techniques observe light emitted by florescence. Nevertheless, for absorption techniques an outer light source is needed and usually a laser is used for this purpose. Here two absorption methods will be presented in more detail.

2.4.3.1 Cavity Ring-Down Spectroscopy

Cavity ring-down spectroscopy is a direct absorption, highly sensitive, versatile technique suitable for the analysis of a wide range of plasma. CRDS belongs to a wide class of cavity-enhanced spectroscopies. A typical experimental setup of the CRDS technique is presented in Fig. 2.14.

In cavity-enhanced absorption spectroscopy, the laser light is guided into an optical cavity with two highly reflective mirrors with a reflectivity greater than 0.999. A light pulse from a tunable pulsed laser enters a cavity and travels back and forth between the mirrors, which increases the absorption length by several orders of magnitude. The signal detected on the other side of a cavity is, therefore, the decaying signal of the subsequent sequence of pulses, each with lower intensity than the previous one, hence the name ring-down. The amplitudes of pulses decay exponentially. When an absorbing medium is introduced into the cavity, the decay time is further reduced. From this an absorption coefficient can be determined. With changing the wavelength of the laser, a spectrum can be obtained.

The absorber density N can be calculated with a simple formula [65]:

$$N = 16\pi^2 \frac{\tau_{21}}{\lambda_{12}^2} \frac{w_1}{w_2} \frac{L}{l} \int \frac{1}{\lambda^2} \left(\frac{1}{\tau(\lambda)} - \frac{1}{\tau_0} \right) d\lambda, \tag{2.12}$$

where τ_{21} is the lifetime of the excited state, λ_{12} is the wavelength of the resonance transition, w_i the statistical weight of state i, L is the cavity

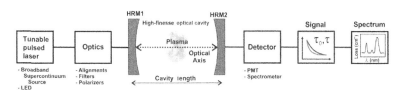

Figure 2.14 Typical experimental setup for CRDS diagnostic of a plasma.

length, l is the length of the absorbing medium, and τ_0 is the laser pulse lifetime of an empty cavity.

2.4.3.2 Two-Photon Absorption Laser-Induced Fluorescence

Laser induced fluorescence (LIF) is a very sensitive technique for the species-selective spatially resolved detection of ground state atoms, which are the most populated state in a low temperature plasma. LIF uses a single photon per excitation of the probed atom or molecule. The use of single-photon transitions limits the available transition energies. The light atoms such as hydrogen, nitrogen, and oxygen have their ground state transition at energies higher than 6.5 eV, which makes the use of VUV laser system necessary. The use of VUV photons has its problems. First, the media is not always transparent enough for VUV photons, which makes the probed volume optically thick. And second, it is experimentally challenging to provide an oxygen free environment for the laser pathway to allow the VUV radiation to reach the plasma. However, problems connected with single-photon excitation can be avoided with the application of a multiphoton excitation scheme. The use of two UV-photons instead of one VUV-photon is, therefore, simpler. In Fig. 2.15 two-photon excitation schemes for hydrogen, nitrogen, and oxygen atoms are presented.

In the oxygen atom example, for a transition from the ground state $2p^4\ ^3P$ to the excited state $3p\ ^3P$ around 11 eV is needed, which corresponds to the energy of two photons with wavelengths of 225.58 nm.

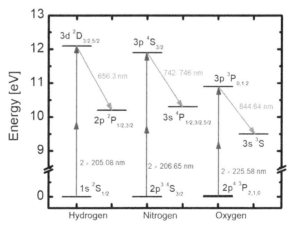

Figure 2.15 A two-photon excitation scheme for hydrogen, nitrogen, and oxygen atoms.

The density of the oxygen atoms in the ground state can then be determined with the use of measurement of the fluorescent light emitted at 844.64 nm (transition from 3p ^{3}P state to 3 s ^{3}S state [66]). In order to get the absolute densities, this method needs to be calibrated for each measured species. For example, for O atoms usually NO titration or calibration with noble gases are used [67].

Even though TALIF is a powerful diagnostic tool for monitoring the ground state atom properties in a plasma environment, it is not used that often because it is a very complex and expensive technique.

2.4.4 Catalytic Probes

Electrical probes enable at least an estimation of the ion density in plasma which is useful information, but major reactive gaseous particles suitable for tailoring surface properties of polymer materials are neutral, especially atoms. Charged particles are almost absent in the afterglow, therefore, an electrical probe is obsolete for characterization of nonequilibrium gas in the afterglow chamber (Fig. 2.3). A similar, but complementary, tool which enables measurements of the neutral reactive particles is a catalytic probe.

The origin of CP is calorimetry. It is definitely the simplest technique for plasma characterization. A piece of metal is mounted into plasma, kept at a floating potential, and the energy flux is determined from the rise of the metal temperature after igniting plasma. Then the density of atoms in the vicinity of the metal is calculated using several assumptions which are usually not correct. These assumptions are: the energy flux comes only from surface association of atoms to parent molecules and the accommodation of energy is complete. The first prediction is obviously wrong for glowing plasma but almost true in the afterglow. The second one is always wrong because the accommodation depends on the surface finish, recombination coefficient (Tables 2.3 and 2.5) as well as temperature of the catalytic material. A sophisticated version of a catalytic probe operates at a constant temperature of the catalytic tip and is protected from other sources of energy fluxes [68]. Such a probe is schematically presented in Fig. 2.16.

CP take advantage of the heterogeneous surface recombination of atoms to parent molecules. The recombination coefficient (Tables 2.3, 2.5, 2.8) depends on the type of atoms and a solid material which is called the catalyzer. The energy flux onto the surface of the catalyzer is [43]:

Figure 2.16 A sophisticated version of a catalytic probe.

Table 2.8 Recombination coefficients for nitrogen atoms on selected surfaces at room temperature

Material	Recombination probability	References
Al	$<10^{-2}$	[69]
Co	0.43	[70]
Cu	0.07	[69]
Fe	0.021	[71]
Pt	0.31	[70]
Pyrex glass	10^{-5}	[72]
Quartz	10^{-4}	[73]
Stainless steel	0.07	[74]
Teflon	$2.9\ 10^{-5}$	[75]

$$P = \frac{1}{4}nv\frac{W_{\mathrm{D}}}{2}\gamma A, \qquad (2.13)$$

where n is the atom density in the vicinity of the catalytic tip, v is the average thermal velocity of atoms $v = \sqrt{8k_{\mathrm{B}}T/\pi m}$, W_D is a dissociation energy, γ is a recombination coefficient for atoms, and A is the area of the catalyst. The recombination coefficient (γ) depends also on the surface morphology as well as temperature. Typically, the recombination coefficient increases with the increasing temperature, but the temperature dependence is often complex; therefore, it is preferable to keep the temperature constant during measurements. One method is heating of the catalytic tip by laser radiation. The catalyst is typically heated by laser to approximately 1000K. As long as the gas in the vicinity of the tip is in thermal equilibrium (negligible concentration of atoms) no additional heating occurs. Once molecular gas is dissociated, there is a rather high concentration of atoms and they associate to parent molecules releasing the dissociation energy on the surface of the catalytic tip. The association of atoms on the catalyst surface represents an additional heating thus the temperature starts rising when gas is transformed to the nonequilibrium state and plasma is ignited. The temperature is measured by fast electronics connected to the laser. As soon as additional heating due to surface association of atoms appears, the heating by laser is suppressed in order to keep the catalyst temperature constant. The power difference is measured and transformed to the atom density using Eq. (2.13). The response time of the probe depends on adjustment of the electronics and is typically well below a second, therefore, almost real-time measurements are possible. Fig. 2.17 represents a measurement of the H-atom density in an empty reactor powered by a RF generator of variable power. The discharge is electrodeless inductively coupled using a helicon coil. The power was chosen in such a way to enable transformation of a discharge mode from E to H. The H-atom density is larger in the H than in the E-mode because of the more efficient heating of electrons using the induced electric field inside the coil.

As shown in Fig. 2.16 the catalytic probe is placed in the cooled housing to suppress other channels for heat dissipation on the surface of the catalytic tip, thus, overcoming the deficiency of a simple calorimetry. Namely, if a tip is placed into the glowing plasma without the housing it is heated also by radiation arising from plasma, bombardment with positive ions, surface neutralization of charged particles, and relaxation of metastables of high excitation energy. The role of the housing is, thus, to allow for neutralization of almost all charged particles (the probability for such a reaction on any surface is almost 100%) screening radiation and allows for the relaxation of metastables of high excitation energy

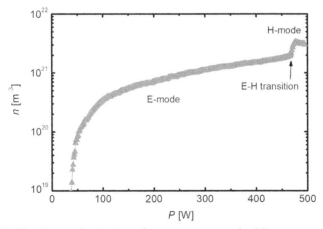

Figure 2.17 The H-atom density in a glass reactor versus the RF power.

(the probability is also very high). The metastables of low excitation energy still persist but they are supposed not to interact extensively with the catalyst. The housing is cooled to suppress any changes in kinetic or rotational gas temperature because after igniting the discharge, the inner walls of plasma reactor warm up in a time scale of seconds and the neutral gas temperature adapts to this temperature. The gas in the vicinity of the catalytic tip, therefore, resembles afterglow conditions. Obviously, the atom density at the tip is somewhat lower than in unperturbed plasma, therefore, calibration in an afterglow (or by any other technique such as TALIF [50]) is necessary. The housing is obviously obsolete if the measurements are performed in the afterglow chamber.

The density of neutral atoms in a plasma reactor depends on the quantity and type of materials immersed into the plasma. The recombination of atoms does not occur only on the catalytic tip or surfaces of plasma or afterglow chambers, but also on the surfaces of materials treated by the nonequilibrium state of gas. Worst cases represent hollow objects made from a material with a high recombination coefficient. Fig. 2.18A represents a schematic of the experimental setup and Fig. 2.18B the O–atom density versus the position of the probe outside and inside a copper tube. Copper has a large recombination coefficient, therefore, it represents a sink for O atoms. Plasma was sustained in a large glass cylinder by an inductively coupled RF plasma. The pressure was 100 Pa. This is the optimal pressure for sustaining extremely nonequilibrium gaseous plasma with high maximal density of neutral atoms. The powers were 500, 700,

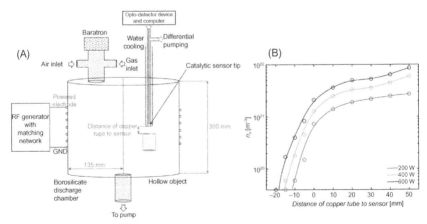

Figure 2.18 The setup for measuring O-atom density inside a hollow object (A) and the O-atoms density versus probe position (B).

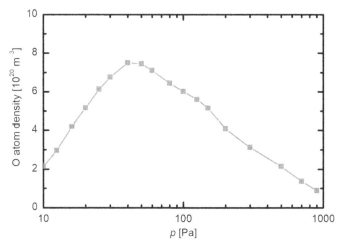

Figure 2.19 The O-atoms density in an afterglow chamber.

and 800 W. The O-atom density 5 cm away from the hollow object is several 10^{21} m^{-3}. As the probe moves toward the object, the O-atoms' density slowly decreases because of heterogeneous surface association to parent molecules. The gradient is rather small until the probe comes 1 cm away from the copper tube. The gradient, however, increases inside the tube (negative values of distance in Fig. 2.18) and eventually the O-atom density becomes immeasurably small approximately 2 cm inside the cylinder. Such effects are rarely addressed in scientific literature but should be taken into account when designing reactors for the treatment of catalytic materials [68,76].

Finally, we schematically present the O-atom density in an afterglow chamber in Fig. 2.2. A MW discharge in the surfatron mode was applied for sustaining plasma in the small chamber (glass tube of inner diameter 6 mm. The behavior of the O-atom density versus the pressure at fixed discharge power of 150 W is presented in Fig. 2.19. There is a well-expressed maximum in the O-atom density that appears at the pressure of 50 Pa inside the afterglow chamber. The pressure in the plasma chamber is larger in order to assure subsonic drift of nonequilibrium gas from the discharge to the afterglow chamber. The maximum is explained by several mechanisms. First, the atom density in plasma increases with the increasing pressure because the density of molecular gas leaked into the system increases. Therefore, there are more molecules available for dissociation. Second, the loss of atoms on surfaces decreases with the increasing pressure, thus, both effects allow for increasing the O-atom density in the afterglow chamber at rather low pressures. After reaching 80 Pa, the O-atom density decreases with the further increase of pressure for the following reasons: first, the plasma in the discharge tube shrinks to a smaller volume thus dissociation is less efficient, and second, the gas-phase recombination is not negligible anymore because the probability for three-body collisions increases as the square of the pressure. Such maxima always appear, but are shifted to different pressures depending on particular conditions. As a general rule, the maximum shifts to higher pressures as the discharge power is increased.

2.5 CONCLUSIONS

The chemical reactivity of molecular gases is increased significantly by excitation, dissociation, and/or ionization of molecules which occupy the lowest states in a thermal equilibrium gas at room temperature. The reason for excitation of molecules to energetic states while keeping the gas temperature close to room temperature is a negligible transfer of an electron energy to a translation energy of gaseous molecules. Such poor gas heating is the consequence of a huge difference in the mass between an electron and a molecule. Free electrons are heated in an electric field, gain substantial energy, and diffuse in the plasma volume where they cause the formation of chemically reactive species such as radicals and ions. The charged particles neutralize quickly after turning off the electric field due to extensive surface reactions. The neutral reactive particles like atoms in the ground state persist long after the glowing plasma vanishes, because the probability for heterogeneous surface recombination could be as low as 10^{-5}. The flowing afterglow is, therefore, rich in neutral reactants but the concentration of charged particles is negligible. Therefore, it is particularly useful for tailoring surface properties of delicate materials. Nonequilibrium state of gas is usually created in a gaseous plasma where the main parameters are the electron temperature and density. The simplest method for the estimation of these parameters is a floating electrical probe. Density of neutral reactive species should be determined with other techniques. The optical emission and absorption techniques represent almost nondestructive methods for measuring densities of such reactants, whereas a catalytic probe is particularly suitable for determining the density of neutral atoms in the ground state.

REFERENCES

[1] NIST, 2017. NIST atomic spectra database. https://www.nist.gov/pml/atomic-spectra-database (accessed October 2017).

[2] N.E. Smallwarren, L.Y. Chowchiu, Lifetime of metastable $^3P_2$ and $^3P_0$ states of rare-gas atoms, Phys. Rev. A. 11 (6) (1975) 1777−1783.

[3] A. Vesel, XPS study of surface modification of different polymer materials by oxygen plasma treatment, Informacije Midem-J. Microelectr. Electr. Compon. Mater. 38 (4) (2008) 257−265.

[4] A. Vesel, M. Mozetic, New developments in surface functionalization of polymers using controlled plasma treatments, J. Phys. D-Appl. Phys. 50 (29) (2017) 293001.

[5] A. Vesel, Modification of polystyrene with a highly reactive cold oxygen plasma, Surf. Coat. Technol. 205 (2) (2010) 490−497.

[6] R.W.B. Pearse, A.G. Gaydon, The Identification of Molecular Spectra, Chapman and Hall, London, 1984.

[7] R.M. Badger, A.C. Wright, R.F. Whitlock, Absolute intensities of the discrete and continuous absorption bands of oxygen gas at 1.26 and 1.065 μ and the radiative lifetime of the $1\Delta g$ state of oxygen, J. Chem. Phys. 43 (12) (1965) 4345−4350.

[8] M. Mozetic, Extremely non-equilibrium oxygen plasma for direct synthesis of metal oxide nanowires on metallic substrates, J. Phys. D-Appl. Phys. 44 (17) (2011) 174028.

[9] P. Natalis, J.E. Collin, J. Delwiche, G. Caprace, M.J. Hubin, Accurate ionization energy values for the transitions O_2^+, $X^2\Pi_g$ ($v' = 0$-20) ← O_2, $X^3\Sigma_g^-$ ($v'' = 0$) and molecular constants of oxygen ground ionic state, determined by Ne(I) (73.6 nm) photoelectron spectroscopy, J. Electr. Spectros. Related Phenomena 17 (3) (1979) 205−207.

[10] S. Wickramanayaka, S. Meikle, T. Kobayashi, N. Hosokawa, Y. Hatanaka, Measurements of catalytic efficiency of surfaces for the removal of atomic oxygen using NO^*_2 continuum, J. Vacuum Sci. Technol. A-Vacuum Surf. Films 9 (6) (1991) 2999−3002.

[11] C. Guyon, S. Cavadias, J. Amouroux, Heat and mass transfer phenomenon from an oxygen plasma to a semiconductor surface, Surf. Coat. Technol. 142 (Suppl. C) (2001) 959−963.

[12] P. Cauquot, S.C. Cavadias, J. Amouroux, Thermal energy accommodation from oxygen atoms recombination on metallic surfaces, J. Thermophys. Heat Transfer 12 (2) (1998) 206−213.

[13] I. Sorli, R. Rocak, Determination of atomic oxygen density with a nickel catalytic probe, J. Vacuum Sci. Technol. A-Vacuum Surf. Films 18 (2) (2000) 338−342.

[14] G.A. Melin, R.J. Madix, Energy accommodation during oxygen atom recombination on metal surfaces, Trans. Faraday Soc. 67 (0) (1971) 198−211.

[15] S. Wickramanayaka, N. Hosokawa, Y. Hatanaka, Variation of the recombination coefficient of atomic oxygen on Pyrex glass with applied RF power, Jpn. J. Appl. Phys. 30 (11R) (1991) 2897.

[16] S. Krongelb, M.W.P. Strandberg, Use of paramagnetic-resonance techniques in the study of atomic oxygen recombinations, J. Chem. Phys. 31 (5) (1959) 1196−1210.

[17] M. Mozeti, A. Zalar, Recombination of neutral oxygen atoms on stainless steel surface, Appl. Surf. Sci. 158 (3) (2000) 263−267.

[18] M. Mozetic, A. Vesel, U. Cvelbar, A. Ricard, An iron catalytic probe for determination of the O-atom density in an Ar/O_2 afterglow, Plasma Chem. Plasma Process. 26 (2) (2006) 103−117.

[19] R. Zaplotnik, A. Vesel, M. Mozetic, A powerful remote source of O atoms for the removal of hydrogenated carbon deposits, J. Fusion Energy 32 (1) (2013) 78−87.

[20] A. Doliska, A. Vesel, M. Kolar, K. Stana-Kleinschek, M. Mozetic, Interaction between model poly(ethylene terephthalate) thin films and weakly ionised oxygen plasma, Surf. Interf. Analysis 44 (1) (2012) 56−61.

[21] A. Vesel, M. Kolar, A. Doliska, K. Stana-Kleinschek, M. Mozetic, Etching of polyethylene terephthalate thin films by neutral oxygen atoms in the late flowing afterglow of oxygen plasma, Surf. Interf. Analysis 44 (13) (2012) 1563−1571.

[22] C.E. Johnson, Lifetime of the c³ Pi_u metastable state of H_2, D_2, and HD, Phys. Rev. A. 5 (3) (1972) 1026−1030.

[23] W. Lichten, Lifetime measurements of metastable states in molecular nitrogen, J. Chem. Phys. 26 (2) (1957) 306−313.

[24] A.D. Tserepi, T.A. Miller, Two-photon absorption laser-induced fluorescence of H atoms: a probe for heterogeneous processes in hydrogen plasmas, J. Appl. Phys. 75 (11) (1994) 7231−7236.

[25] B.J. Wood, H. Wise, Diffusion and heterogeneous reaction. II. Catalytic activity of solids for hydrogen-atom recombination, J. Chem. Phys. 29 (6) (1958) 1416−1417.

[26] B.J. Wood, H. Wise, The kinetics of hydrogen atom recombination on Pyrex glass and fused quartz, J. Phys. Chem. 66 (6) (1962) 1049−1053.

[27] M. Mozetič, M. Drobnič, A. Zalar, Recombination of neutral hydrogen atoms on AISI 304 stainless steel surface, Appl. Surf. Sci. 144 (Suppl. C) (1999) 399−403.

[28] S.M. Zyryanov, A.S. Kovalev, D.V. Lopaev, E.M. Malykhin, A.T. Rakhimov, T.V. Rakhimova, et al., Loss of hydrogen atoms in H_2 plasma on the surfaces of materials used in EUV lithography, Plasma Phys. Rep. 37 (10) (2011) 881.

[29] I. Shkurenkov, D. Burnette, R.W. Lempert, V.I. Adamovich, Kinetics of excited states and radicals in a nanosecond pulse discharge and afterglow in nitrogen and air, Plasma Sour. Sci. Technol. 23 (6) (2014) 065003.

[30] H. Yoshimine, H. Toshio, S. Koichi, Lifetime of molecular nitrogen at metastable $A^3\Sigma u$ + state in afterglow of inductively-coupled nitrogen plasma, Jpn. J. Appl. Phys. 51 (12R) (2012) 126301.

[31] B. Mutel, J. Grimblot, O. Dessaux, P. Goudmand, XPS investigations of nitrogen-plasma-treated polypropylene in a reactor coupled to the spectrometer, Surf. Interf. Analysis 30 (1) (2000) 401−406.

[32] M. Kolar, M. Mozetic, K. Stana-Kleinschek, M. Frohlich, B. Turk, A. Vesel, Covalent binding of heparin to functionalized PET materials for improved haemocompatibility, Materials 8 (4) (2015) 1526−1544.

[33] M.-J. Wang, Y.-I. Chang, F. Poncin-Epaillard, Effects of the addition of hydrogen in the nitrogen cold plasma: The surface modification of polystyrene, Langmuir 19 (20) (2003) 8325−8330.

[34] W.W. Stoffels, E. Stoffels, K. Tachibana, Polymerization of fluorocarbons in reactive ion etching plasmas, J. Vacuum Sci. Technol. A-Vacuum Surf. Films 16 (1) (1998) 87−95.

[35] X.F. Tang, X.G. Zhou, M.M. Wu, Z. Gao, S.L. Liu, F.Y. Liu, et al., Dissociation limit and dissociation dynamic of CF_4^+: application of threshold photoelectron-photoion coincidence velocity imaging, J. Chem. Phys. 138 (9) (2013) 094306.

[36] Y. Iriyama, H. Yasuda, Fundamental aspect and behavior of saturated fluorocarbons in glow-discharge in absence of potential source of hydrogen, J. Polym. Sci. Part A Polym. Chem. 30 (8) (1992) 1731−1739.

[37] M. Strobel, S. Corn, C.S. Lyons, G.A. Korba, Surface modification of polypropylene with CF_4, CF_3H, CF_3Cl, and CF_3Br plasmas, J. Polym. Sci. Part A Polym. Chem. 23 (4) (1985) 1125−1135.

[38] F.D. Egitto, F. Emmi, R.S. Horwath, V. Vukanovic, Plasma etching of organic materials. I. Polyimide in $O_2−CF_4$, J. Vacuum Sci. Technol. B-Microelectr. Process. Phenomena 3 (3) (1985) 893−904.

[39] F. Fanelli, F. Fracassi, R. d'Agostino, Deposition and etching of fluorocarbon thin films in atmospheric pressure DBDs fed with $Ar-CF_4-H_2$ and $Ar-CF_4-O_2$ mixtures, Surf. Coat. Technol. 204 (11) (2010) 1779–1784.

[40] J.L. Jauberteau, I. Jauberteau, Plasma parameters deduced from cylindrical probe measurements: determination of the electron density at the ion saturation current, Plasma Sour. Sci. Technol. 17 (1) (2008) 015019.

[41] M. Stanojevic, M. Cercek, T. Gyergyek, Experimental study of planar Langmuir probe characteristics in electron current-carrying magnetized plasma, Contribut. Plasma Phys. 39 (3) (1999) 197–222.

[42] D. Babič, I. Poberaj, M. Mozetič, Fiber optic catalytic probe for weakly ionized oxygen plasma characterization, Rev. Sci. Instr. 72 (11) (2001) 4110–4114.

[43] R. Zaplotnik, A. Vesel, M. Mozetic, A fiber optic catalytic sensor for neutral atom measurements in oxygen plasma, Sensors 12 (4) (2012) 3857–3867.

[44] P. Vašina, V. Kudrle, A. Tálský, P. Botoš, M. Mrázková, M. Meško, Simultaneous measurement of N and O densities in plasma afterglow by means of NO titration, Plasma Sour. Sci. Technol. 13 (4) (2004) 668.

[45] J.A. Ferreira, F.L. Tabares, Characterization of minority species in reactive plasmas by cryotrap-assisted mass spectrometry, Plasma Sour. Sci. Technol. 18 (3) (2009) 034019.

[46] Z. Kregar, M. Biscan, S. Milosevic, A. Vesel, Monitoring oxygen plasma treatment of polypropylene with optical emission spectroscopy, IEEE Trans. Plasma Sci. 39 (5) (2011) 1239–1246.

[47] Z. Kregar, N. Krstulovic, N.G. Vukelic, S. Milosevic, Space and time resolved optical emission spectroscopy characterization of inductively coupled RF water vapour plasma, J. Phys. D-Appl. Phys. 42 (14) (2009) 145201.

[48] M.A. Worsley, S.F. Bent, N.C.M. Fuller, T. Dalton, Characterization of neutral species densities in dual frequency capacitively coupled photoresist ash plasmas by optical emission actinometry, J. Appl. Phys. 100 (8) (2006) 083301.

[49] N. Krstulovic, N. Cutic, S. Milosevic, Modeling of cavity ring-down spectroscopy characterization of laser-induced plasma plume, IEEE Trans. Plasma Sci. 36 (4) (2008) 1130–1131.

[50] F. Gaboriau, U. Cvelbar, M. Mozetic, A. Erradi, B. Rouffet, Comparison of TALIF and catalytic probes for the determination of nitrogen atom density in a nitrogen plasma afterglow, J. Phys. D-Appl. Phys. 42 (5) (2009) 055204.

[51] J. Fengdong, I. Kenji, T. Keigo, K. Hiroyuki, K. Jagath, K. Hiroki, et al., Spatiotemporal behaviors of absolute ,density of atomic oxygen in a planar type of Ar/O_2 non-equilibrium atmospheric-pressure plasma jet, Plasma Sour. Sci. Technol. 23 (2) (2014) 025004.

[52] Y. Xiaoli, S. Koichi, N. Masaaki, Self-absorption-calibrated vacuum ultraviolet absorption spectroscopy for absolute oxygen atomic density measurement, Plasma Sour. Sci. Technol. 24 (5) (2015) 055019.

[53] A.B. Cruden, M.V.V.S. Rao, P.S. Sharma, M. Meyyappan, Fourier-transform infrared and optical emission spectroscopy of $CF_4/O_2/Ar$ mixtures in an inductively coupled plasma, J. Appl. Phys. 93 (9) (2003) 5053–5062.

[54] S. Bhattarai, L.N. Mishra, Theoretical study of spherical Langmuir probe in Maxwellian plasma, Int. J. Phys. 5 (3) (2017) 73–81.

[55] D.N. Ruzic, Electronic Probes for Low Temperature Plasmas, American Vacuum Society, New York City, 1994.

[56] B.M. Annaratone, G.F. Counsell, H. Kawano, J.E. Allen, On the use of double probes in RF discharges, Plasma Sour. Sci. Technol. 1 (4) (1992) 232–241.

[57] M. Mozetic, Characterization of extremely weakly ionized hydrogen plasma with a double Langmuir probe, Mater. Tehnol. 45 (5) (2011) 457−462.

[58] J.W. Coburn, M. Chen, Optical-emission spectroscopy of reactive plasmas - A method for correlating emission intensities to reactive particle density, J. Appl. Phys. 51 (6) (1980) 3134−3136.

[59] J.P. Booth, O. Joubert, J. Pelletier, N. Sadeghi, Oxygen atom actinometry reinvestigated - Comparison with absolute measurements by resonance-absorption at 130 Nm, J. Appl. Phys. 69 (2) (1991) 618−626.

[60] R.E. Walkup, K.L. Saenger, G.S. Selwyn, Studies of atomic oxygen in $O_2 + CF_4$ RF discharges by 2-photon laser-induced fluorescence and optical-emission spectroscopy, J. Chem. Phys. 84 (5) (1986) 2668−2674.

[61] P. Macko, P. Veis, G. Cernogora, Study of oxygen atom recombination on a Pyrex surface at different wall temperatures by means of time-resolved actinometry in a double pulse discharge technique, Plasma Sour. Sci. Technol. 13 (2) (2004) 251−262.

[62] A. Ricard, T. Czerwiec, T. Belmonte, S. Bockel, H. Michel, Detection by emission spectroscopy of active species in plasma-surface processes, Thin. Solid. Films. 341 (1-2) (1999) 1−8.

[63] SpectralFit, 2012. Specair. http://www.specair-radiation.net/, (accessed October 2017).

[64] SRI-International, 2016. LIFBASE spectroscopy tool. https://www.sri.com/engage/products-solutions/lifbase, (accessed October 2017).

[65] R. Zaplotnik, M. Biscan, N. Krstulovic, D. Popovic, S. Milosevic, Cavity ring-down spectroscopy for atmospheric pressure plasma jet analysis, Plasma Sour. Sci. Technol. 24 (5) (2015) 015026.

[66] L.F. Dimauro, R.A. Gottscho, T.A. Miller, 2-Photon laser-induced fluorescence monitoring of O-atoms in a plasma-etching environment, J. Appl. Phys. 56 (7) (1984) 2007−2011.

[67] A. Goehlich, T. Kawetzki, H.F. Dobele, On absolute calibration with xenon of laser diagnostic methods based on two-photon absorption, J. Chem. Phys. 108 (22) (1998) 9362−9370.

[68] G. Primc, A. Vesel, G. Dolanc, D. Vrancic, M. Mozetic, Recombination of oxygen atoms along a glass tube loaded with a copper sample, Vacuum 138 (2017) 224−229.

[69] R.A. Hartunian, W.P. Thompson, S. Safron, Measurements of catalytic efficiency of silver for oxygen atoms and the $O−O_2$ diffusion coefficient, J. Chem. Phys. 43 (11) (1965) 4003−4006.

[70] B. Halpern, D.E. Rosner, Chemical energy accommodation at catalyst surfaces. Flow reactor studies of the association of nitrogen atoms on metals at high temperatures, J. Chem. Soc. Faraday Trans. 1: Phys. Chem. Condensed Phases 74 (0) (1978) 1883−1912.

[71] M. Mozetic, U. Cvelbar, A. Vesel, A. Ricard, D. Babic, I. Poberaj, A diagnostic method for real-time measurements of the density of nitrogen atoms in the postglow of an $Ar−N_2$ discharge using a catalytic probe, J. Appl. Phys. 97 (10) (2005) 103308.

[72] A. Ricard, A. Popescu, A. Baltog, C.P. Lungu, Influence of the wall material on N atom density in a downstream nitrogen plasma, Vide-Sci. Technique Et Applications 52 (280) (1996) 248−254.

[73] T. Marshall, Surface recombination of nitrogen atoms upon quartz, J. Chem. Phys. 37 (10) (1962) 2501−2502.

[74] H. Singh, J.W. Coburn, D.B. Graves, Recombination coefficients of O and N radicals on stainless steel, J. Appl. Phys. 88 (6) (2000) 3748−3755.
[75] R.A. Young, Pressure dependence of the absolute catalytic efficiency of surfaces for removal of atomic nitrogen, J. Chem. Phys. 34 (4) (1961) 1292−1294.
[76] U. Cvelbar, M. Mozetic, A. Ricard, Characterization of oxygen plasma with a fiber optic catalytic probe and determination of recombination coefficients, IEEE Trans. Plasma Sci. 33 (2) (2005) 834−837.

CHAPTER 3

Plasma Assisted Polymer Synthesis and Processing

Shrikaant Kulkarni
Department of Chemical Engineering, Vishwakarma Institute of Technology, Pune, Maharashtra, India

3.1 INTRODUCTION

This chapter introduces the effect of plasma on polymer behavior, plasma technology, processing behavior of the polymer, plasma deposition of polymer films, and ageing of polymers as well as future trends given the unique aspects of the subject and the use of plasma technology in the polymer industry.

The influence of plasma technology has made inroads into almost every human activity during the past decade or so. Some of the key areas of plasma technology, applications, and areas of operation are shown in Table 3.1. Despite having widespread applications, the subject of plasma has not yet been explored to its potential and has been looked upon as a complex and impenetrable area. The aspects of plasmas that make the subject different from many other areas of physics and engineering are dealt with in this chapter.

Plasma in its simplest form, is constituted by the two elementary particles that make up an atom, that is, electrons and ions. Over 99% of the universe is believed to be made up of plasma, as against condensed matter (solids, liquids, and gases) such as comets, planets, and cold stars. The term plasma was first used by Langmuir in 1927 and the genesis of the word lies in the Greek word meaning to shape or to mold and its semblance with biological plasma, which is an electrolyte. The self-regulating behavior of plasma is in contrast to the apparently random behavior of fluids [1].

The science of plasma embodies many forms like space plasmas, kinetic plasmas, and technological plasmas, and is governed by enormous variations of parameters such as pressure, distance, and energy. One method depending on the relative value of the ion temperature Ti to the electron temperature Te of distinguishing different areas of plasma technology that is often used is as hot or cold plasmas as shown in Table 3.2.

Non-Thermal Plasma Technology for Polymeric Materials
DOI: https://doi.org/10.1016/B978-0-12-813152-7.00003-2
67

Table 3.1 Type of plasma technology, areas of operation and applications

Nonequilibrium plasmas		Equilibrium plasmas (high current)
Low pressure	**Atmospheric**	
Electron and ion beams-based processes	Sterilization of medical devices, Surgical applications	Circuit breakers and contactors
Mercury vapor lamp	Surface treatment using barrier and partial discharges	Arc gas heaters and electric discharge led fuel flames
Intense electron beams	Diesel engine emissions treatment, biomedical applications	Electricity generation, and space exploration
Surface modification (glow discharge)	Surface processing to enhance adhesion strength, hydrophobicity, wettability, etc.	
Inductively coupled plasma (ICP)	Lines	Streamlined plasma (torch)
Cold-cathode and electrode-less discharge lamps	Surface and atmospheric pressure glow discharges	Arc discharges and discharge lamps operating at high pressure
	Surface modification of polymer films	Ion lasers
Fabrication of electronic components	Discharges	Arc welding
Glow discharge diode	Corona discharges	Gas welding
Magnetron	Corona discharges on power	Arc welding (submerged)

Table 3.2 Temperature and pressure range for hot and cold plasmas

Cold plasmas		Hot plasmas
Thermal	**Nonthermal**	
$T < 2 \times 10^4$ K Arcs at pressure of the order of 100 kPa	Temperature (T) approx. 300 K Glow and arc discharge at low pressure of the order of 100 Pa	Temperature (T) > 10^6 K Plasma (kinetic and fusion type)

Although widely and conveniently used to describe individual areas, they accentuate the differences and exhibit the anomalous behavior at several thousand degrees Kelvin being described as cold and is not always helpful. Other common nomenclatures used are glow, corona, arc, and beams. These kinds of definitions, however, often confuse and create hurdles in the way of newcomers in the field or even to those already participating in it. The subject of plasma is better described as a potential energy continuum primarily of the electrons T_e and ions T_i and the electron number density, n_e [2,3].

The reason for plasmas unique characteristics and relevance to high-energy processes is apparent from Fig. 3.1, where the electron temperature, T_e, is dependent upon electron number density for different plasma processes. The relation between Energy and temperature with the Boltzmann constant, k_B is given as:

$$1/2\ mu^2 = k_B\,T$$

where $k_B = 1.38 \times 10^{-23}$ J/K [2]. In a cold plasma such as a neon lamp, the kinetic energy equals the electron energy and, although the mean electron temperature may be far greater by many a times than room temperature, the hot electron density and thermal mass is too small and, therefore, the rise in average temperature is consequently small. The potential and energy of most plasma processes are also too large in magnitude vis-à-vis other processes, for example, the energy of molecules at room temperature is about 0.025 eV and at 4000 K it is 0.35 eV [4,5]. Table 3.2 shows the temperature and pressure ranges of hot and cold plasmas.

Plasma parameters at the atomic level normally considered are particle morphology, length and time scales, particle number densities, kind and magnitude of forces between particles, etc., which interrelate with each other. Atomic particles are vulnerable to resonance and the difference in resonance frequencies between electrons and ions due to their mass difference influence the rheological properties of a plasma and is influenced by static and fluctuating magnetic field strength, as well as current and gas pressure conditions.

Fig. 3.1 shows the prerequisites for plasma processing leading to different applications.

3.2 EFFECT OF PLASMA ON FIBERS AND POLYMERS

Textile materials subjected to plasma treatments undergo major chemical and physical changes in surface chemistry, for example, deposition of

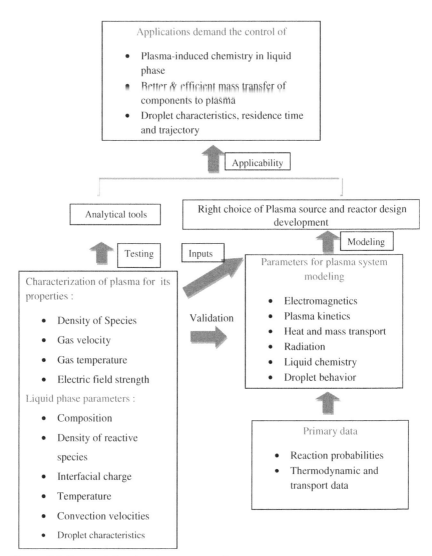

Figure 3.1 Prerequisites for plasma processing.

layers on surface, change in its morphology, and physical property profile. Plasmas generate a huge number of free radicals by dissociating molecules by frequent electron collisions and photochemically induced processes. New chemical moieties are formed in the fiber polymer surface due to weakening followed by rupturing of the chemical bonds leading to a profound effect on surface chemistry, topography, and the specific surface area of the fibers. Plasma treatment brings about changes in functionality and formation of new functional groups, such as $-OH$, $-CKO$, $-COOH$, of the polymer surfaces, which reflects on wettability grafting in polymers thereby imparting lipophobicity on treatment. Adhesion behavior, subsequently reflects on coating, bonding, and printing of materials. The poor adhesion tendency of polymeric materials is ascribed to its low surface energy [6]. Wet-chemical surface treatment methods for adhesion strengthening at times fail miserably in complying with environmental and safety regulations. However, one established technique presents a host of solutions to the problems posed by wet-chemical methods. Energized particles and photons are generated on exposing the substrate surface to plasma by following free-radical chemistry which is governed by prominent factors namely, the substrate, gas chemistry, the reactor design, and the operating parameters [7,8].

The four major effects are:
- surface preparation
- ablation or etching
- crosslinking of near-surface molecules
- maneuvering the surface-chemical structure

All these processes, in isolation or in synergistic combination, affect adhesion behavior. Plasma treatment can enhance the bond strength of polymers with other substrates. The improvement in adhesion behavior is attributed mainly to the increase in wettability and the surface chemistry of the polymer. Wettability is quantified as the increase or decrease in contact angles for specific liquids. Plasma induced polarity is responsible for increasing the surface free energy, γ, of the polymers which decreases the contact angle, θ, and provides better adhesion strength [9].

Plasmas offer unique, effective, and efficient surface engineering tools by virtue of their unique physical, chemical, and thermal property profile. It is a low temperature, nondestructive, dry, eco-friendly, and distinct treatment which holds much promise in frontier research areas. The advantages of an industrially viable plasma processing system vis-a-vis conventional textile processing are summarized in Table 3.3.

3.2.1 Plasma Finishing of Textiles

An extensive literature review and patent survey shows the availability of excellent state-of-the art method descriptions of plasma technology. Many research articles and patents have been granted in the field of plasma treatment for range of materials like fibers, polymers, nonwovens,

Table 3.3 Industrially amenable plasma processing vis-a-vis conventional wet chemistry

Criterion	Plasma processing	Conventional wet chemistry
Medium	No wet chemistry but treatment by excited gas phase	Aqueous phase based
Energy	Electricity-only free and mobile electrons energized (<1% of system mass)	Heat entire system to raise temperature of whole mass
Reaction type	Complex involving many simultaneous processes	Simpler, well established
Reaction locality	Highly specific, no effect on bulk properties	Bulk of the material generally affected
Potential for new, innovative processes	Huge potential, field under continuous exploration	Very low, not much amenable to technology
Equipment availability and development	Experimental, industrial and laboratory prototypes, rapid industrial development underway	Mature, but slow evolution
Energy consumption level	Low	High
Water consumption level	Abysmally low	High

coated fabrics, filter media, and composites, etc., to improve their functional and performance behavior [10−15]. The potential use of plasma treatments lies in the following types of functionalization:

- Antifelting or shrink-resistance of woolen fabrics
- Increasing hydrophilicity for enhancing wetting, dyeing, and adhesion
- Increasing hydrophobicity of water and oil-repellent textiles
- Facilitating and promoting the removal of sizing agents
- Scouring of cotton, viscose, polyester, and nylon fabrics
- Deposition of silver particles as antibacterial fabrics
- Disinfecting medical textiles at ambient temperatures
- Improving the adhesion strength between different materials by reducing interfacial tension
- Manufacturing polymers with oleophobic, hydrophobic and stain-resistant surfaces and color stability

- Functionalization of polymer fabric on either side
- Applying a flame-retardant and crosslinked silicone coating
- Acquiring durable antistatic properties
- Enhancing electro-conductivity

3.2.2 Plasma Systems for Laboratory and Industrial Scale Processing of Polymers and Fibers

Plasma systems used for polymer and textile treatments were low pressure systems designed and confined to a batch process was the only option till recently. This type of plasma machinery also exists for "air-to-air" or "cassette-to-cassette" batch−continuous in-line systems for, say, fabric which is wound from one batch to another inside a vacuum chamber [16].

There is a huge demand, however, for plasma systems which can be used for continuous on-line but economically viable processing of fabrics and films for finishing and coating applications. Both the low-pressure and the atmospheric-pressure plasma techniques would be very handy for said purpose. There have been some commercial low-pressure and atmospheric-pressure plasma systems serving both on-line and in-line applications to functionalize the polymers and fabrics [17].

3.2.3 Industrial Objectives of Plasma Systems

The industrial objectives of these plasma systems have been to:
- produce plasma reactors and fabric-handling systems able to process continuous, roll-to-roll lengths and widths ∼5 m or more of fabrics
- process at speeds of >10 m/min
- process at, or near, ambient temperature
- obtain uniform, deep-penetration, and surface modification of polymers
- be capable of handling large enough fabrics with wide electrode spacing.

Although plasma technology has proven capability and flexibility in the laboratory environment and holds much market potential, the commercial potential for processing at an industrial scale is yet to be tapped [18].

Incumbent plasma systems in the European market are:
- Low-pressure and atmospheric-pressure plasma systems capable of in-line and batch treatment
- On-line and continuous treatment systems like Dow Corning Plasma Systems (Ireland)

Manufacturing companies offer standardized and customized, specially designed portable laboratory plasma machines. Some companies even offer proprietary liquid precursor delivery/deposition systems [19,20].

3.2.4 Cost Considerations

The cost incurred towards energy, chemicals, and cooling water used in plasma-based treatments is far less compared with conventional wet chemistry-based ones. However, installation cost is far more for the former than the latter. The projected utilization rates and number of cycles are the deciding factors for the sustenance of the systems offered by the companies [21].

3.3 THE PLASMA TECHNOLOGY FOR THE POLYMER INDUSTRY

Low-pressure plasmas have primarily found applications in microelectronic industries since its inception in 1960s. However, in the 1980s the application areas were extended to surface treatments, especially in the fields of metals and polymers. It is used for treating a range of surfaces like plastics, polymers and resins, paper and board, metals, ceramics and inorganics, and biomaterials. Properties strengthened by plasma technology are wettability, adhesion, biocompability, protection and antiwear, sterilization, and chemical affinity or inertness. The technology offers various technical and economically viable options. Plasma technology is, therefore, a distinct but effective engineering tool for functionalization related applications. It is found to be a flexible and versatile technology [22−27].

Since 1980s much research work has been carried out to evolve the applications of low-pressure plasma technology for a host of materials with excellent results in terms of functionalization of the treated surfaces. Many of the commercially used low-pressure plasma machines have been in use for both batch and in-line processing of polymeric fiber-based materials for a decade or so. In spite of their advantages both on industrial and laboratory scale, the impact of plasma processing on a mass scale has not been encouraging. This may be due to discrete relevant applied research, inordinate delay in the design and development of suitable industrial plasma systems, the late focus on developing in-line atmospheric pressure plasma systems, and less awareness and exposure of the successes and failures of the systems [28].

Due to globalization of host of polymer and textile industries across the globe have been exposed to huge challenges. Therefore, there has been a paradigm shift to high functional, value added and technical materials as a prerequisite for their sustainable growth to take place and growing environmental and energy-saving concerns which will require dry and clean processes which are less water and energy intensive, unlike wet-chemical ones. Plasma technology, has huge commercial potential for developing new functionalities in fibers. Advantages of plasma technology has been recognized by manufacturers to design and develop machines using low and atmospheric plasma processes on an industrial scale for the functionalization of fibers in recent years [13,14,29−31].

3.3.1 Plasmas—Effective Surface Engineering Tools

The electromagnetic power into a process gas volume generates the plasma medium as a dynamic mix of ions, electrons, neutrons, photons, free radicals, meta-stable excited species, and molecular and polymeric fragments at ambient temperature. This leads to the surface functionalization without compromising their properties in bulk. These species have the ability to migrate under the influence of electromagnetic fields and diffusion gradients, etc., on the substrates soaked in, or moved through, the plasma. A sequence of generic surface processes including surface activation by bond breaking to create reactive sites, grafting of chemically active species and functional groups, material volatilization and removal (etching), ionization of surface contaminants/layers (cleaning/scouring), and deposition of conformal coatings occur [15,32−37]. Plasmas are acknowledged as unique and effective surface engineering tools due to their:

- Unparalleled range of physical, chemical, and thermal properties allowing the manipulating and truncating of the surface properties with extraordinary precision.
- Prevents destruction of the sample from taking place at low temperatures

3.3.2 Power Requirement for Plasma Reactors

Different types of power supply to generate the plasma in the reactor are as shown in Table 3.4:

Table 3.4 Different types of power supply to generate the plasma

Frequency	Range
Low frequency	50−460 kHz
Radio frequency	13.50 or 27.10 MHz
Microwave	910 MHz or 2.50 GHz

Size of the reactor as well as the kind of treatment desired influences the power required for the reactor which ranges from 10 to 5000 W [38].

3.3.3 Low Pressure Plasmas and Its Use

The low-pressure plasma process is a mature technology designed and developed for the microelectronics industry. However, the requirements of the microelectronics industry are different from those of the textile industry. Low pressure technology is used by many companies for functionalization of textile fibers and nonwovens. To achieve this, the gas is fed to the vacuum vessel and subjected to ionization by a high frequency generator. The vacuum vessel is pumped down to a pressure range of 10^{-2} to 10^{-3} mbar using high vacuum pumps. The advantage of the low-pressure plasma technique is that it is reproducible and controllable [39,40].

3.3.4 Atmospheric Pressure Plasmas

The most common forms of atmospheric pressure plasmas are described below.

3.3.4.1 Corona Treatment Technique

Corona treatment is a highly established and most widely used plasma process. It uses a bright filament from a sharp and high-voltage electrode. It is advantageous due to its operability at atmospheric pressure, with the air as a reagent. The corona treatment system confirms to the requirements of the manufacturing of the textile industry in terms of width and speed, but constrained by the type of plasma generated. It doesn't bring about the desired variation in surface functionality in textiles and nonwovens, affects only loose fibers and unable to diffuse to any greater depth into yarn or woven fabric. Corona systems also rely heavily on very small inter-electrode spacing (≈ 1 mm), which don't meet the requirements of a rapid, uniform treatment [41−44].

3.3.5 Dielectric Barrier Discharge or Silent Discharge

The dielectric barrier discharge is provided with an insulating (dielectric) cover over one or both of the electrodes and works at high voltage ranging between low frequency AC to 100 kHz. It produces nonthermal plasma with a large number of random arcs formed in between the electrodes. However, these microdischarges are not uniform and bring about nonuniform surface treatments [45].

3.3.6 Atmospheric Pressure Glow Discharge

Glow discharge is a uniform, homogeneous, and stable discharge usually generated by applying radio frequency voltage across two parallel-plate electrodes in the presence of inert gases like He, Ar, and N_2, etc. Atmospheric pressure glow discharge offers an alternative homogeneous cold-plasma source and is similar to vacuum cold-plasma operating at atmospheric pressure [46].

3.3.7 Applications of Cold Plasmas

Cold plasmas are used for applications like polymerization (gaseous monomers), grafting of polymers, deposition of polymers, on suitable substrate with the right choice of gas and process parameters, and plasma liquid deposition in vaporized form can be done using cold plasmas [47,48].

3.3.8 Plasma Technology for the Textile Fiber Industry

Plasma technology is a potential tool to introduce new functionalities in textiles products, like water and stain repellence, hydrophilicity, and conductivity, biocompatibility apart from new or improved mechanical, optical, and tribological properties [49].

The plasma state is a dynamic mix of anions and cations (quasi-neutrality), excited states, free radicals, metastables, and ultraviolet (VUV) radiation in the vacuum region. Surface treatments of materials can be carried out by nonequilibrated plasmas, with the application of electric fields. A mean temperature of the order of about 100,000 K is attained. The atoms and molecules ionize and the degree of ionization ranges between 10^4 and 10^6. The highly reactive particles and radiation gives rise to nano-scaled interaction with substrate material surfaces due to chain scission and crosslinking reactions, radical formation, etching or deposition without compromising the bulk properties of the materials. Textiles and fibers differ in length (micrometer to nanometer range) such as filament or inter-fiber distances, offering higher specific surface areas unlike other materials [50−52].

The nano-scale effect has been used to the advantage of microfibers, with the improvement in their fineness, for example, wicking and softness. Nano-fibers having even higher specific areas are in use, for example, air filtration. The deposition of nano-scaled coatings beginning with few nanometers thickness to cover the entire textile surface can develop a high surface functionality, for example, for adhesion or wettability improvement, while using a very little of the material. Nano particles embedded in a surface coating are used for drug release and photo catalytically induced or antimicrobial interfaces [53]. Nano-porous coatings possess dyeability, antifogging, or nonfouling properties. Plasma technology can, thus, be used to functionalize, etching, and deposition of the substrate surfaces. However, the commercialization on an industrial scale depends on the feasibility of scale-up and economic aspects of the plasma processes [54,55].

3.3.9 Plasma Polymerization

Plasma polymerization has enabled the modification of surface chemistry with widespread industrial applications. Nano-scaled modification of fabrics and fibers has added high value to these products due to the acquisition of properties like water/stain repellence, hydrophilicity, dyeability, abrasion resistance, and biocompatibility keeping the bulk properties unperturbed. Moreover, plasma polymerization is eco-benign, with minimum consumption of material and energy compared to wet-chemical processes [56].

Plasma polymerization processes are radical-induced governed by the composite parameter power input/gas flow W/F, which depends on the macroscopic kinetics concept. This approach assumes that the particle in the active plasma region forms an excited state, such as a radical, and forms a stable product such as a deposition in the passive region on recombination [57]. The correlation between a monomer-dependent activation energy E_a and the mass deposition rates R_m is:

$$\frac{R_m}{F} = G \exp\left(-\frac{E_a}{W/F} \right)$$

as with a reactor-dependent geometrical factor G. An Arrhenius-type straight line plot of above equation is obtained, where its slope is E_a, which validates the adopted approach [58]. Energy below the activation energy, produce films which show swelling behavior, for example, polyethylene oxide-like coatings [59].

3.3.10 Fabrics

Low-pressure plasma polymerization processes have been scaled up to industrial level. Fluorine-free hydrophobic siloxane coatings are obtained from organo-silicone based discharges. Polymer-based siloxane coatings having low surface energies and can be synthesized, for example, with hexamethyldisiloxane (HMDSO) as a monomer. While the hydrophobic properties of plasma polymer of HMDSO coatings reduce, the crosslinking and consequently the mechanical properties, improve [60]. The plasma polymerization impart the hydrophilic character for better wicking, adhesion, and dyeability, leading to regio-selectivity in fibers used for blood dialysis [61].

3.3.11 Fibers

Siloxane-based coatings on fibers are deposited using a HMDSO discharge in a continuous process to obtain a hydrophobic, water-repellent yarn. The continuous process involves each segment of the fiber passing through different domains of the plasma chamber along the given axis and at different radial positions. The deposition with all-side treatment of the fibers is obtained by variations in deposition process parameters for a continuous plasma polymerization process [62]. However, the film chemistry (retention of methyl groups and crosslinking) using a HMDSO discharge depends on the amount of energy invested, which is also influenced by pressure sans the deposition position within the plasma chamber Thus, the continuous process can be used for the deposition of nano-scaled functional plasma polymers on the fiber with given geometry [63].

3.3.12 Scale-up

The retention of functional groups and the degree of crosslinking of plasma-polymerized films is dependent on the amount of energy invested (J/cm^3) within the active plasma region and the degree of interaction among energetic particles during film growth. Radio frequency plasma excitation enables well-defined deposition conditions like plasma expansion and power coupling. The influence of energetic particles can be enhanced by reducing either the pressure or increasing the degree of asymmetry. The smaller RF electrodes cause higher bias potentials [64].

Different reactor geometries with HMDSO discharges can be used for the scale-up to industrial level. For symmetric reactors, the activation energy for HMDSO discharges is in tune with the requirement, unlike asymmetric reactors wherein a shift toward lower energy input was obtained [65,66].

3.3.13 Plasma Copolymerization

The surface properties of biomaterials (e.g., chemistry, wettability, and morphology) are of pivotal importance nurturing their interaction with the surrounding environment and the success of their application in medical devices and the failure of these devices may take place due to bacterial colonization [67].

One of the highly effective antibacterial methods currently is the design of a silver-containing coating. Pathogens containing sulfhydryl functional groups are killed by the monovalent silver. Silver has been proved lethal in killing over 650 pathogens and is a potential threat to gram-negative bacteria, like *Pseudomonas aeruginosa*, as well as gram-positive bacteria, fungi, protozoa, and certain viruses. However, the cytotoxicity of Ag is still under investigation for certain medical applications. It is, therefore, considered that Ag/plasma polymer films may exhibit resistance in bacterial infection by preventing the bacterial colonization on biomaterial surfaces. Such Ag/plasma polymer nano-composite materials, consisting of nano-scaled metal clusters as a filler reinforced within the plasma—polymer matrix, can be deposited using a mixed plasma polymerization/sputtering process. The deposition of such composites, demand a silver cathode, a suitable monomer to develop the desired material properties of the matrix like nonfouling, adhesion to cell and hydrophilicity, etc., as well as a suitable reactor design [68].

The plasma deposition of nano-Ag/plasma polymer films involves the deposition of a plasma polymer matrix and the sputtering of silver atoms. During the discharge favoring ion bombardment of the cathode, lower pressure (0.1 mbar) and higher input power are required for the sputtering of silver. However, such conditions may lead to a rise in monomer fragmentation and reduced functionality of the plasma polymer matrix. Concentration of Ag nano-particles and matrix characteristics of the films can be maneuvered to meet the requirements of the specific applications by controlling the deposition process parameters like power input, pressure, and Ar feed ratio [69]; for example, Ag/poly(ethylene-oxide) (Ag/PEO-like) coatings, which combine the nonfouling properties of PEO that deactivates the proteins and the germicidal properties of silver. The nonfouling nature of plasma-deposited PEO-like coatings is attributed to complete remediation of the adhesion tendency towards four different *P. aeruginosa* bacterial strains when deposited on medical grade PVC. Increasing the Ag-content results in a commensurate reduction of the bacterial colonization.

Other than silver, some other nano-composites used in biotechnology are:

- Au-containing, Teflon like, and quartz-like coatings where the Au nano-particles selectively bind to cancer cells.
- TiO_2 nano-particles as a UV-absorbing, or odor-repellent layer on textiles.

This technique having metal nano-cluster/plasma polymers provides for an in situ immobilization of nano-particles during plasma deposition which prevents the direct handling of the nano-particles and has all the advantages of plasma processing [70].

3.3.14 Plasma Technology: Limitations and Challenges in Polymer and Fiber Processing

Although plasma technology is a highly sought-after, versatile and environmentally friendly technology for a range of textile functionalization, initiatives still have not gained much commercial momentum due to some limitations, such as:

- Discrete relevant applied research.
- Development of plasma systems at a slow pace suiting the needs of industries.
- Belated focus on developing on-line atmospheric pressure plasma systems.
- Less public awareness about the successes and failures of industrial systems.

However, although plasma systems manufacturers undertake quite interesting product development, the work is primarily carried out in a strictly confidential manner and the successes or failures of these activities are not public. More exposure regarding the results and experience will pay off in gathering the momentum in the commercial application of plasma technology [71,72].

3.4 MATERIALS AND METHODS

All types of materials can be subjected to treatment by plasma in principle. However, the factors like the special geometry, manufacturing residuals (sizes) and additives, as well as the water content, have to be considered for treatment at nano-scale using plasma technology—for example, in the case of textile products, the structure of the fiber which includes dimensions like filament, inter-fiber distance, and contact area, etc. Treatment conditions depend on different incident angles of reactive particles and shadowing effects influenced by surface areas which further affect the homogeneity in treatment. The factors like the mean free path (pressure), the lifetime (reaction probability), and the width of the plasma sheath surrounding the substrates determine the nature and type of active particles that reach out to the substrate surfaces for their treatment. Manufacturing residuals (leftovers) on the substrate surfaces, might demand surface cleaning before plasma functionalization is undertaken. Moreover, desorption from substrate surfaces, etching, or sputtering products might substantially affect the composition of the plasma gas [73].

3.4.1 Pressure Range

The pressure range chosen for substrate surface treatments not only strongly affects the functionalization, but also determines the plasma equipment required—for example, for the textile industry, mainly atmospheric pressure plasma systems are preferred to enable a continuous process. Corona discharge treatment in air, used on a commercial scale since the 1950s, is used for surface activation of a range of materials like polymers, textiles, and other flexible substrates with a simple geometry. Corona discharges with high voltage and frequency range (10−20 kHz) consisting of filamentary microdischarges generating hot spots and, thereby, spoiling textile surfaces, however, are heterogeneous in themselves.

The other methods like atmospheric pressure dielectric barrier discharges (DBDs), characterized by high process velocity and encapsulation require a given gas composition as an advanced plasma treatment technique. Low pressure discharges contain energetic species, radicals with longer life, and radiation in the VUV, oxygen and hydroxyl radicals with long life are able to diffuse into the open structures to the greater depth in membranes and fibers [74].

Plasma process can be optimized by variation in pressure. Moderate pressure plasmas (100−1000 Pa) gave optimum cleaning and process environments. Plasma polymerization can best be regulated by using low pressure plasmas with a pressure range of 1−0100 Pa, ascribed to well-defined plasma zones. Sputtering processes are carried out in the end at the pressure range of 0.5—5 Pa to prevent in-flight scattering collisions between background gas and the sputtered atoms [75].

3.4.2 Plasma Reactors

A vacuum chamber with a winding inside is required in the design of a reactor for plasma polymerization in low-pressure plasma semicontinuous treatment for a host of fabrics, nonwovens, membranes, and papers. Large-sized reactors for low-pressure plasma processes are readily available. Plasma is excited at a frequency 40 kHz. Other sources like radio frequency (RF) or microwave (MW) are also available. Reactors typically designed with widths up to 120 cm are loaded with textile materials, closed, evacuated, and run in a semicontinuous process. One major advantage of fiber processing is the possibility to perform unwinding (before) and winding (after) plasma treatment in air by guiding the fibers into the low-pressure plasma region with the help of a differentially pumped sealing system. A modular set-up consisting of coupled plasma chambers facilitates steps such as plasma pretreatment (cleaning), deposition, and posttreatment. The plasma reach and process velocity can be enhanced by moving the fibers many times through plasma zones. Furthermore, different treatment times for different steps like cleaning, deposition, and posttreatment are used. Therefore, both semicontinuous and continuous web and fiber coats are available as pilot-plant reactors for a commercial scale-up initiative [74,75].

The fibers are wound and unwound in air and are moved through a sealing system into the vacuum chamber at a velocity of 100 m/min with the fiber diameters (100 μ to 1 mm). Automatically controlled pumping systems (rotary, roots, and turbo molecular) facilitate a fast evacuation of the reactor chambers. Moreover, batch reactors with variation in geometries and excitation sources (RF, MW, and magnetron sputtering) are employed for the optimization of plasma processes [76].

3.5 PLASMA PROCESSING OF POLYMERS

Traditional wet-chemical processes demand solvents, oxidants such as chromates or permanganates, strong acids or bases for etching, etc., have a huge impact on the environment and are hazards. Further, these processes lack uniformity, diffusivity, and reproducibility. In contrast, the plasma processing technique is a versatile and dry process which can tailor polymers in different forms, namely webs, fibers, and particles, etc., to modify their surface behavioral properties without compromising natural bulk properties. The technological application areas of plasma processed polymers cover the automobile, microelectronics, decoration, packaging, bio-analytical and biomedical industries, among others. A wide range of plasma devices, both for laboratory and industry-scale experiments, have been proposed for the surface treatment of polymers. Corona discharge treatments in air which were used over the past five decades in preference for polymer film production and processing on industrial scale. They have the advantage of operating at atmospheric pressure in air. DBD is used wherein either the work piece rolls or high voltage electrode is used as a dielectric barrier in corona-based treatments [76].

Different configurations of electrodes are in use to treat polymer foils, fabrics, and 3D substrates. Electrode assemblies having several parallel knife edges are used for treating foils with a width of <10 m at a speed of 1−10 m/s, which demands a discharge power of 1−100 kW, and operating frequencies of 10−70 kHz. DBDs using pulsed sources are developed for more uniform treatments. A broad range of atmospheric pressure discharges have been successfully used for various applications for a range of substrates. However, most of the early research shows use of low-pressure cold plasmas by employing direct current, RF, or MW based power sources with or without an additional electric or magnetic field. Such plasmas are characterized by a pressures range of 10−1000 Pa. The commercial plasma systems operate at low frequency (40−450 kHz), radio frequency (13.56 or 27.12 MHz), or MW (915 or 2.45 GHz). The first study on polymers at low pressure was reported in their study entitled "Attachment of amino groups to polymer surfaces by radiofrequency plasmas" by Hollahan and Stafford (1969), wherein they used plasmas of ammonia and mixtures of nitrogen and hydrogen in an inductively coupled reactor to adsorb heparin to polypropylene which was plasma treated with the treatment time range of 3−50 h [77].

Although the chemistry inside various plasma reactors is governed by electron collision processes, even with similar electron energy distribution environments, one does not get the reproducibility in terms of final surface modification outcomes. More specific factors like the electrode configuration, reactor geometries, the external properties, the position of the electrodes or substrate in the reactor, and also individual specificity of the processes should be taken into account for better results. This demands to monitor continuously the plasma parameters given a polymer substrate for the selective application. There is no universal discharge applicable to varied substrates since the chemical nature of the polymer influences plasma—polymer surface exposures.

3.6 PLASMA DEPOSITION OF FLUOROCARBON FILMS

The first studies using tetrafluoroethylene (CF_4) or fluorocarbon and other such feeds-based plasmas were used from the 1970s in microelectronics for dry etching of silicon, SiO_2, and other materials. Since then, areas like plasma and surface diagnostics, deposition kinetics, and fluorocarbon coatings applications have been explored.

Choice of the monomer is important, as it is the precursor of reactive fragments and the film in the plasma. Volatile aliphatic, cyclic and aromatic, fluorocarbon compounds are the right kind of monomers for plasma-based fluorocarbon coatings. These fluorocarbon compounds are able to generate a high density of CFx radicals and are responsible for etching and deposition and, therefore, should be preferred in plasma-enhanced chemical vapor deposition (PE-CVD).

Table 3.5 shows some of the most utilized monomers. The choice of the gas feed in plasma processing techniques is governed by the environmental effect. In the recent past, research efforts have been aimed at the right choice of monomers generating low emission of greenhouse gases in the exhaust to reduce carbon footprint While designing and developing innovative PE-CVD processes in microelectronics industry, it has become indispensable. There is growing evidence that most fluorocarbons-based processes have a global warming potential more than that of CO_2 [78].

3.6.1 Deposition of Nanostructured Thin Films Using TFE Discharges

In continuous discharge (CD) processes the substrate is directly exposed to discharge leading to the direct interaction of active species generated

Table 3.5 Representative list of fluorocarbon monomers utilized

Monomer utilized	Structural formula
Perfluoroalkanes (PFA)	C_nF_{2n+2}
Tetrafluoroethylene (TFE)	$CF_2 = CF_2$
Hexafluoropropylene (HFP)	$CF_2 = CF-CF_3$
Hexafluoropropylene oxide (HFPO)	

by the plasma with the substrate surface. CD deposition kinetics is influenced by both chemical reactions brought about by neutral atoms and cationic species on their interaction. Coatings with variable composition and crosslinking are obtained, while the structure and composition of the precursor (monomer) is no more in existence.

D'Agostino et al. obtained a F/C ratio of two for a very thin film deposited from a $C_3F_6H_2$ discharge. Chemically active species found in fluorocarbon plasmas are CFx radicals, F and C neutral atoms, ions generated by fragmentation of the monomer, and heavier species of recombination reactions (between different fragments and monomeric molecules). All active species are not found in the same excitation states, degree, and range of distribution. It is governed by the process parameters of the plasma discharge. The distribution of active species and its close intimacy reflects upon the effectiveness of the interactions with the substrate in addition to the nature, temperature, frequency, and intensity of ion bombardment and position in the reactor. It is further responsible in regulating the structure and composition of the deposit or the etched/fluorinated layer. Therefore, in CD fluorocarbon plasmas, the fluorine to radicals F/CFx density ratio in the plasma plays a major role in defining the coating of fluoropolymers. Optical emission spectroscopy is a diagnostic tool to probe the species generated in a plasma in a nonintrusive manner. Actinometric optical emission spectroscopy is a tool used to analyze the variation in density of emitting species with variation in the experimental variables [78].

3.6.2 Deposition of Fluorocarbon Films by Modulated Glow Discharges

D'Agostino et al. studied the decomposition trend of tetrafluoroethylene (C_2F_4), and many other halogenated hydrocarbons, with a modulated RF discharge. C_2F_4 shows the high polymerization behavior, and the lowest F/CFx ratio in the analyzed feed compared with other fluorocarbon plasmas. The various energy-exchange processes are found to occur in the discharge. [79].

3.7 AGEING OF PLASMA-TREATED SURFACES

The long-term stability of a plasma-treated surfaces is of vital importance if the sample is not preserved under controlled environmental conditions or deposited with immediate effect for the specific application. It has been investigated by earlier researchers that polymer surfaces with modified chemistry are vulnerable to aging effects when exposed to a nonpolar medium like air [80−82]. The wettability of substrate surface may reduce by the plasma treatment with time. This process called ageing might be attributed to a combination of factors, including:

- A thermodynamically driven reorientation of polar moieties away from the surface into the subsurface.
- A reaction of the surface with the atmospheric agents like oxygen, water vapor, and CO_2.

The surface reorientation or reconstruction is commensurate with the interfacial energy difference between the plasma polymer and its surrounding environment. Substantial reduction in the surface density of functional groups may take place [81]. Greisser et al. have presented a quantitative analysis of surface restructuring as immobile and mobile polar functional groups [83]. However the internalization of the mobile polar groups within the bulk of the nonpolar polymer is not a thermodynamically favorable process, while the polar groups stabilize themselves by forming hydrogen-bonded dimers or micromicellar clusters in the subsurface region [81−83].

It is well known that plasma-treated surfaces react with the atmospheric agents, with a loss of desired surface chemistry. Gerenser [83] showed that on exposing plasma-treated PS to air, 4%−5% of the nitrogen was found to be lost.

3.8 CONCLUSIONS

Plasma-based systems provide the best option for the surface treatment of substrates irrespective of substrate porosity, area, size, and topography, and it is of immense interest in the key area of the processing of polymeric substrates. The conventional systems, for example, had limitations in terms of the maximum substrate thickness and discharge gap between electrode and counter electrode. Systems carrying indirect or remote plasmas are capable of diffusing easily through the dense and thick polymeric substrate surfaces and uniformly modify their three-dimensional network structures. A challenge is posed in the design of a compatible plasma system given the wide surface area (working width) of polymers [81−84].

Low-pressure plasma processes have dominated even where there are very slight economic benefits, as plasma treatments are characterized by unique and distinct capabilities and are eco-benign as well. Plasma-based processes are, therefore, preferred over many traditional wet-chemical processes. The approach to coat nano-scaled deposits without compromising bulk properties is of great interest.

Plasma processing of polymers for acquiring better water and oil phobicity is feasible and is being applied to a range of products that are either in the advanced stages of development or already in the market, and is cost effective for a mass-scale (industrial) process. Further, optimization and automation in the equipment is required to make it affordable. In addition, to broaden the horizons of the commercial applications, the processing volumetric capacity of the equipment will have to be scaled-up.

The low-pressure plasma technique is of special importance for imparting a high degree of lipophobicity to polymeric products and has come off age due to its value-added performance behavior and its advantages in terms of processing and environmental characteristics.

Plasma treatment is characterized by its capacity of selective modification of the exterior of a wide range of substrate surfaces and has many manifestations in the form of improved dyeing, printing, shrink proofing, and consequent chemical finishing. Plasma treatments are therefore looked upon as the most innovative and sought-after approach as a substitute to chlorination stages in finishing and upgrading incumbent processes with special reference to a synergy between economical–ecological optimization.

3.9 FUTURE TRENDS

Controlled drug delivery is an area which is gaining momentum in the field of biotechnology, where plasma processing provides for the synthesis of novel substrates with tremendous control on polymer crosslinking and release rates on swelling. The development of multilayer, gradient reservoir systems which are difficult to obtain by conventional wet-chemical techniques has become possible by plasma processing techniques. Smart textiles that allow wound healing at a faster pace, check infection, or give immediate feedback in tune with physiological responses will be a great achievement in the quest of product developments [82,83]. Hydrophobic and oleophobic coatings are still the most sought-after ones, since wet-chemical treatments are bereft of sufficient rapidity in washing. Thus, crosslinked plasma coatings should further be explored for identifying their applications in the broader perspective. The same holds for hydrophilic coatings with endurance, where state-of-the-art nano-scaled functional plasma coatings that are characterized by reduced ageing and enhanced washing rapidity are being developed. Hence, plasma treatments deal mainly with hydrophobic/oleophobic, hydrophilic functional, antistatic and conductive, antimicrobial and medicinal coatings, as well as multifunctional surfaces in tandem with textile trends.

A further trend will demand strong, thermally stable, and highly abrasion-resistant textiles, where plasma treatments can make a substantial contribution. The comfort of functional textiles from the user's point of view becomes a priority. For example, integrating electronics into clothing is an innovative concept, which opens up new horizons in the form of a host of multifunctional, wearable electro-textiles for sensing/monitoring body functions, delivering communication facilities, data transfer, individual environment control, and many other applications. The true potential of plasma technology for wool finishing processes gave a fillip to the development of state-of-the-art machinery providing for a treatment at ambient conditions for various fiber complexions. Thus the integration of plasma technology with the wool industry is a big step forward in the further development toward mass-scaled machinery, providing for a cost-effective treatment with great regards to a high material throughput, highly efficient but tailored material for accomplishing special effects [25,85].

Plasma technology will emerge as an established and cost-effective industrial solution provider technique on a wide scale [86].

REFERENCES

[1] N. Abidi, E. Hequet, Cotton fabric graft copolymerization using microwave plasma. I. Universal attenuated total reflectance−FTIR study, J. Appl. Polym. Sci. 93 (2004) 145.

[2] N. Abidi, E. Hequet, Polymer Grafting and Crosslinking, J. Appl. Polym. Sci. 98 (2005) 896.

[3] G. Akovali, N. Dilsiz, Plasma polymerization of hexamethyldisiloxane in the gas phase. I. Polymerization studies, Polym. Eng Sci. 30 (8) (1990) 485.

[4] L. Andreozzi, V. Castelvetro, G. Ciardelli, L. Corsi, M. Faetti, E. Fatarella, et al., Free radical generation upon plasma treatment of cotton fibers and their initiation efficiency in surface-graft polymerization, J. Colloid. Interface. Sci. 289 (2005) 455.

[5] R. Barni, C. Riccardi, E. Selli, M.R. Massafra, B. Marcandalli, F. Orsini, et al., Characterization of a low pressure supersonic plasma jet, Plasma, Process. Polym. 2 (2005) 64.

[6] R. Benerito, T. Ward, D. Soignet, O. Hinojosa, Modifications of Cotton Cellulose Surfaces by Use of Radiofrequency Cold Plasmas and Characterization of Surface Changes by ESCA, Text Res. J. 51 (1981) 224.

[7] N.V. Bhat, Y.N. Benjamin, Surface Resistivity Behavior of Plasma Treated and Plasma Grafted Cotton and Polyester Fabrics, Text Res. J. 69 (1) (1999) 38.

[8]   E.J. Blanchard, E.E. Graves, R.M. Reinhardt, Dyeable durable-press cationic cotton, Col. Ann. (1999) 55.
[9]   Z. Cai, Y. Qiu, C. Zhang, Y. Hwang, M. McCord, Preliminary Investigation of Atmospheric Pres-sure Plasma-Aided Desizing for Cotton Fabrics, Text Res. J. 73 (8) (2003) 670.
[10] P. Chaivan, N. Pasaja, D. Boonyawan, P. Suanpoot, T. Vilaithong, Environmentally-Friendly RF Plasma Treatment of Thai Silk Fabrics with Chitosan for Durable Antibacterial Property, Surf Coat Tech. 193 (2005) 356.
[11] M. Charbonnier, M. Alami, M. Romand, Metallization of Polymers 2, J. Electrochem. Soc. 143 (2) (1996) 472.
[12] J. Chen, Surface modification of textiles by plasma treatments, J. Appl. Polym. Sci. 42 (1991) 2035.
[13] Y. Hiraguchi, K. Nagahashi, T. Shibayama, T. Hayashi, T. Yano, K. Kushiro, et al., Effect of the distribution of adsorbed proteins on cellular adhesion behaviors using surfaces of nanoscale phase-reversed amphiphilic block copolymers, Acta Biomater. 10 (2014) 2988−2995.
[14] B.B. Shotorbani, E. Alizadeh, R. Salehi, A. Barzegar, Adhesion of mesenchymal stem cells to biomimetic polymers: a review, Mater. Sci. Eng. C. 71 (2017) 1192−1200.
[15] Z.-Y. Qiu, C. Chen, X.-M. Wang, I.-S. Lee, Advances in the surface modification techniques of bone-related implants for last 10 years, Regen. Biomater. 1 (2014) 67−79.
[16] J. Chen, Atmospheric pressure plasma treatments, J. Appl. Polym. Sci. 62 (1996) 1325.
[17] L.E. Cruz-Barba, S. Manolache, F. Denes, Polymer Surface Modification: Relevance to Adhesion, Volume 5, Langmuir 18 (2002) 9393.
[18] R. D'Agostino, F. Cramarossa, S. Benedictis, Diagnostics and decomposition mechanism in radio-frequency discharges of fluorocarbons utilized for plasma etching or polymerization, Plasma Chem. Plasma Process. 2 (1982) 213.
[19] F. Denes, et al., Surface functionalization of polymers under cold plasma conditions - A mechanistic approach, J. Photopol. Sci. Tech. 10 (1) (1997) 91−112.
[20] R. Ganapathy, et al., Primary amine implantation onto polypropylene surfaces from melamine and urea RF plasmas, J. Photopolym. Sci. Tech. 9 (2) (1966) 181.
[21] J. Goodman, The formation of thin polymer films in the gas discharge, J. Polym. Sci. 44 (1960) 551.
[22] J. Hautojarvi, S. Laaksonen, Plasma treatments of fibers and textiles, Text Res. J. 70 (5) (2000) 100−107.
[23] F. Hochart, R. De Jaeger, J. Levalois-Grützmacher, Graft-polymerization of a hydrophobic monomer onto PAN textile by low-pressure plasma treatments, Surf. Coat Tech. 165 (2003) 201.
[24] H. Höcker, H. Thomas, A. Küsters, J. Herrling, Dyeing of plasma treated wool, Melliand Textilberichte 75 (1994) 506.
[25] L.J. Gerenser, in: M. Strobel, C. Lyons, K.L. Mittal (Eds.), Plasma Surface Modification of Polymers, VSP, Utrecht, 1994, pp. 43−64.
[26] X. Wang, Overview on biocompatibilies of implantable biomaterials, in: R. Pignatello (Ed.), Advances Biomaterials Science and Biomedical Applications, InTech, 2013.
[27] M. Saini, Implant biomaterials: a comprehensive review, World J. Clin. Cases 3 (2015) 52.
[28] H. Höcker, Plasma treatment of textile fibers, Pure App. Chem. 74 (3) (2002) 423.
[29] I. Holme, Proceedings of 10th International Wool Textile Research Conference, Aachen, (2000).

[30] M. Kamel, B.M. Youssef, G.M. Shokry, Plasma Technologies for Textiles, J. Soc. Dyers Color 114 (1998) 101.

[31] M.F. Maitz, Applications of synthetic polymers in clinical medicine, Biosurf. Biotribol. 1 (2015) 161—176.

[32] C.W. Kan, K. Chan, C.W.M. Yuen, M.H. Miao, The Impact of Nitrogen Plasma Treatment upon the Physical-Chemical and Dyeing Properties of Wool Fabric, J. Mater. Process. Technol. 83 (1998) 180.

[33] J. Kang, M. Sramadi, Plasma Technologies for Textiles, AATCC Rev. 4 (11) (2004) 29.

[34] M.S. Kim, T.J. Kang, Plasma Technologies for Textiles, Text Res. J. 72 (2) (2002) 113.

[35] K.T.M. Tran, T.D. Nguyen, Lithography-based methods to manufacture biomaterials at small scales, J. Sci. Adv. Mater. Devices. 2 (2017) 1—14.

[36] T. Zhou, Y. Zhu, X. Li, X. Liu, K.W.K. Yeung, S. Wu, et al., Surface functionalization of biomaterials by radical polymerization, Prog. Mater. Sci. 83 (2016) 191—235.

[37] Y. Zhao, K.W.K. Yeung, P.K. Chu, Functionalization of biomedical materials using plasma and related technologies, Appl. Surf. Sci. 310 (2014) 11—18.

[38] S. Kobayashi, T. Wakida, S. Niu, S. Hazama, C. Doi, Y. Sasaki, Plasma Technologies for Textiles, J. Soc. Dyers Color 111 (1995) 111.

[39] H. Lee, et al., Plasma Technologies for Textiles, Surf. Coat Tech. 468 (2001) 142—144.

[40] J. Liao, C. Chen, Y. Wu, C. Weng, Modification of Alkanethiolate Self-Assembled Monolayers by Free Radical-Dominant Plasma, Plasma Chem. Plasma Process. 25 (3) (2004) 255.

[41] M.R. Massafra, G.M. Colonna, B. Marcandalli, E. Occhiello, Proceedings of the 17th IFVTCC Congress, Wien, (1996) June 6—7.

[42] M.G. McCord, Y.J. Hwang, Y. Qiu, L.K. Hughes, M.A. Bourham, Surface analysis of cotton fabrics fluorinated in radio-frequency plasma, J. Appl. Polym. Sci. 88 (2003) 2038.

[43] R. Molina, P. Erra, L. Julià, E. Bertran, Plasma Technologies for Textiles, Text Res. J. 73 (11) (2003) 955.

[44] M. Möller, 2006. Proceedings of the 3rd International Conference of Textile Research Division, Cairo. K.W. Oh, S.H. Kim, E.A. Kim, J. Appl. Polym. Sci. 81 (2001) 684.

[45] T. Öktem, H. Ayham, N. Seventekin, E. Piskin, Plasma Technologies for Textiles, J. Soc. Dyers Color 115 (1999) 274.

[46] T. Okuno, T. Yasuda, H. Yasuda, Effect of Crystallinity of PET and Nylon 66 Fibers on Plasma Etching and Dyeability Characteristics, Text Res. A new approach for dyeability of cotton fabrics by different plasma polymerisation methods, J. 62 (8) (1992) 474.

[47] E. Özdogan, R. Saber, H. Ayahn, N. Seventekin, J. Soc. Dyers Color 118 (2002) 100.

[48] J. Park, I. Henins, H.W. Hermann, An atmospheric pressure plasma source, Appl. Phys. Lett. 76 (2000) 288.

[49] A. Raffaele-Addamo, C. Riccardi, E. Selli, R. Barni, M. Piselli, G. Poletti, et al., Plasma Technology for Deposition and Surface Modification, Surf. Coat Tech. 174—175 (2003) 886.

[50] A. Raffaele-Addamo, E. Selli, R. Barni, C. Riccardi, F. Orsini, G. Poletti, et al., Plasma Technologies for Textiles, Appl. Surf. Sci. 252 (2006) 2265—2275.

[51] W. Rakowski R. Osella O. Demuth Proceedings of the 17th IFVTCC Congress, Wien, (1996) June 6—7.

[52] M. Ghaffari, S. Moztarzadeh, F. Rahmanian, A. Yazdanpanah, A. Ramedani, D. Mills, et al., Nanobiomaterials for bionic eye: vision of the future, in:

A. Grumezescu (Ed.), Engineering of Nanobiomaterials, first ed, Elsevier, 2016, pp. 257–285.

[53] E.C. Ranger, W.C.A. Bento, M.E. Kayama, W.H. Schreiner, N.C. Cruz, Plasma Technologies for Textiles, Surf. Interface Anal. 35 (2003) 179.

[54] C. Riccardi, R. Barni, M. Fontanesi, B. Marcandalli, M. Massafra, E. Selli, et al., Waterproof and Water Repellent Textiles and Clothing, Plasma Sour. Sci. Technol 10 (2001) 92.

[55] A.M. Sarmadi, Y.A. Kwon, Plasma Technologies for Textiles, Text Chem. Color 93 (1993) 25.

[56] E. Selli, G. Mazzone, C. Oliva, F. Martini, C. Riccardi, R. Barni, et al., 300 Plasma technologies for textiles, J. Mater. Chem. 11 (2001) 1985.

[57] E. Selli, C. Riccardi, M.R. Massafra, B. Marcandalli, Plasma Technologies for Textiles, Macromol. Chem. Phys. 202 (2001) 1672.

[58] L. Shi, Plasma Technologies for Textiles, Reactive Funct. Pol. 45 (2000) 85.

[59] S. Sigurdsson, R. Shishoo, Surface properties of polymers treated with tetrafluoro-methane plasma, J. Appl. Polym. Sci. 66 (1997) 1591.

[60] M. Šimor, J. Ráhel, M. Černák, Y. Imahori, M. Štefečka, M. Kando, Surface Behavior of Polyamide 6 Modified by Barrier Plasma in Oxygen and Nitrogen, Surf. Coat Tech. 172 (2003) 1.

[61] A. Sparavigna, Plasma technologies for textiles, Recent Res. Devel Appl. Phys. 5 (2002) 203–222.

[62] R.B. Stone Jr, J.R. Barrett Jr, Lignocellulosic Fibers and Wood Handbook: Renewable Materials, Textile Bull 88 (1962) 65.

[63] Y.C. Tyan, J. Liao, Y. Wu, R. Klauser, Assessment and characterization of degradation effect for the varied degrees of ultra-violet radiation onto the collagen-bonded polypropylene non-woven fabric surfaces, Biomaterials. 23 (2002) 67.

[64] Y.C. Tyan, J. Liao, S.P. Lin, Surface properties and in vitro analyses of immobilized chitosan onto polypropylene non-woven fabric surface using antenna-coupling microwave plasma, J. Mater. Sci. Mater. Med. 14 (2003) 775.

[65] G.W. Urbanczyk, B. Lipp-Symonowicz, S. Kowalska, A Novel Green Treatment for Textiles: Plasma Treatment as a Sustainable, Melliand Textilberichte 11 (1983) 838.

[66] T. Wakida, K. Takeda, I. Tanaka, T. Takagishi, Free Radicals in Cellulose Fibers Treated with Low Temperature Plasma, Text Res. J. 59 (1989) 49.

[67] T. Ward, O. Jung, R. Benerito, Characterization and use of radio frequency plasma-activated natural polymers, J. Appl. Polym. Sci. 23 (1979) 1987.

[68] J. Yip, K. Chan, K.M. Sin, K.S. Lau, Sustainable Apparel: Production, Processing and Recycling, J. Mat. Proces. Technol. 123 (1) (2002) 5.

[69] N.S. Yoon, Y.J. Lim, M. Tahara, T. Takagishi, Mechanical and dyeing properties of wool and cotton fabrics treated with low temperature plasma and enzyme, Text Res. J. 66 (1996) 329.

[70] T. Yuranova, A.G. Rincon, A. Bozzi, S. Parra, C. Pulgarin, P. Albers, et al., Antibacterial textiles prepared by RF-plasma and vacuum UV mediated deposition of silver, J. Photochem. Photobiol. A: Chem. 161 (1) (2003) 27.

[71] J. Zhang, P. France, A. Radomyselskiy, S. Datta, J. Zhao, W. van Ooji, Hydrophobic cotton fabric coated by a thin nanoparticulate plasma film, J. Appl. Polym. Sci. 88 (2003) 1473.

[72] H. Zubaidi, T. Hirotsu, Graft plomerization of hydrophilic monomers onto textile fibres treated by glow discharge plasma, J. Appl. Polym. Sci. 61 (1996) 1579.

[73] E. Kissa, Functional finishes Part B, in: M. Lewin, S.B. Sello (Eds.), Handbook of Fibre Science, Vol. II, Marcel Dekker, New York, 1984.

[74] A. Grill, Cold Plasma Materials Fabrication: From Fundamentals to Applications, IEEE Press, Piscataway, NJ, 1994.

[75] H. Biederman, Plasma Polymer Films, Imperial College Press. Papers, London, 2004.

[76] S.R. Coulson, S.A. Brewer, C.R. Willis, J.P.S. Badyal, GB Pat. Patent: WO 98, (1997) 58117.

[77] S.R. Coulson, I.S. Woodward, J.P.S. Badyal, S.A. Brewer, C.R. Willis, Plasmachemical functionalization of solid surfaces with low surface energy perfluoro-carbon chains, Langmuir 16 (2000) 6287−6293.

[78] S.R. Coulson, I.S. Woodward, J.P.S. Badyal, S.A. Brewer, C.R. Willis, Super-repellent composite fluoropolymer surfaces, J. Phys. Chem. B 104 (2000) 8836−8840.

[79] S.R. Coulson, I.S. Woodward, J.P.S. Badyal, S.A. Brewer, C.R. Willis, Ultralow sur-face energy plasma polymer films, Chem. Mater. 104 (2000) 2031−2038.

[80] M. Morra, E. Occhiello, F. Garbassi, in: J.J. Pireaux, P. Bertrand, J.L. Brédas (Eds.), Polymer−Solid Interfaces, Galliard Ltd, Great Yarmouth, Norfolk, Great Britain, 1991, pp. 407−428.

[81] H.J. Griesser, Y. Da, A.E. Hughes, T.R. Gengenbach, A.W.H. Mau, Langmuir 7 (1991) 2484−2491.

[82] K.S. Siow, L. Britcher, S. Kumar, H.J. Griesser, Plasma methods for the generation of chemically reactive surfaces for biomolecule immobilization and cell colonization - A review, Plasma. Process. Polym. 3 (2006) 392−418.

[83] R.C. Chatelier, X. Xie, T.R. Gengenbach, H.J. Greisser, Quantitative analysis of polymer surface restructuring, Langmuir 11 (1995) 2576.

[84] C.I. Justino, A.C. Duarte, T.A. Rocha-Santos, Critical overview on the application of sensors and biosensors for clinical analysis, TrAC Trends Anal. Chem. 85 (2016) 36−60.

[85] L. Wang, B. Sun, K.S. Ziemer, G.A. Barabino, R.L. Carrier, Chemical and physical modifications to poly(dimethylsiloxane) surfaces affect adhesion of Caco-2 cells, J. Biomed. Mater. Res. A. 93 (2010) 1260−1271.

[86] M.R. Wertheimer, L. Martinu, J.E. Klember-Sapieha, in: K.L. Mittal, A. Pizzi (Eds.), Adhesion Promotion, Techniques, Marcel Dekker, New York, 1999, p. 139.

FURTHER READING

E. Kissa, Fluorinated Surfactants and Repellent, Marcel Dekker, New York, 2001.

CHAPTER 4

Plasma Assisted Polymer Modifications

Gomathi Nageswaran[1], Lavanya Jothi[1] and Saravanakumar Jagannathan[2]

[1]Department of Chemistry, Indian Institute of Space Science and Technology, Thiruvananthapuram, Kerala, India
[2]JN-UTM Cardiovascular Engineering Centre, Faculty of Biosciences and Medical Engineering, Universiti Teknologi Malaysia, Johor Bahru, Malaysia

4.1 INTRODUCTION

Polymers have attracted considerable attention from various industries, including automotive, aerospace, medical, building, and packaging, etc., due to their excellent properties such as low density, high flexibility, elasticity, conductivity, strength, ease of processing, low bulk material cost, and processing cost. Easy and cost effective processes allow tailoring of polymer properties to fit the needs of their application and make them appealing for use in advanced application fields. Despite the excellent properties of polymeric materials, surface properties are of paramount importance in designing material for various applications. Polymers are usually selected for a given application primarily because of their desirable bulk properties such as thermal stability, mechanical strength, solvent resistance, ease of processability, and cost, etc. For instance, the following are the materials selected for various applications for their favorable bulk properties: (1) material used for hip joint selected for their mechanical strength as they must perform under high stress for many years, (2) materials for vascular prosthesis selected as they are compliant and strong, and (3) polymeric membranes selected for their low affinity to separated molecules and their ability to withstand the harsh cleaning environment. However biological responses to these biomaterials and the performance of these polymeric membranes are influenced by surface properties of the material. The selected polymers for various biomedical applications and separation processes, etc., often have poor surface characteristics that are less than optimum for the intended application due to its inertness. As it is quite difficult to find a material to meet the needs of both the desired

Non-Thermal Plasma Technology for Polymeric Materials
DOI: https://doi.org/10.1016/B978-0-12-813152-7.00004-4

bulk and surface properties, it is common practice to fabricate a material with the required bulk properties for the intended application and subject it to a special treatment to enhance the surface properties subsequently according to the requirement. Surface modification techniques striving to retain favorable bulk properties of polymers involve modification of the outermost layer of a polymer by inserting some functional groups and changing the surface morphology in order to cater to specific needs. It is usually performed to improve its interaction with the biological environment to increase its bio/blood compatibility, barrier properties, wettability, printability, and adhesive properties, while striving to retain favorable bulk properties of the polymer. Among the various surface modification techniques available, plasma treatment is the most suitable method due to its advantages, such as controlled surface modification to a depth of few μm to Å, formation of multilayer films, deposition of highly crosslinked polymers irrespective of geometries, eco-friendly dry method, and low temperature, etc. Various active species generated in plasma are sufficiently energetic to break the covalent bonds of polymers. Surface modification of a polymer by plasma treatment is a well-established method used in various applications including textiles, biomedical applications, and membrane separation, etc. [1−3].

This chapter will give an introductory overview of various plasma treatment methods such as plasma etching, plasma activation and functionalization, and plasma-induced grafting, etc., which cause the surface modification of polymers. Insight into the interaction between the plasma species and polymer surface considering various plasma gases, such as argon, argon-oxygen, and nitrogen and helium, effects of types of polymer surfaces especially polypropylene and polycarbonate as plasma sources are discussed in detail. Surface changes caused by plasma treatment on polymers are reversible to some extent. Hence the ageing effect of plasma-modified polymers is also discussed in this chapter. Various applications of plasma-modified polymers are explored in the final section.

4.2 PLASMA TREATMENT

Plasma, a fourth state of matter, often thought of as a subset of gases, differ from gases and behave differently since it is not simply made up of atoms, but consists of electrons and positively charged ions that are formed when some or all of the electrons are stripped away. The degree of ionization of gases determines the type of plasma such as cold plasma or thermal plasma. A low degree of ionization leads to the formation of cold plasma while nearly full ionization leads to the formation of thermal plasma. Cold plasma is a collection of various species including electrons, positively charged ions, excited atoms, metastables, reactive radicals, and photons and can be used for the surface modification of polymers. This partially ionized gas consists of an equal number of positive ions and electrons and thus maintains electro neutrality and has a very low degree of ionization in the range of $10^{-3}-10^{-4}$. So, the cold plasma consists mostly of neutrals such as excited atoms or excited molecules and metastables. Cold plasma is used to cause various changes on polymer surface such as cleaning, etching, functionalization, grafting, and polymerization.

4.2.1 Plasma Cleaning

Plasma cleaning of a polymer surface is achieved by removing material by physical sputtering process or chemical reaction to form volatile products such as CO_2, H_2O, and low molecular weight hydrocarbons, as shown in Fig. 4.1. Various contaminants such as weekly bonded surface layers, oxide layers, and finger prints, etc., can be removed from surfaces. Inert gases such as argon and neon are used for plasma cleaning by physical sputtering. Argon gas ionized up on applied voltage is accelerated towards the substrate by the applied electric field and transfers the energy through elastic and inelastic collisions with the substrate surface atoms. The surface atoms that acquired enough energy will leave the surface, resulting in cleaning. Argon plasma generated at a low pressure has neutral

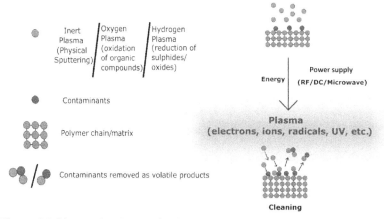

Figure 4.1 Plasma cleaning mechanism.

metastable lives for longer duration of 55.9 s existing in high concentration bombards the polymer surface and causes surface modifications. Due to the longer lifetime of the metastables they can travel long in remote/afterglow plasmas. Low-pressure noble gas plasmas emit vacuum ultraviolet radiation which causes crosslinking in the subsurface layers. Argon gas has limited diffusivity and, therefore, it cannot penetrate deeper into the substrate to reach the subsurface layers. Helium plasma, on the other hand, has a high diffusivity and can diffuse deeper. However, if plasma cleaning is performed for a longer duration then, besides the removal of the first layer of substrate surface, the underlying layers will also be removed which will result in etching or the degradation of the polymer surface. Contaminants such as organic contaminants, oxides, or sulphides can be removed by using reactive gases such as oxygen or hydrogen. Organic contaminants are removed by oxygen plasma through oxidation of the contaminants, while oxides or sulphides can be removed by reducing the contaminants using hydrogen plasma.

4.2.2 Plasma Etching

Plasma etching is a process for removing materials. Depending on the nature of the plasma gas used (inert or reactive gas) etching is accomplished either by physical or chemical etching, as depicted in Fig. 4.2. In physical etching, ion bombardment through directional momentum transfer causes physical sputtering of atoms. It has the advantage of anisotropic etching, but has the limitation of poor selectivity. Chemical etching

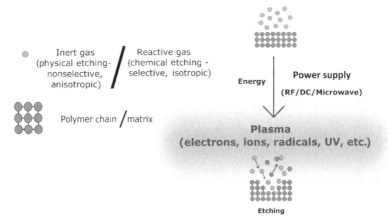

Figure 4.2 Plasma etching mechanism.

involves transport of reactive species through diffusion followed by reaction with the substrate atom. Chemical etching is isotropic due to the diffusion of reactive species occurring in all directions. However, it has very good selectivity. It is believed that etching at higher pressure is due to the neutrals since the ion energies are low, while etching at lower pressure is due to the ions. However, neutrals are responsible for almost all etching for a wide range of pressure (0.13–0.66 Pa) except in the case of reactive ion etching where the ions react with, and remove, the substrate material. Selectivity, uniformity, and directionality of etching and etching rate are the parameters used to evaluate the quality of the plasma etching process. Etching rate depends on the density of the reactive radicals available. Plasma density is determined by the plasma process parameters such as input power, process pressure, and gas flow rate. Excitation frequency affects the discharge characteristics and etching characteristics by changing the spatial distribution of species, electric field across the discharge and electron energy distribution. Evenness of etching across a substrate and the degree of etching rates maintained between the substrates in the same reactor are the measures of uniformity of etching. Unevenness across the substrate is actually caused by depletion of gas phase radicals locally which can be overcome by maintaining uniform plasma density. Etching rate variations between the substrates is caused by poor reactor geometry and nonuniform plasma composition arises from variations in the gas flow or nonuniform discharge current flowing. Plasma etching that results in directional or anisotropic etching are capable of producing a straight wall etching profile.

Oxygen plasmas are commonly used in polymer etching which can simultaneously cause modification of the polymer surface. Atomic oxygen free radicals are the predominant species in the plasma region initiating the etching process. Etching rate is highly influenced by the type of etching gas used. Etching rates are strongly influenced by the chemical reactivity of the plasma species generated, which varies with the different gases used to generate plasma. In addition to the chemical reactivity of the plasma species, the type of polymers also greatly influences the etching rate. For instance, when exposed to oxygen plasma different polymers such as poly(vinyl acetate), poly(methyl methacrylate), low density polyethylene, high density polyethylene, poly(ethylene terephthalate), poly (butylene terephthalate), polystyrene, and natural rubber exhibit etching rates in a decreasing order. Plasma etching of polymers is commonly used in the removal of polymeric photoresists in integrated circuits.

Anisotropic plasma etching transfers even submicron features into polymer films, which forms high resolution lithographic masks.

In plasma etching it is possible to selectively etch the polymers from a composite. Selective oxygen plasma etching of pigment-polymer composite based on the selective interaction of reactive gaseous particles from oxygen plasma was reported [4]. Inductively coupled oxygen plasma resulted in extremely selective etching of the polymer matrix leaving the particles unaffected. In inductively coupled plasma, the density of neutral particles is large and has a higher order of magnitude (five orders of magnitude) when compared with the density of charged particles. Selective etching was achieved by interaction of neutral plasma radicals which are actually not accelerated by the electric field, therefore, possess very low kinetic energy ($<$0.1 eV). The neutral particles are capable of distinguishing between the different materials. The reaction probability of the neutral particles depends on the binding energy of atoms in solids since even a small difference in binding energy causes a substantial difference in etching probability. The etching is isotropic since the radicals are approaching the surface of the composite from all the directions. In oxygen plasma etching, atomic oxygen initiates the etching process. Plasma etching initiated by atomic oxygen involves abstraction of hydrogen atoms from the polymer surface, in addition to unsaturated moieties and absorption of oxygen dissociation energy [5]. Abstraction by atomic oxygen for saturated or unsaturated polymers does not result in bond-weakening, but addition of oxygen to the unsaturated molecule leads to the formation of saturated radical resulting in weakening the $C-C$ bond. Etching rate of polymers by oxygen plasma can be increased by increasing the density of atomic oxygen in the plasma. Addition of N_2 or N_2O increases the concentration of atomic oxygen in plasma. The additive gases do not change the surface mechanism since they do not alter the activation energy associated with etching. Activation energy associated with etching by atomic oxygen in plasma after addition of N_2 or N_2O gases remains same as the activation energy of etching by atomic oxygen species generated in pure oxygen plasma without any additive gases. Lower etching rate of pure oxygen plasma can also be increased through addition of fluorine containing gases, such as CF_4, $C2F_6$, SF_6, and NF_3, to the oxygen gas feed. The density of atomic oxygen species is increased by addition of fluorine-containing gases through electron induced dissociation process. The major role of fluorine-containing gases in increasing the atomic oxygen density is through increased number density of electrons or the energy of electrons. Though the increased etching rate with the addition of fluorine-containing gases is through the enhancement of oxygen atom density, fluorine can also take part in the reactions to generate polymer radical sites through abstraction or additional reactions with unsaturated polymer. Enhancement of the etching rate of pure oxygen plasma can be obtained either by elevating the temperature or with the assistance of ion bombardment. Besides oxygen atom, various other energetic species in plasma, such as ions, metastables, and photons, can also result in etching. The energy of ions range between 20 eV and several hundred eV, which is sufficient to induce polymer chain rupture while that of metastables are several times greater than the typical polymer bond energies. VU and VUV radiations possess very high energy which is greater than the ionization potential of polymer. Concentration of various energetic species vary with the type of gas used. Noble gases emit strongly VUV and the metastables, which results in increased bondability of polymers through the process called CASING (crosslinking by activated species of inert gases). Enhancement of etch rate is high with ion/photon bombardment and elevated temperatures compared to atomic oxygen due to the relatively small rate constant for the reaction of atomic oxygen with polymers to form radicals. To increase the reaction rate of atomic oxygen with polymers, the flux of atomic oxygen has to be increased. Oxygen atoms form hydroxyl groups on reacting with saturated hydrocarbons, while forming polymer radicals and excited intermediates with unsaturated hydrocarbons and phenols with aromatic hydrocarbons. Molecular oxygen, though, does not initiate etching of polymers, it reacts with polymer radical sites (formed during etching initiated by ion bombardment) and leads to scission of polymer backbone chains. Therefore, the plasma parameters should be maintained in such a way to increase the flux of atomic oxygen, increase the dissociation of oxygen molecule, and decrease the recombinative loss of atomic oxygen.

4.2.3 Plasma Functionalization

Plasma functionalization of polymers is a process which neither removes nor deposits atomic or molecular layers, but is done to make chemically inactive polymers active through the attachment of special, most polar, and chemically reactive groups, such as O-, N-, S-, or halogen-containing groups. The degree and type of functionalization during plasma functionalization are determined by three major components, namely plasma chemistry (influenced by type of plasma gas used), plasma polymer interaction (influenced by both types of plasma and polymer) and postplasma reaction, as illustrated in Fig. 4.3. Attachment of functional groups to the polymer depends on the plasma gas used which results in the attachment of the main element, or a characteristic group, to the carbon radical sites at the polymer surface. This plasma gas-specific functionalization can attach a broad spectrum of functional groups containing mother elements from the plasma gas and oxygen obtained from postplasma reactions. Specific functionalization of polymers is achieved by plasma species which is influenced by the nature of the plasma gas. Plasma treatment of polymers involves fragmentation of plasma precursor gas as a first step and formation of active species through various collision processes. Various energetic plasma species, such as excited, ionized, and fragmented species, on reaching the polymer surface form the plasma gas specific functionalization. Plasma precursor gases used may be either an inert gas or a reactive gas. For functionalization with oxygen or hydroxyl groups, reactive plasma precursor gases such as O_2, O_2/O_3, H_2O, and O_2/H_2O_2 and sometimes even noble gases like Ar, are used [6−8]. Functional group attachment with Ar plasma is mainly due to the postplasma treatment reactions. Nitrogen-containing groups are obtained by N_2, NH_3, and N_2/H_2 plasma. Sometimes the reactive gases may be mixed with argon, an inert gas, to enhance the oxygen and nitrogen

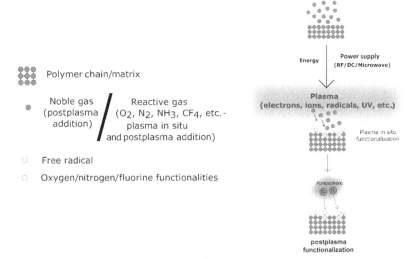

Figure 4.3 Plasma functionalization mechanism.

functionalities through enhanced free radical formation where more functional groups can be attached [9].

Attachment of functional groups is done through hydrogen abstraction initiated by UV in the plasma region followed by attachment of plasma fragments and polymer radicals. In aliphatic hydrocarbon, hydrogen atoms from the backbone or side chains is abstracted and is replaced by a heteroatom, the main element of plasma gas, or by plasma fragments. This substitution process involves recombination and simultaneous chain scission processes. Plasma functionalization sometimes has very poor selectivity which can be marginally improved by optimizing plasma parameters, lowering the wattage, and using pulsed plasma or remote plasma. When the plasma conditions are severe, the electron temperature will be high and will result in increased fragmentation of monomers and the destruction of the polymer substrate surface due to plasma particle bombardment, while low-energy plasma retains the structure and composition of monomer and substrate. Plasma-induced radical formation plays an important role in plasma functionalization as the radicals generated by plasma from monomer fragmentation initiates the reactions. Continuous bombardment of the polymer surface with high energy particles and irradiation with high energy vacuum UV radiation result in a variety of products, degradation, fragmentation, crosslinking, and postplasma oxidation, etc. UV photons can result in photo oxidation when the energy of radiation (with a wavelength of 200—400 nm) is in the range of binding energies in polymers. Photo oxidation reactions can be made more specific by applying the specific wavelength of the radiation. When the plasma-activated polymer is exposed to the atmosphere, oxygen from air initiates auto-oxidation.

4.2.3.1 Functionalization With Oxygen-Containing Groups

Plasma treatment for 2 s of polyolefin (polyethylene or polypropylene) in continuous wave radiofrequency discharge with 100 W power is sufficient to form a nearly saturated level of oxygen functional groups at the surface. Exposure of polyolefin to plasma after 2 s results in only a marginal increase in the concentration of oxygen functional groups in the sub surface layers that are not accessible to XPS analysis. Penetration of reactive oxygen species deeper into the subsurface layers of polymer and the resulting postplasma auto-oxidation of these plasma irradiated subsurface layers are beyond the accessibility region for XPS analysis. Further exposure to plasma results in polymer degradation and etching.

Functionalization of polyolefin with oxygen plasma involves: (1) functionalization of uppermost layer to a higher percentage of oxygen-containing groups (18%−24%) during an exposure time of <2 s; (2) penetration of oxygen reactive species to deeper into the layers beyond XPS analysis depth; and (3) degradation, etching and formation of oxygen functionalities initiated by far-ultraviolet radiation of plasma [10]. The type of plasma influences the plasma functionalization. Functionalization of polyethylene with oxygen-containing groups is faster with atmospheric pressure oxygen plasma when compared to low pressure oxygen plasma which produces double bonds and crosslinking. Plasma jet results in deposition of traces of metal oxides, also as a result of the sputtering of the electrode material.

4.2.3.2 Functionalization With Nitrogen-Containing Groups

Attachment of amino groups are important for biomedical applications as amino acids play an important role in nature. Ammonia plasma is commonly used to attach amino groups onto polymer surfaces such as polycarbonate, polyvinyl chloride, polytetrafluoroethylene, and polyethylene, etc. Ammonia plasma treatment of polymers involves a recombination reaction between NH_2 radicals formed by hydrogen abstraction of NH_3 and carbon radicals formed simultaneously by hydrogen abstraction from the polymer backbone. Dissociation of ammonia in ammonia plasma is not only due to the primary mechanisms such as heavy particle collision and electron collision, but also due to VUV/UV radiation which is more predominant especially in microwave discharge. Continuous discharge of the nitrogen or ammonia plasma sometimes leads to significant weight loss due to etching of the surfaces induced by the energetic species present in plasma, which can be overcome by pulsed plasma. Limitations such as electrical charging and arcing of continuous plasma are overcome by providing an off-time during each pulse in pulsed DC plasma, which also enhances the plasma power during each cycle. Surface damage induced by surface charging and etch profile distortion are prevented by pulsed discharge. Suitable duty cycle helps in minimizing the heat load on the substrate surface. Nitrogen and nitrogen plus hydrogen plasmas created in continuous wave plasma introduces oxygen-containing groups besides nitrogen-containing groups simultaneously during plasma treatment or during post plasma reactions upon exposure to atmosphere. Formation of nitrogen atoms is mainly through electron impact dissociation of N_2 $(e + N_2 \rightarrow e + 2N)$ which increases strongly with 1% H_2 admixture with

N_2 discharge in both DC and microwave plasma, whereas 10% H_2 admixture causes destruction of atomic nitrogen. Hydrogen atoms are produced through various kinetic paths associated with electron impact on hydrogen and collisions with metastable/excited nitrogen molecules or with N_2^+. Hydrogen atoms reduce the nitrogen functionalities and leads to the formation of amino groups. Use of pulsed DC instead of continuous wave RF plasma minimizes the introduction of oxygen groups. The rate of attachment of nitrogen groups by exposure to plasma is slower than that of oxygen groups by exposure to oxygen plasma. Primary amino groups play a significant role in biomedical applications to immobilize biomolecules. Nitrogen plasma in both microwave and RF plasma can introduce primary amino groups to polymer, but at a lower concentration which can be increased by introducing hydrogen gas at a very low concentration (1%) into the nitrogen discharge or by using ammonia plasma. Presence of nitrogen increases the concentration of nitrogen atoms. Changing plasma parameters, type of plasma (continuous wave RF plasma or DC plasma), and different gases such as ammonia or N_2/H_2 plasma, etc., did not significantly change the concentration of primary amino group attached to the pentafluorobenzaldehyde [11]. Among the various intermediate radicals such as $NH_2,$: $NH^{\cdot}$, $N^{\cdot}_{\cdot}$, and $\dot{H}$, primary amino (NH_2) groups are the major dissociation products of ammonia and $N_2 + H_2$ plasma. Excitation of ammonia molecules with an electron is the primary step resulting in the formation of intermediate species such as NH and NH_2. Favia et al. achieved attachment of selective nitrogen functionalities by tuning the relative densities of the reactive species, especially NH radicals [12]. Formation of NH_2 radicals is influenced by discharge power. Plasma power influences the type of radical forming which subsequently influences the plasma polymer surface interaction. McCurdy et al. reported that the formation of NH_2 in ammonia RF plasma increases rapidly with increasing RF power up to 100 W and decreasing further at higher RF power due to the dissociation of the radicals themselves [13]. Therefore, at high power >120 W concentration of: NH is high which does not decrease much at higher power. Steen et al. reported that at high RF powers, ions enhance NH_2 surface generation but suppress the reactions leading to the formation of NH [14]. N_2 molecules which interact with only the free sites of the surface become more reactive when they are reduced by a hydrogen atom to form NH radicals. Increase in amino selectivity in remote plasma is achieved with hydrogen admixtures with ammonia plasma and by a hydrogen plasma posttreatment. Exposure of polymers to discharges of various nitrogen-containing gases such as N_2, NH_3, and $N_2 + H_2$ introduces various nitrogen functionalities including primary, secondary, and tertiary amino groups, as well as imino, nitrile, and amido groups. Incorporation of these functional groups is not only due to the direct plasma polymer interaction, but also due to the postplasma reactions upon exposure to the atmosphere, as shown in Fig. 4.3.

4.2.3.3 Functionalization With Fluorine-Containing Groups

Oxygen-containing plasmas are used to increase the wettability of the polymer, whereas fluorine-containing polymers are used to increase its hydrophobicity. Among the various fluorine-containing gases, such as SiF_4, BF_3, NF_3, XeF_2, and $ClCF_3$, the most commonly used gases for fluorination are CF_4 and SF_6. CF_4 and SF_6 cannot react with the polymer in the neutral form. Therefore, the precursors on dissociation produce fluorine atoms which react with the polymer and result in fluorination through substitution with hydrogen atoms or recombination reactions with carbon radicals. CF_4 plasma, a nonpolymerizable gas, produces mainly fluorine atoms and ions and CF_x species at low concentrations. The fluorine radicals result in surface fluorination rather than etching or deposition [15]. Fluorine atoms produced on dissociation of these fluorine-containing gases lead to the fluorination of the polymer backbone by replacing hydrogen atoms from the polymer backbone, which is an exothermic, self-accelerating, chain-propagation process. Since the fluorination of the polymer backbone is a self-accelerating process, it does not need additional plasma energy. Plasma energy is required only to disrupt the fluorine bonds. However, the excess energy associated with the fluorination process results in poor selectivity towards different polymer units and functional groups. Duration of fluorination process is much longer than plasma oxidation due to the smaller size of the fluorine atoms which can easily migrate. Rotational and diffusional migration of fluorinated moieties into the polymer near the surface region subsequently reacts and results in loss of hydrophobicity. Fluorination depth is limited to a few nanometers in the case of plasma fluorination, unlike chemical fluorination which goes up to a depth of about 2 μm. The difference in the penetration depth of fluorine atoms is due to the difference in the diffusivity of fluorine atoms at atmospheric pressure and under vacuum conditions. The chemical nature of a polymer surface influences the degree of surface fluorination. Degree of fluorination achieved by plasma fluorination is higher than that of chemical fluorination. Co-introduction of oxygen with the fluorine-containing precursor gas or postplasma reactions on exposure to the atmosphere results in an auto-oxidation process. Electron acceleration upon power supply to CF_4 gas leads to the formation of fluorine atoms at a high concentration with a smaller concentration of other species, such as CF, CF_2, and CF_3 radicals at low concentrations. Besides the high ratio of F:C in CF_4, vacuum UV radiation also gives rise to excited fluorine atoms. Excited fluorine atoms lead to the formation of various fluorine functionalities, such as $-CF$, $-CF_2$, and $-CF_3$. Plasma fluorination makes the polymer surface hydrophilic through fluorination or perfluorination and hydrophobic through oxy-fluorination. Hydrogen abstraction by fluorine atoms is the initial step which is simultaneously ruptured by vacuum UV radiation. Fluorine-containing gases, such as tetrafluoromethane, are also used as a dopant gas along with molecular oxygen to enhance the production of atomic oxygen. Atomic fluorine helps in enhancing etching via hydrogen abstraction. Hydrogen abstraction by fluorine is more efficient than by oxygen as the F−H bond is stronger than O−H bond.

4.2.4 Plasma Surface Activation and Grafting

Plasma surface activation and grafting, or plasma induced grafting, is a two-step process of plasma surface activation followed by the incorporation of functional groups and reactive sites to the surface activated polymer, as illustrated in Fig. 4.4. Plasma surface activation is achieved by exposing the polymer to a noble gas to form the free radicals on the polymer surface which can initiate the grafting process and subsequently exposes the plasma-activated polymer surface to the suitable reaction environment for the incorporation of functional groups. Free radicals are created in the polymer chains on exposure to plasma through the inelastic collision between electrons in the plasma and the polymer surface. Free radical density on the surface of the plasma-activated polymer is crucial as this determines the degree of grafting. The low density of free radicals results in incomplete grafting, while high free radical density results in loss of reactivity due to recombination/termination reactions as the radicals are in close proximity and easily accessible to each other.

Exposure of the plasma-activated polymer surface to the suitable reaction environment for grafting, can be achieved through the following processes: (1) direct contact of plasma-activated surface with monomers from gas phase; (2) immersing the plasma-activated surface directly into a solution with a reactive agent; or (3) by exposure to

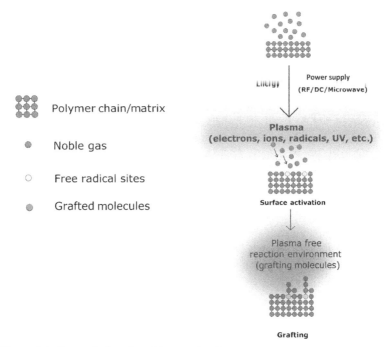

Figure 4.4 Plasma induced grafting mechanism.

oxygen. The following are the different grafting reaction environment reported after plasma activation of the polymer surface. Plasma induced vapor phase grafting polymerization (PIVPGP) of acrylic acid on cotton fabrics was reported to enhance silver nanoparticle loading on the fabric which exhibited excellent laundry durability. During PIVPGP, oxygen plasma was used to activate the cotton fabrics in the first step and subsequently, surface activated cotton fabric was exposed to acrylic acid from the vapor phase which was supplied to the same chamber without plasma generation [3]. Shourgashti et al. used oxygen plasma to activate the surface of polyurethane film and exposed the plasma induced polyurethane for subsequent grafting reaction to the solution of vinyl siloxy terminated polydimethysiloxane polymer [16].

The plasma-activated polymer, thus exposed to the reaction environment, can be grafted by the following three different ways:

1. Functional groups available on the polymer surface will react with the reactive groups or the end groups of molecules, oligomers, and polymers available in the reaction environment.
2. Grafting onto carbon radical sites.
3. Grafting onto postplasma formed radicals.

Plasma-induced grafting is different from plasma polymerization where the monomer itself is used as the plasma gas leading to polymerization. Grafting can be accomplished by either directly using the monomer as a plasma gas or indirectly by activating the surface by noble gas plasma followed by exposure to the monomer. When the monomer is directly used as plasma gas, care must be taken to retain the structure of the monomer by controlling the plasma process parameters such as low power, high pressure to obtain a low mean free path and shorter residence time of the monomer. Plasma activation can also be achieved by using atmospheric pressure plasma which reduces the cost of the vacuum system significantly. Plasma activation is usually accomplished by using noble gases, such as argon, helium, or nitrogen, for a duration of a few seconds which is sufficient to form free radicals on the polymer surface. The free radicals formed on the surface of the polymer stay alive for several days. Plasma forms the free radicals only on the surface. Therefore, the grafting is limited to the near-surface layer only. Plasma grafting was earlier performed to improve the wettability and dyeability of acrylic acid and acrylic amide for fibers and textiles. Plasma grafting is highly beneficial for biomedical applications due to the homogeneous finish of the surface with chemical groups and their distribution. Various groups generated on polymer surface are carboxyl, amine, hydroxyl, and thiol, among others.

4.2.5 Plasma Polymerization

Plasma polymerization is a special means of preparing an ultrathin film of unique polymers retaining many of the bulk properties of the polymer such as permeability, electric volume resistivity, and dielectric constant, etc., which is true for the films as long as their thickness is above the critical value of $0.05-0.1$ μm. Plasma polymerization may be plasma induced polymerization or plasma state polymerization, as illustrated in Fig. 4.5. Plasma-induced polymerization is similar to plasma activation and grafting where, instead of the attachment of functional groups, polymerization occurs. Plasma creates free radicals on the surface and they initiate polymerization. Therefore, in this process plasma is considered as a radiation source which generates free radicals. The free radicals formed on the surface of the polymers exposed to the plasma initiate graft polymerization in a similar way to preirradiation grafting. The polymerization reaction occurs through direct addition of degassed monomers under vacuum conditions. The necessary chemical reaction for polymerization in

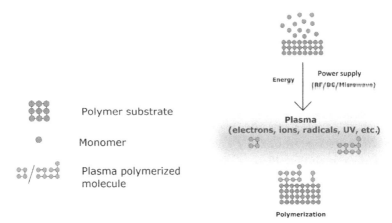

Figure 4.5 Plasma polymerization mechanism.

plasma-induced polymerization is considered to be the conventional molecular polymerization that occurs without plasma. Therefore, it differs from plasma state polymerization in which polymerization occurs only under plasma conditions as the monomer vapor is in the plasma state. The nature of the substrate polymer and the monomer that is to be polymerized influence the efficiency of graft polymerization in conventional graft polymerization. The same monomer which grafts well onto one polymer substrate may not graft at all onto another type of polymer in conventional grafting, especially when it is initiated by chemical means. Efficiency of graft polymerization depends on the free radical yield of polymer substrate and monomer. Plasma polymerization is independent from the nature of the polymer substrate. Similar deposition during plasma polymerization can be obtained on different substrates such as glass, organic polymers, and metals. Therefore, the choice of material does not restrict surface modification by plasma polymerization.

4.3 PLASMA-POLYMER INTERACTIONS

Electrons receive energy from the applied electric energy and are accelerated by the applied electric field until they gain sufficient energy to cause electron ejection or bond breaking resulting in the formation of positive ions, molecular fragments, or free radicals respectively. Collision processes during plasma generation in the gaseous phase between the species may be elastic or inelastic which is determined by the changes in internal energies of the particles. Neutral particles have kinetic and different forms of potential energy such as excitation, ionization, and dissociation. Elastic collision is a process where there is no internal energy change. When an electron collides with any atom, transfer of energy is negligible due to the great difference in the masses of the colliding particles. During this process, the electron only changes its momentum without significantly changing its kinetic energy. Elastic collisions restrict the velocity of electrons in the direction of the field. Inelastic collisions can create different types of new particles. Different types of reactions are responsible for the generation of new particles, such as electron-induced, photo-induced, ion-induced, and neutral particle-induced reactions. Classification of various collision processes between different particles are depicted in Fig. 4.6. During the collision process, atoms or molecules are excited. The excited atoms lose their energy through a de-excitation process in a short time ($\leq 10^{-8}$ s) or they may collide with other particles or on the wall of the chamber if they have lifetime of $\geq 10^{-3}$ s. Ionization happens when sufficient energy is provided to strip away the loosely bound electron from the atom. Plasma, consisting of equal numbers of positive ions and electrons, gains a positive charge as the electrons escape to the surroundings due to their higher mobility than that of ions. Positive ions accelerated

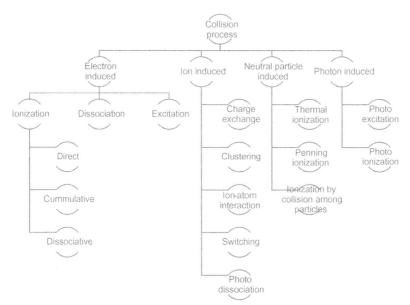

Figure 4.6 Various collisions processes forming plasma species interacting with polymers.

from the plasma towards the substrate surface bombard the surface. At low pressure, the energy is dissipated during ion bombardment while at high pressure it is dissipated as heat due to the collision between particles. The energy of plasma arrives from various components, such as highly reactive free radicals, UV radiation, and photos produced upon relaxation of electronically excited plasma molecules. The following are the effects of ion bombardment on the surface [17]:

1. Increasing threshold energy.
2. Desorption of adsorbed gases.
3. Activation of chemical reactions between the adsorbed gaseous molecules or with the surface of the substrate.
4. Momentum transfer with the surface atoms and their removal.
5. Crystallographic disorder of the top-most layers or deeper layers due to intermixing of atoms or ion implantation respectively.
6. Anisotropic etching which proceeds only in the direction of ion arrival.

Highly reactive, neutral free radicals with dangling bonds that diffuse out of plasma may result in the following effects:

1. Chemisorption resulting in a surface layer containing the chemisorbed gases.
2. Continued chemisorption resulting in chemical vapor deposition.
3. Etching.

Particle matter interaction occurring in the substrate depends on the energy distribution of neutral and metastable species, positive and negative ions, electrons and photons, particle dynamics and the nature of the collisions in the substrate region. Plasma sheath, a positively charged layer, is formed near the surface of the solid substrate where plasma meets the substrate surface due to the larger thermal velocity of electrons than the thermal velocity of ions. Ions accelerated with a sheath field collide with liquid or solid polymer surfaces resulting in several physical events, such as elastic collisions, recombination of ions, and orientation of dipole groups constituting a polymer surface. Trapping of ions by CH_2 groups and charging of the polymer surface which stops when the electric field produced by the charged polymer surface attains the sheath field [18].

Bombardment of heavier particles on the surface of the polymer causes bond-breaking and reorganization of $C-C$ and $C-H$ bonds. Free bonds formed on the surface from hydrogen desorption may recombine with the neighboring bonds or with the free radicals from the gaseous phase which produces crosslinked structures. Hydrogen abstraction, carbonate cleavage, dearomatization, ether bond breakage, double bond formation, and hydroxyl group formation have been reported as the major surface chemistry changes observed in polycarbonates treated with various plasmas, such as argon, argon-oxygen, nitrogen, and helium [19,20,21,22]. Argon-oxygen and nitrogen plasma treated polycarbonate introduced aromatic aldehyde and amide group. Similarly, hydrogen abstraction, double bond formation, and incorporation of hydroxyl and carbonyl groups were the major surface chemistry changes observed in the case of polypropylene. NH_2^+, NH stretching vibration with nitrogen plasma and $C-O-C$ stretching with helium plasma were also observed in polypropylene. Reactive oxygen species causing chemical etching resulted in the emission products such as CO^+ and CO_2 which were identified using optical emission spectrum from the argon-oxygen emission spectrum. Two noble gases, namely argon and helium, result in different etching rates and etching behavior. Argon plasma caused more homogeneous etching while helium plasma produced more enhanced, deeper, and localized etching. During plasma polymerization, the deposition rate decreases exponentially with temperature increases. Charged particles, UV radiation, and excited and chemically active neutrals also significantly contribute to the deposition rate. However, it is very difficult to say which plasma component plays a predominant role in the deposition rate.

In general when cold plasma interacts with a polymer surface it results in changes in the surface properties such as surface energy, wettability, adhesion, surface electric resistance, dielectric loss tangent, dielectric permittivity, catalytic activity, tribological parameters, gas absorption, and permeability characteristics [23]. Plasma surface modification of polymers is done for various applications, such as for the efficient painting of textiles, printing on synthetic wrapping materials, enhancing the biocompatibility of polymeric material, and to increase the adhesive properties of polymers, etc. Both low pressure and high pressure or atmospheric pressure plasma result in various surface modifications such as chemical functionalization by the former and surface cleaning or changing wettability by the later. All the major plasma components, including electrons, ions, excited/metastable particles, atoms, free radicals and UV radiation, cause surface modifications to produce free radicals on the polymer surface, unsaturated organic compounds, crosslinking between the polymer chains, and volatile products—mostly hydrogen. Formation of free radicals on the polymer surface is due to the impact and UV irradiation.

4.4 INFLUENCE OF THE TYPE OF POLYMER

Physical and chemical modifications of polymers, induced by energetic species of plasma such as ions, free radicals, electrons, excited atoms or molecules, metastables and UV not only depend on the plasma gas used, but also on the type of polymer being treated. The influence of the type of polymer on plasma surface modification is discussed in detail with reference to two polymers—polypropylene and polycarbonate. The effect of RF plasma surface modification of polyethylene and polycarbonate and various physico-chemical changes occurring during plasma treatment studied by Nageswaran et al. are depicted in Tables 4.1 and 4.2 [20−25]. It was performed for different plasma gases such as argon, oxygen, a mixture of nitrogen and argon, and helium plasma to study the effect of various plasma parameters, such as RF power, pressure, gas-flow rate, and treatment time on surface functionalization measured as surface energy and etching measured as weight loss. Increased surface energy was reported for both polycarbonate and polypropylene. Among the two components of surface energy, including the polar component corresponding to the hydrophilicity and the dispersion component corresponding to crosslinking, polarity increased significantly with both polycarbonate and polypropylene. Presence of oxygen-containing groups make polycarbonate relatively polar compared to polypropylene. But after plasma treatment at optimized conditions, the polarity of polypropylene was more than that of polycarbonate due to the etching effect of active species which might have broken oxygen containing groups. However, plasma-treated polycarbonate exhibited increased polarity compared to the polarity of untreated polycarbonate. Among the various plasma-treated polycarbonates, nitrogen plasma resulted in higher surface energy of 42.79 mN/m, followed by argon, argon-oxygen, and helium plasma treated polycarbonate. In case of polypropylene, the highest surface energy of 38.69 mN/m was found with argon plasma, followed by helium, nitrogen, and argon-oxygen plasma treated polypropylene. The increase in polarity of polycarbonate with various plasma treatments were all in the same range (5.14−5.86 mN/m) which was slightly wider in the case of polypropylene (5.41−8.46 mN/m). The etching rate of oxygen-containing polymers are generally higher compared to hydrocarbon polymers which means that the etching rate of polycarbonate will be higher compared to that of polypropylene and, thus, the polar component of polypropylene is higher than that of polycarbonate. So, it can be

Table 4.1 Effect of process variables on surface energy and weight loss of various plasma treated polypropylene

Variables	Argon plasma		Argon-oxygen plasma		Nitrogen plasma		Helium plasma	
	Surface energy	Weight loss	Surface energy	Weight loss	Surface energy	Weight loss	Surface energy	Weight loss
Power	Increasing	Increasing	Not significant	Decreasing up to 65 W then increasing	Increasing up to 110 W then decreasing	Increasing except at 110 W	Decreasing up to 110 W then increasing	Increasing
Pressure	Increasing up to 175 mTorr then decreasing	Increasing	Not significant	Decreasing	Decreasing up to 125 mTorr then increasing	Increasing except at 150 mTorr	Decreasing	Decreasing
Flowrate/ O2%	No change	Decreasing	Not significant	Increasing up to 16% O$_2$ then decreasing	Increasing	Increasing except at 12 sccm	No change	Increasing
Treatment time	Increasing	Increasing	Not significant	Increasing	Increasing up to 6 min then decreasing	Increasing except at 6 min	Increasing	Increasing

Table 4.2 Effect of process variables on surface energy and weight loss of various plasma treated polycarbonate

Variables	Argon plasma		Argon-oxygen plasma		Nitrogen plasma		Helium plasma	
	Surface energy	Weight loss	Surface energy	Weight loss	Surface energy	Weight loss	Surface energy	Weight loss
Power	Decreasing	Increasing	Decreasing	Increasing	Increasing up to 65 W then decreasing	Increasing	Decreasing up to 110 W then increasing	Increasing
Pressure	Increasing	Increasing	No change	Decreasing	No change	No change	Increasing	Increasing up to 16.7 Pa then decreasing
Flowrate/ O2%	No change	Increasing	No change	Decreasing	Decreasing	Decreasing up to 15 sccm then increasing	No change	No change
Treatment time	No change	Increasing	Decreasing	Increasing	Decreasing up to 6 min then increasing	Increasing	Decreasing	Increasing

concluded that functionalization was predominant in polypropylene whereas etching was predominant in polycarbonate. However, the plasma process variables also influence the degree of functionalization and etching. Since plasma polymer interaction is a complex process it is difficult to generalize the trends of the effect of process variables on surface energy and weight loss. Highest weight loss in the case of polycarbonate was found with nitrogen plasma and then by argon–oxygen plasma due to reactive species like atomic oxygen and nitrogen. For both polymers, the etching rate by argon plasma and, hence, the weight loss, were less. Higher weight loss of polypropylene with helium and nitrogen are due to the higher penetrating power of helium metastables, or excited atoms, and nitrogen plasma. Various active species responsible for functionalization as well as etching in argon and helium plasma were identified from an optical emission spectrum as excited atom and ion. Argon-oxygen and nitrogen plasma produced a patterned etching in polycarbonate at a specific combination of 110 W, 20% O_2, 16.7 Pa and 6 min and 110 W, 19.99 Pa, 25 sccm and 6 min respectively resulted, as shown in Fig. 4.7 [22]. Microdepression caused by multiple interactions during ion sputtering which is subsequently surrounded by the dispersed deposited material, produced patterned etching. Formation of microdepressions is influenced by the physical properties of the polymer as well as plasma-processing parameters. When the micro discharges interact with the polymer surface it results in the patterning of a ripple formation around the striking point. No pattern formation was observed with polypropylene. Oxygen-containing polymers degrade more compared to hydrocarbon polymers. High energy ions in argon plasma and vacuum UV radiations in helium and nitrogen plasma promoting crosslinking also degrade hydrocarbon polymers, such as polypropylene.

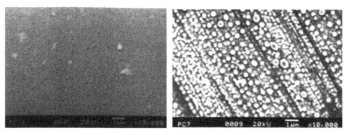

Figure 4.7 Scanning electron micrograph of untreated and argon-oxygen plasma-treated polycarbonate [22].

The effect of process variables on surface energy and weight loss of polypropylene and polycarbonate are depicted in Tables 4.1 and 4.2 respectively with the process conditions of power being $20-200$ W, pressure $100-200$ mTorr, flow rate $4-20$ sccm or $4\%-20\%$ O_2, and a treatment time of $2-10$ min for all the different conditions. It is very difficult to generalize the trend of theses process variables due to the complexity of the interaction of plasma species with polymer surfaces. An increase in surface energy with increasing power is due to the large number of active species in argon plasma. In nitrogen plasma, surface energy of polypropylene increased up to 110 W and then decreased due to more predominant effect of weight loss than functionalization. In helium plasma, surface energy initially decreased due to the cleaning effect and then increased due to the radical formation. Reactive oxygen species form argon-oxygen plasma caused an increase in surface energy initially due to the incorporation of polar groups during which the decrease in weight loss was reduced. Weight loss increased at a higher power level due to the etching effect. Higher density of plasma species at high pressure and decreased mean free path at low pressure cause positive and negative effects on surface energy. Similarly, increased active species at lower flow rates with sufficient ionization and poor ionization due to gas impedance at higher flow rates cause an increase or decrease in surface energy respectively. Increased surface energy with increasing treatment time due to more polar groups being incorporated in the case of argon and helium plasma treated polypropylene and increased surface energy up to 6 min treatment time and then decreased surface energy due to increased etching in the case of nitrogen plasma treated polypropylene, were observed. The effect of plasma process conditions on surface energy and weight loss of polycarbonate caused slightly different behavior when compared to polypropylene. Argon and argon-oxygen plasma resulted in increased surface energy at low power due to polar group incorporation and a decrease in surface energy due to etching. Nitrogen plasma resulted in increased surface energy, which reached saturation level at 65 W. But in helium plasma, decreased surface energy up to 110 W and then increased surface energy, was observed. The effect of pressure on surface energy in the case of noble gases was found to increase with increasing pressure due to the more active species with higher ion density that interacts with the polymer surface, while there was no change in the case of argon and argon-oxygen plasma. Surface energy of polycarbonates was not influenced by flow rate of argon and helium gases, but flow rate of nitrogen gas decreased surface energy due to poor ionization at higher flow rate.

4.5 PLASMA SOURCES

The desired plasma surface modification of polymers, which depends on the purpose of application, is obtained using different discharges of plasma. Cold plasma used for polymer surface modification is generally obtained by certain types of gas discharges at low pressure.

Glow discharge is a type of plasma that is generated when DC voltage is applied to the two electrodes inside a glass tube containing precursor gas in the pressure range of $0.1-10$ Torr. The type of plasma gas used and the process pressure to be maintained determines the minimum voltage required to maintain the discharge. There are different types of glow discharges such as hollow cathode discharge, penning glow discharge, and magnetron discharge used for sputter deposition. Glow discharge has potential application in surface modification of polymers. Atmospheric pressure glow discharge of pure nitrogen, nitrogen-hydrogen, or nitrogen-ammonia mixtures for surface modification of polyethylene and polypropylene were reported to increase their surface energy with only 10% hydrophobic recovery [26,27] reported increased wettability of various polymers, such as low-density polyethylene, polypropylene, polyethylene terephthalate, and polytetrafluoroethylene and textile DynemaSB21 and DynemaSB51 by atmospheric pressure argon glow discharge.

High frequency discharge or RF discharges are widely applied for surface modification and film formation due to the large volume of stable plasma. High frequency discharge operated at a frequency >300 kHz is supplied to the load by means of an external electrode which reduces the effect of the electrodes and is called electrodeless discharge. Depending on the type of the coupling of the electrode to the load it may be classified as: (1) inductively coupled plasma, or (2) capacitively coupled plasma. RF discharges are generally operated at a frequency of 13.56 MHz, allowed by the international communication authority. Plasma characteristics such as density and composition influence the characteristics of the film deposited. Nonthermal RF discharge is subdivided into two types, namely low-pressure discharges and moderate-pressure discharges. Electron relaxation length in low-pressure discharge is comparable with the discharge size, unlike under moderate pressure where it is much smaller. Electron energy distribution function is determined by the electric field distribution in the entire region in the low-pressure discharge. In moderate pressure discharge it is determined by the local electric field.

Modification of surface physical and chemical properties without affecting bulk properties through RF plasma modification is advantageous for the design, development, and manufacture of biocompatible polymers. The RF plasma surface modification of polymers' biomedical applications has been the topic of extensive investigations pertaining to a wide range of applications. RF nitrogen plasma modification of polypropylene investigated at different process conditions for enhancement of bio and blood compatibility was found to be effective in bringing about significant changes in the surface properties, including surface chemistry and surface morphology [25].

4.6 AGEING OF PLASMA-MODIFIED POLYMERS

Surface modifications obtained by plasma treatment are permanent which changes with time and storage conditions due to the reorientation of molecules. Various polymers, including polycarbonate, polyethylene, polypropylene, and polystyrene, etc., when treated with argon, helium, or nitrogen plasma are reported to cause etching, crosslinking, and activation of polymer surfaces [28]. They result in improved wetting and friction properties of polymers and activation of polymer surfaces. However, a serious problem of potential ageing of treatment effects is associated with plasma surface modification. Various studies reported hydrophobicity recovery of plasma-treated polymers over time [29–31]. The degree of hydrophobicity recovery of plasma-treated polymers varies with the type of polymer and plasma gas used. Ageing effects of nitrogen plasma treated polycarbonate is less compared to other plasma treatments including argon, oxygen, and mixtures of nitrogen. Composition of plasma gas influences the ageing effect, as reported by Ren et al. [32] Ageing effect of helium plasma treated ultrahigh modulus of polyethylene was brought down by adding 1% of oxygen which is more efficient in bringing down the hydrophobic recovery when compared to the mixture of helium plus 2% oxygen, which showed accelerated ageing. Hydrophobic recovery is due to the reorientation of polar groups introduced during plasma treatment and the diffusion of the bulk molecules to the surface [33]. Hydrophobicity of plasma treated polydimethylsiloxane (PDMS) is reported to recover due to the diffusive burial of polar groups in the bulk and condensation of silanol groups formed by plasma treatment and consequent crosslinking on the surface of plasma modified PDMS [34]. Decreasing the molecular mobility through crosslinking to prevent the reorientation of molecules in the modified surface layer is one of the possible ways to retard hydrophobic recovery reported by Behnisch et al. [31] Saturation of functional groups in the surface layer by repeated oxygen plasma treatment is another way to hinder hydrophobic recovery.

4.7 APPLICATIONS OF PLASMA-MODIFIED POLYMERS

Plasma surface modification of polymers has major applications in the enhancement of the adhesive properties of polymers. The effect of argon and oxygen plasma treatment on the adhesive properties of polyethylene has been studied [35]. Plasma treatment enhances wettability by increasing surface energy. Increased surface roughness through etching or ion bombardment results in an anchoring effect. Both the increased wettability and increased surface roughness improve the adhesive property of the polymer. Various oxygen-containing functional groups, such as carbonyl, carboxyl, and hydroxyl groups, are introduced with both oxygen and argon plasma. However, the additional anchor effect obtained in argon plasma due to the increased surface roughness makes it more efficient than oxygen plasma. Argon plasma produces a large number of free radicals which causes increased chain scission resulting in increased chain mobility due to decreased crystallinity. Incorporation of oxygen containing polar functional groups is reported to be more efficient with argon plasma which also results in a smoother surface due to homogeneous etching [36].

Polymers used in various applications, including packaging, aircraft, appliances, biomedical, and automotive industries, though, have superior bulk properties that require suitable surface properties, for instance low friction coefficient and hydrophilicity. Especially polymers used for biomedical devices are selected for their desirable mechanical strength or stability in the body. However, their poor surface properties lead to problems associated with surface-induced thrombosis and poor cell adhesion. Plasma surface modification, a dry, environmentally friendly process, is limited to a depth of few microns, does not leave residues on the polymer surface, and does not generate any chemical waste polluting the environment. Improved wettability, printability, bondability, and biocompatibility are achieved by nitrogen plasma surface modification of the polymer surface [37]. Nitrogen plasma treated polypropylene undergoes surface chemistry with the introduction of new functional groups such as ($-NH_2$), imine (CH NH), imine (CH NH), cyano ($-CN$), or nitrile, as well as oxygen-containing functional groups and morphological changes due to the increased surface roughness [25].

Plasma treatment of polymers has wide applications due to the improvement of adhesive properties. Plasma-treated polypropylene and polycarbonate using various plasma gases such as argon, argon oxygen, nitrogen, and helium were found to increase its adhesive property according to a mechanical interlocking mechanism through increased surface roughness which increase anchoring sites and remove weak boundary layers. In addition to the incorporation of polar groups and increased surface roughness, strengthening of weak boundary layers through crosslinking also resulted in improved adhesive property in the case of helium plasma. Plasma surface modification of polymers has potential application in biomedical applications. Increased hydrophilicity or hydrophobicity helps in enhancing the performance of the biomaterial. Unaffected cell morphology and increased cell adhesion and cell spreading, as shown in Figs. 4.8 and 4.9, indicating enhanced biocompatibility in plasma-treated polymers was due to increased protein adsorption. Protein adsorption to the surface happens first, even when a foreign substance is in contact with the biological environment, which can also enhance blood compatibility indicated by the reduced platelet adhesion, as shown in Figs. 4.10 and 4.11. Various polar groups such as hydroxyl, carbonyl, amide, amine, and aldehyde on various polymers can help in enhancing bio- and blood-compatibility of plasma-treated polymers. Increased surface roughness and hydrophilicity help in enhancing bio- and blood-compatibility.

The antireflective and/or water repellent property of polymers through nanostructuring of polymers using plasma treatment has been reported [38]. Antireflective and water-repellent properties are essential in various applications such as architecture, photovoltaic, and automotive.

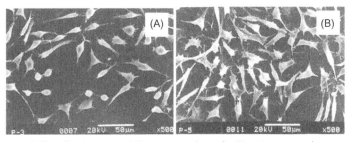

Figure 4.8 Cell adhesion on (A) untreated and (B) nitrogen plasma treated polypropylene.

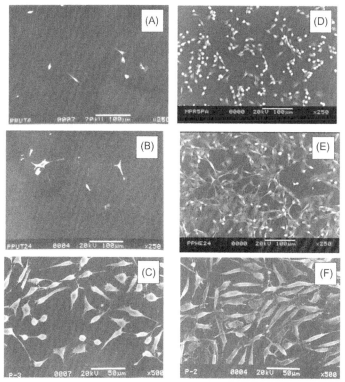

Figure 4.9 Cell adhesion on polypropylene control after (A) 4 h, (B) 24 h, (C) 72 h, and helium plasma treated polypropylene after (D) 4 h, (E) 24 h, and (F) 72 h.

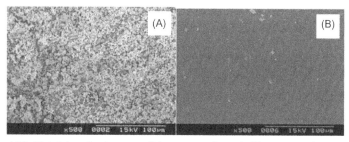

Figure 4.10 Platelet adhesion on (A) untreated and (B) argon plasma treated polypropylene.

Role to roll dual magnetic plasma etching is reported to increase the anti-reflective and water-repellent properties. Spatially varied properties on the surface of a material arising from the gradient surface has potential application in the biomedical field. Wettability gradient surface on a

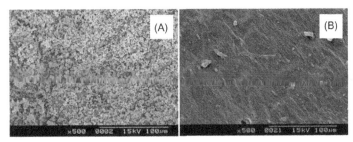

Figure 4.11 Platelet adhesion on (A) untreated and (B) nitrogen plasma treated polypropylene.

polypropylene surface was obtained by using O_2 plasma followed by SF_6 plasma treatment under a masked surface [39]. Increasing the power of radiofrequency corona discharge apparatus with a knife-type electrode formed a wettable and rough gradient polyethylene surface [40]. Surface roughness gradient on polyethylene terephthalate using dielectric barrier discharge plasma apparatus with asymmetric electrode configuration which results in inhomogeneous distribution of active species leads to the formation of three different zones, namely the central zone, diffuse zone, and boundary zone [41]. Surface roughness in these zones are different. Conductive polymer nanostructures exhibiting unique electrical properties have great importance in the new developments for field emission displays and heat dissipation. Enhanced surface interaction of polymer nanostructures has the possible application in gas sensing. Microwave plasma produced at electron cyclotron resonance specially designed to allow the operation at low precursor pressure and high plasma homogeneity was used for synthesis of polyaniline and the surface was further modified by argon plasma to create a nanodot textured surface by controlling various plasma parameters [42].

4.8 CONCLUSION

The basic concepts of plasma surface modification of polymers have been summarized in this chapter. It was shown that various plasma treatments of polymers, such as etching, activation, functionalization, plasma-induced grafting, and plasma polymerization, can result in the enhancement of wettability, adhesive property, biocompatibility, and printability, etc. The role of the active species causing surface modification and their interaction with the polymer surface has been discussed, which is crucial in determining the nature and degree of surface modification. The effect of the type of plasma and polymer on physic-chemical properties of plasma-treated polymer are discussed with reference to polypropylene and polycarbonate. It is our opinion that plasma surface modification of polymers has been, and will continue to be, a vital and state-of-the-art technique to enhance the success of modern surface engineering of materials.

REFERENCES

[1] Z.-P. Zhao, A.-S. Zhang, X.-L. Wang, P. Lu, H.-Y. Ma, Controllable modification of polymer membranes by LDDLT plasma flow: grafting acidic ILs into PPF membrane for catalytic performance, J. Memb. Sci. (2018). Available from: https://doi.org/10.1016/J.MEMSCI.2018.02.044.

[2] K. Orihara, A. Hikichi, T. Arita, H. Muguruma, Y. Yoshimi, Heparin molecularly imprinted polymer thin film on gold electrode by plasma-induced graft polymerization for label-free biosensor, J. Pharm. Biomed. Anal. 151 (2018) 324−330. Available from: https://doi.org/10.1016/J.JPBA.2018.01.012.

[3] C.X. Wang, Y. Ren, J.C. Lv, Q.Q. Zhou, Z.P. Ma, Z.M. Qi, et al., In situ synthesis of silver nanoparticles on the cotton fabrics modified by plasma induced vapor phase graft polymerization of acrylic acid for durable multifunction, Appl. Surf. Sci. 396 (2017) 1840−1848. Available from: https://doi.org/10.1016/J.APSUSC.2016.11.173.

[4] U. Cvelbar, M. Mozetic, M. Klanjsek-Gunde, Selective oxygen plasma etching of coatings, IEEE Trans. Plasma Sci. 33 (2) (2005) 236−237. Available from: https://doi.org/10.1109/TPS.2005.845345.

[5] F.D. Egitto, Plasma etching and modification of organic polymers, Pure Appl. Chem. 62 (9) (1990) 1699−1708. Retrieved from http://citeseerx.ist.psu.edu/viewdoc/download;jsessionid = ADE7DA6D560139E8CD9C8A37C8DC5AD4?doi = 10.1.1.503.330&rep = rep1&type = pdf.

[6] V. Andre, F. Tchoubineh, F. Arefi, J. Amouroux, Surface treatment of PP films by a non equilibrium low pressure plasma of NH3, N2, Ar, Plasma-Surface Interactions and Processing of Materials, Springer Netherlands, Dordrecht, 1990, pp. 507−510. Available from: https://doi.org/10.1007/978-94-009-1946-4_34.

[7] H. Wittrich, J. Gähde, J. Friedrich, E. Schlosser, G. Kaiser, 1972. Verfahren zur Herstellung von verstärkten Polymerformstoffen. DD-WP, 106052.

[8] G. Nageswaran, S. Neogi, Surface modification of polypropylene using argon plasma: statistical optimization of the process variables, Appl. Surf. Sci. 255 (17) (2009) 7590−7600. Available from: https://doi.org/10.1016/J.APSUSC.2009.04.034.

[9] F. Khelifa, S. Ershov, Y. Habibi, R. Snyders, P. Dubois, Free-radical-induced grafting from plasma polymer surfaces, Chem. Rev. 116 (6) (2016) 3975−4005. Available from: https://doi.org/10.1021/acs.chemrev.5b00634.

[10] G. Kühn, S. Weidner, R. Decker, A. Ghode, J. Friedrich, Selective surface functionalization of polyolefins by plasma treatment followed by chemical reduction, Surf. Coat. Technol. 116−119 (1999) 796−801. Available from: https://doi.org/10.1016/S0257-8972(99)00232-7.

[11] A. Meyer-Plath, 2002. Grafting of Amino and Nitrogen Groups on Polymers by Means of Plasma Functionalisation. Ernst-Moritz-Arndt University of Greifswald.

[12] P. Favia, M.V. Stendardo, R. d'Agostino, Selective grafting of amine groups on polyethylene by means of NH3 − H2 RF glow discharges, Plasmas Polym. 1 (2) (1996) 91−112. Available from: https://doi.org/10.1007/BF02532821.

[13] P.R. McCurdy, C.I. Butoi, K.L. Williams, E.R. Fisher, 1999. Surface Interactions of NH2 Radicals in NH3 Plasmas. https://doi.org/10.1021/JP9909558

[14] M.L. Steen, K.R. Kull, E.R. Fisher, Comparison of surface interactions for NH and NH2 on polymer and metal substrates during NH3 plasma processing, J. Appl. Phys. 92 (1) (2002) 55−63. Available from: https://doi.org/10.1063/1.1486038.

[15] J. Hopkins, J.P.S. Badyal, Nonequilibrium glow discharge fluorination of polymer surfaces, J. Phys. Chem. 99 (12) (1995) 4261−4264. Available from: https://doi. 10.10 10P1.j1000104056.

[16] Z. Shourgashti, M.T. Khorasani, S.M.E. Khosroshahi, Plasma-induced grafting of polydimethylsiloxane onto polyurethane surface: characterization and in vitro assay, Rad. Phys. Chem. 79 (9) (2010) 947−952. Available from: https://doi.org/ 10.1016/J.RADPHYSCHEM.2010.04.007.

[17] D.L. Smith, H.P. Gillis, T.M. Mayer, Elemental analysis of treated surfaces, in: O. Auciello, D.L. Flamm (Eds.), Plasma Diagnostics - Volume 2, Surface Analysis and Interactions, Academic Press, Inc, 1989.

[18] E. Bormashenko, G. Whyman, V. Multanen, E. Shulzinger, G. Chaniel, Physical mechanisms of interaction of cold plasma with polymer surfaces, J. Colloid. Interface. Sci. 448 (2015) 175−179. Available from: https://doi.org/10.1016/J. JCIS.2015.02.025.

[19] A. Qureshi, S. Shah, S. Pelagade, N.L. Singh, S. Mukherjee, A. Tripathi, et al., Surface modification of polycarbonate by plasma treatment, J. Phys. Conf. Ser. 208 (2010) 012108.

[20] G. Nageswaran, C. Eswaraiah, S. Neogi, Surface modification of polycarbonate by radio-frequency plasma and optimization of the process variables with response surface methodology, J. Appl. Polym. Sci. 114 (3) (2009) 1557−1566. Available from: https://doi.org/10.1002/app.30691.

[21] G. Nageswaran, D. Mishra, T.K. Maiti, S. Neogi, Helium plasma treatment to improve biocompatibility and blood compatibility of polycarbonate, J. Adhesion Sci. Technol. 24 (13−14) (2010) 2237−2255. Available from: https://doi.org/10.1163/ 016942410X511088.

[22] G. Nageswaran, S. Neogi, Investigation on argon−oxygen plasma induced blood compatibility of polycarbonate and polypropylene, J. Adhesion Sci. Technol. 23 (13−14) (2009) 1811−1826. Available from: https://doi.org/10.1163/ 016942409X12459095670476.

[23] F. Alexander, Plasma Chemistry, Cambridge University Press, Cambridge, 2008.

[24] G. Nageswaran, D. Mishra, T.K. Maiti, S. Neogi, Enhanced cell adhesion to helium plasma-treated polypropylene, J. Adhesion Sci. Technol. 23 (13−14) (2009) 1861−1874. Available from: https://doi.org/10.1163/016942409X12459095670593.

[25] G. Nageswaran, R. Rajasekar, R. Rajesh Babu, D. Mishra, S. Neogi, Development of bio/blood compatible polypropylene through low pressure nitrogen plasma surface modification, Mater. Sci. Eng. C 32 (7) (2012) 1767−1778. Available from: https://doi.org/10.1016/J.MSEC.2012.04.034.

[26] M. Šíra, D. Trunec, P. Sťahel, V. Buršíková, Z. Navrátil, J. Buršík, Surface modification of polyethylene and polypropylene in atmospheric pressure glow discharge, J. Phys. D. Appl. Phys. 38 (4) (2005) 621−627. Available from: https://doi.org/ 10.1088/0022-3727/38/4/015.

[27] R.B. Tyata, D.P. Subedi, A. Huczko, C.S. Wong, Surface modification of polymers and textiles by atmospheric pressure argon glow discharge, Int. J. Sci. Eng. Appl. Sci. 2 (6) (2016) 474−481. Retrieved from www.ijseas.com.

[28] D. Hegemann, H. Brunner, C. Oehr, Plasma treatment of polymers for surface and adhesion improvement, Nucl. Instr. Methods Phys. Res. Sect. B 208 (2003) 281−286. Available from: https://doi.org/10.1016/S0168-583X(03) 00644-X.

[29] R.A. Lawton, C.R. Price, A.F. Runge, W.J. Doherty III, S.S. Saavedra, Air plasma treatment of submicron thick PDMS polymer films: effect of oxidation time and storage conditions, Colloids Surfaces A: Physicochem. Eng. Aspects 253 (1−3) (2005)   213−215.   Available   from: https://doi.org/10.1016/J. COLSURFA.2004.11.010.

[30] E. Occhiello, M. Morra, P. Cinquina, F. Garbassi, Hydrophobic recovery of oxygen-plasma- treated polystyrene, Polymer (Guildf). 33 (14) (1992) 3007−3015. Available from: https://doi.org/10.1016/0032-3861(92)90088-E.

[31] J. Behnisch, A. Holländer, H. Zimmermann, Factors influencing the hydrophobic recovery of oxygen-plasma-treated polyethylene, Surf. Coat. Technol. 59 (1−3) (1993) 356−358. Available from: https://doi.org/10.1016/0257-8972(93)90112-2.

[32] Y. Ren, C. Wang, Y. Qiu, Aging of surface properties of ultra high modulus poly-ethylene fibers treated with He/O2 atmospheric pressure plasma jet, Surf. Coat. Technol. 202 (12) (2008) 2670−2676. Available from: https://doi.org/10.1016/J. SURFCOAT.2007.09.043.

[33] I. Banik, K.S. Kim, Y.I.L. Yun, D.H. Kim, C.M. Ryu, C.S. Park, et al., A closer look into the behavior of oxygen plasma-treated high-density polyethylene, Polymer (Guildf). 44 (4) (2003) 1163−1170. Available from: https://doi.org/10.1016/S0032-3861(02)00847-9.

[34] M. Morra, E. Occhiello, R. Marola, F. Garbassi, P. Humphrey, D. Johnson, On the aging of oxygen plasma-treated polydimethylsiloxane surfaces, J. Colloid. Interface. Sci. 137 (1) (1990) 11−24. Available from: https://doi.org/10.1016/0021-9797(90) 90038-P.

[35] M. Ataeefard, S. Moradian, M. Mirabedini, M. Ebrahimi, S. Asiaban, Investigating the effect of power/time in the wettability of Ar and O2 gas plasma-treated low-density polyethylene, Prog. Org. Coat. 64 (4) (2009) 482−488. Available from: https://doi.org/10.1016/J.PORGCOAT.2008.08.011.

[36] O.-J. Kwon, S.-W. Myung, C.-S. Lee, H.-S. Choi, Comparison of the surface char-acteristics of polypropylene films treated by Ar and mixed gas (Ar/O2) atmospheric pressure plasma, J. Colloid. Interface. Sci. 295 (2) (2006) 409−416. Available from: https://doi.org/10.1016/J.JCIS.2005.11.007.

[37] X. Hua, T. Zhang, J. Ren, Z. Zhang, Z. Ji, X. Jiang, et al., A facile approach to modify polypropylene flakes combining O2-plasma treatment and graft polymeriza-tion of l-lactic acid, Colloids Surfaces A: Physicochem. Eng. Aspects 369 (2010) 128−135. Available from: https://doi.org/10.1016/j.colsurfa.2010.08.009.

[38] C. Steiner, J. Fichtner, J. Fahlteich, Nanostructuring of polymer surfaces by magne-tron plasma treatment, Surf. Coat. Technol. (2017). Available from: https://doi.org/ 10.1016/J.SURFCOAT.2017.09.023.

[39] D. Mangindaan, W.-H. Kuo, Y.-L. Wang, M.-J. Wang, Experimental and numerical modeling of the controllable wettability gradient on poly(propylene) created by SF6 plasma, Plasma Process. Polym. 7 (9−10) (2010) 754−765. Available from: https:// doi.org/10.1002/ppap.201000021.

[40] H. Ahn, I. Lee, H. Lee, M. Kim, Cellular behavior of human adipose-derived stem cells on wettable gradient polyethylene surfaces, Int. J. Mol. Sci. 15 (2) (2014) 2075−2086. Available from: https://doi.org/10.3390/ijms15022075.

[41] M. Gao, L. Sun, Y. Guo, J. Shi, J. Zhang, Modification of polyethylene terephthalate (PET) films surface with gradient roughness and homogenous surface chemistry by dielectric barrier discharge plasma, Chem. Phys. Lett. 689 (2017) 179−184. Available from: https://doi.org/10.1016/J.CPLETT.2017.10.009.

[42] A. Zaitsev, A. Lacoste, F. Poncin-Epaillard, A. Bès, D. Debarnot, Nanotexturing of plasma-polymer thin films using argon plasma treatment, Surf. Coat. Technol. 330 (2017)   196−203.   Available   from: https://doi.org/10.1016/J. SURFCOAT.2017.10.010.

CHAPTER 5

Plasma-Induced Polymeric Coatings

Venu Anand[1,2], Rajesh Thomas[1,3], K.H. Thulasi Raman[1] and Mohan Rao Gowravaram[1]

[1]Plasma Processing Laboratory, Department of Instrumentation and Applied Physics, Indian Institute of Science, Bengaluru, Karnataka, India
[2]Department of Electronics and Communication Engineering, National Institute of Technology Calicut, Kozhikode, Kerala, India
[3]International Iberian Nanotechnology Laboratory, Braga, Portugal

5.1 INTRODUCTION

The formation of thin film coatings of polymers in glow discharge systems was recognized in the 19th century. However, they were considered as by-products of electrical discharges and hence received little attention. Only the period after 1950s saw their renaissance, particularly as practical ways to make special coatings on metals. Since then, much research has been done in plasma-induced polymers and interesting as well as industry-relevant applications have been identified for polymers deposited using plasmas because of their flawless thin film structure, excellent adhesion to the substrate, chemical inertness, and low dielectric constant.

The processes that convert a gaseous phase reactant into a solid film essentially proceed due to the influence of the radicals generated in the high energy plasma state. Often, structurally, the polymer formed in plasma does not resemble the conventional polymer due to the existence of an abundant amount of various reactive species which can generate several reaction pathways. Hence the plasma deposited films are mostly amorphous and crosslinked compared to their conventional polymers, as depicted in Fig. 5.1.

Plasma-based techniques, particularly nonthermal plasmas, enable one to have a controlled process of depositing polymers on any substrate for a variety of applications. The highly crosslinked structure is associated with good mechanical resistance [2,3]. Moreover, plasma polymers exhibit strong adhesion to a wide variety of substrates (ceramic, metals, etc.), and

Non-Thermal Plasma Technology for Polymeric Materials
DOI: https://doi.org/10.1016/B978-0-12-813152-7.00005-6
129

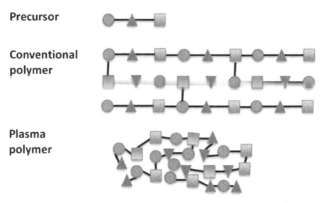

Figure 5.1 Structural difference between polymers deposited using conventional and plasma methods. *Reprinted with permission from F. Khelifa, S. Ershov, Y. Habibi, R. Snyders, P. Dubois, Free-radical-induced grafting from plasma polymer surfaces, Chem. Rev. 116 (2016) 3975−4005. Copyright (2016) American Chemical Society.*

the deposition technique allows even complex geometrical shapes to be covered by thin films with tunable thicknesses extending to hundreds of nanometers [4,5]. These characteristics have been exploited in producing superhydrophobic surfaces, barrier layers for the corrosion protection of metals, for controlled drug delivery, in food packaging, and bio-sensing applications, etc. [6−15].

An additional advantage of the process, discussed later, is the incorporation of a secondary material into the polymer during the polymerization process which will qualitatively enhance the functionality of the coating. In simplistic terms, plasma-induced polymerization is caused by direct or indirect energy transfer from the plasma [16,17]. Indirect energy transfer is through the absorption process of UV-vis radiation emitted by the species in the plasma due to excitation and relaxation processes. Direct energy transfer is through the plasma power which controls the energy of the species causing the polymerization process.

In this chapter the significance of plasma-induced polymer coatings is presented with more emphasis on the process aspects, general methods of characterization of these polymers and application of such polymers in super-hard coatings. In the first section, a clear distinction is made between plasma-polymerized coatings and plasma-induced polymerized coatings based on the mechanism of the process. The processing parameters that greatly influence the properties of such coatings are considered. A definition for the rate of deposition based on these process parameters has been established. This section ends with the most commonly used monomer materials in plasma-induced polymerizations systems.

Inorganic hard coatings are widely used in many industries and these are basically transition metal nitrides and carbides and their combination. The hardness of these compounds ranges from 20 to 40 GPa. The hardness and stiffness (elastic modulus) are completely governed by the nature of the bond (ionic or covalent) and distance between the two atoms, among others. The hardness of the super-hard coatings is above 40 GPa, and this hardness is achieved using nanosize and composite effects—these coatings are widely used in industries. Polymer-based coatings are also gaining momentum to replace some of the inorganic hard coatings. At present, polymer coatings have lower hardness compared with inorganic coatings and the hardness range of polymer coatings is 1−3 GPa. The reason for lower hardness in the polymer coatings is the predominant contribution from sp2 hybridization. To enhance the polymer coating hardness, the polymer should be rich in sp3 hybridization and foreign materials should be incorporate into the polymer matrix, which has higher degree of covalent and ionic nature.

Plasma-enhanced metal organic chemical vapor deposition (MOCVD) and physical or chemical sputtering of a solid metal target along with the plasma polymerization are the two strategies of obtaining the nanoparticle or nanocomposites embedded polymer films. These polymers can be utilized for many applications such as optical coatings, protective films, and biomedical materials, etc. Details of plasma techniques used to produce nanocomposites especially by using microwave and radio frequency (RF) plasmas are discussed later. Various characterization tools employed for studying different properties of plasma-induced polymer coatings are discussed in detail. The chemical structure and nature of the polymer films can be analyzed using optical and electrical methods. Optical methods such as Raman, Fourier Transform Infrared and UV visible absorption, fluorescence spectroscopy, and ellipsometry are used to study polymer properties. Electrical properties of polymer films are studied by their response to imposed electric fields of various strengths and frequencies.

We review the mechanical properties of existing polymer coatings and the science behind higher hardness and stiffness of different polymer coating materials and the mechanical methods of polymers and the failure mechanism of polymer coatings will be discussed. It is interesting to see the increase in hardness of monolithic polymer coatings by incorporating new materials and enhancing the polymer network stiffness. The probable process methods to incorporate the foreign materials and network stiffening methods using plasma process will be discussed.

5.2 PLASMA-INDUCED POLYMERIZATION VERSUS PLASMA POLYMERIZATION

An organic vapor can polymerize in the plasma state under favorable process conditions. This is termed glow discharge polymerization [18—??]. This process is essentially electron-driven, in which energetic electrons in the plasma bombard the organic vapor molecules to create a variety of radicals, ions, and excited molecules. Usually, the process is operated under low-pressure conditions and in the presence of an inert gas like argon or helium, which is used to initiate and sustain the plasma, while the organic vapor is introduced into the system as the reactant. Since plasma contains electrons of both very low and very high kinetic energies, many chemical reactions which might be thermally unfavorable are possible in the plasma state.

Depending on the mechanism with which the reactions proceed, glow discharge polymerization can be categorized into two categories: *plasma polymerization* and *plasma-induced polymerization*. The subtle differences between the two processes are:

1. Plasma-induced polymerization (molecular polymerization) is similar to the conventional polymerization process in which the product polymer retains the structure of its starting material (monomer), whereas, in plasma polymerization the product will have an entirely different structure and the starting material serves only as a source of radicals for the polymerization to proceed, as shown in Fig. 5.2 [23].

2. In the case of plasma-induced polymerization, the starting material must contain polymerizable structures such as olefinic double bonds, triple bonds, or cyclic structures.

3. Induced polymerization is a more direct process which proceeds as single-step reactions, whereas plasma polymerization has multiple steps involving intermediate species.

4. Plasma-induced polymerization does not create any gaseous phase product, whereas plasma polymerization proceeds through the creation of intermediates in the gaseous phase, which might also be nonpolymerizable.

5. Since no by-products are formed in plasma-induced polymerization, its kinetics are not disturbed by the reaction between the solid product and reactive gaseous intermediates. However, this can significantly affect the reaction rates in the plasma polymerization processes.

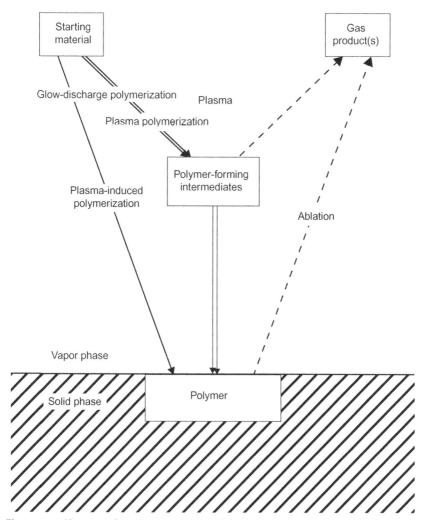

Figure 5.2 Plasma polymerization versus plasma-induced polymerization. *Reprinted from H. Yasuda, T. Hsu, Plasma polymerization investigated by the comparison of hydrocarbons and perfluorocarbons, Surf. Sci. 76 (1978) 232–241, Copyright (1978), with permission from Elsevier.*

5.3 MECHANISM OF POLYMER FORMATION

The reactions in plasma state are mostly driven by collisions due to energetic electrons. In the presence of organic vapors, ionization due to electron collision is an unlikely event, since the ionization energies are typically in the order of more than 10 eV. A more likely scenario is the dissociation of the hydrocarbon chains to produce large-sized activated fragments and short-sized reactive radicals, which can crosslink to form a polymer.

The unpaired electron in free radicals makes it highly reactive towards other radicals and unsaturated bonds in molecules. This is similar to conventional free radical polymerization, but the difference is related to the species initiating the polymerization process. The starting site is often random in plasma polymerization because of the diversity of precursor fragmentation routes, whereas in conventional free-radical polymerization, the building block (the monomer to be polymerized) is always the same. However, similar to conventional radical polymerization, the reactions of initiation, propagation, termination, and reinitiation of radicals (with each other and molecules) are principally applied for a description of the plasma polymer growth mechanism.

The mechanism of polymer formation in glow discharge polymerization process is depicted in Table 5.1 [24].

Where i and j are the numbers of repeating units; $i = j = 1$ for the starting material and S represents a reactive species, which can be either an ion of any charge, an excited molecule, or a free radical. S needn't necessarily retain the structure of the starting material.

As depicted in Table 5.1, while plasma-induced polymerization is similar to the conventional polymerization process, consisting of propagation and termination of chain reactions, in plasma polymerization, initiation and reinitiation of intermediate species are mandatory for the reaction to sustain. These intermediates are created mostly by the electron impact dissociation of hydrocarbons in the plasma state. In most glow discharge plasma polymerization reactions, both plasma polymerization and plasma-induced polymerization reactions exist at varying degrees. The nature of

Table 5.1 Mechanism of polymer formation in plasma polymerization versus plasma-induced polymerization [24]

Plasma polymerization	Plasma induced polymerization
Initiation/reinitiation	Propagation/termination
$Si \rightarrow Si^*$	$S^* + S \rightarrow SS^*$
$Sj \rightarrow Sj^*$	$Si^* + S \rightarrow Si^* + 1$
	$Si^* + Sj^* \rightarrow Si - Sj$
Propagation/termination	
$Si^* + Sj^* \rightarrow Si - Sj$	
$Si^* + Sj \rightarrow Si - Sj$	

the starting materials and the discharge process conditions dictate the dominating process.

Apart from these reactions, one mechanism which is often overlooked, but has a huge impact on both the structure of the final product as well as the polymerization rate, is the etching (ablation) of the product polymer. Etching can happen due to both ion impact from plasma as well as due to the reactions between the product polymer and the reactive species in the product gas. Often the latter is more aggressive than the former. Models proposed by Lam et al., Poll et al., and Yasuda are often used to describe the kinetics of these processes [25–27]. Kay [28] shows that even though polymerizable intermediates can be formed in CF_4 plasma, the presence of highly reactive F atoms etch away the product polymer. By adding H_2 to the reactor, this etching rate can be lowered as hydrogen depletes the plasma of fluorine atoms by forming hydrogen fluoride (HF). Oxygen is another element which has a similar effect as fluorine and the rate of polymerization with hydrocarbons containing these elements will be low [27]. Presence of nitrogen, on the other hand, creates more nitrogen-based functional groups in the product polymer [29]. Thus, the polymer formation and its properties are dictated by the competitive processes of plasma polymerization/plasma-induced polymerization versus the ablation, which is usually referred as competitive ablation and polymerization (CAP) [30].

5.4 FACTORS THAT INFLUENCE POLYMER FORMATION

5.4.1 Reactor Geometry

The reactors for glow discharge polymerization are similar in design to conventional plasma-enhanced chemical vapor deposition systems. However, certain geometrical factors contribute to the polymer formation on the substrate. One such factor is the *bypass ratio* which represents the fraction of the feed gas that does not traverse the discharge zone and, hence, will not actively participate in the polymer formation process. The higher the bypass ratio, the less polymer will be formed and this factor depends on the ratio of the volume of the discharge to the total volume of the reactor. Another factor is the relative position of the substrate with respect to the discharge zone. Direct exposure of the substrate to the plasma might do more harm because of the etching phenomenon in CAP limited systems. Instead, an after-glow configuration is mostly used, in which the active species reach the substrate surface from the plasma zone.

In this case, the polymer formation rate is low, but it has the added advantage that the density and energy of ions that strike the substrate surface can be controlled for minimal damage. Also, this configuration is best-suited for plasma-induced polymerization because the after-glow region will be devoid of high energy charged particles and this favors the binding of already formed monomer units. Taking all these aspects into consideration, a generalized geometrical similarity parameter has been defined as [31]:

$$S = \frac{W}{F}\bigg|_{dep} = \frac{W}{F}\frac{d_{act}V_{gas}}{d_{gas}V_{dis}} \qquad (5.1)$$

where d_{act} is the length of the active plasma zone; d_{gas}, the (mean) distance between gas inlet and deposition area; V_{gas}, the volume occupied by the gas; V_{dis}, the volume occupied by the gas discharge; and S is the specific energy consumed within the active plasma zone, which determines the measured deposition rate ($S = W/F$).

5.4.2 Power Supply

The power supplied to the plasma becomes an important parameter because the product polymer is highly temperature sensitive and a fraction of the input power will be lost in heating the electrodes and the chamber walls due to charged particle bombardment. The two components of the power supply that contributes to the polymer formation rate are discharge power and frequency.

5.4.2.1 Discharge Power

Discharge power represents the effective power density (power/volume) in the glow region of the plasma. The amount of discharge power required to initiate and sustain the polymerization process is different for different monomers. Hence, the power needed to maintain a fixed rate of polymerization depends on the flow rate and on the structure of the starting material, as shown in Fig. 5.3 [32]. So a quantity defined as the power per unit flow rate per unit mass (W/FM) is defined to establish constant process conditions across different polymer depositions. Here M represents the molecular mass of the reactant. The variation of this ratio with flow rate for different gases is shown in Fig. 5.4 [32]. For low flow rates, the ratio is independent of the flow and depends only on the structure (M). Interestingly, the slope of this curve often indicates the H_2 yield in the process. So, for a certain

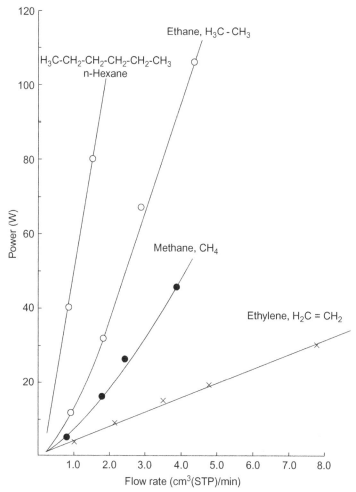

Figure 5.3 Discharge power needed to maintain a constant polymerization rate for different flow rates of several starting materials. *Reprinted from H. Yasuda, T. Hirotsu, Critical evaluation of conditions of plasma polymerization, J. Polym. Sci. Part A: Polym. Chem. 16 (1978) 743–759, Copyright (2003), with permission from John Wiley and Sons.*

polymer formation rate, the discharge power has to be fixed based on the flow rate and the molecular mass of the starting material.

5.4.2.2 Frequency

The reactor may be powered with direct current (DC), low frequency (50 Hz–20 kHz) or high frequency (13.56 MHz, 2.45 GHz) AC power sources. The efficiency of plasma formation is high at higher frequencies

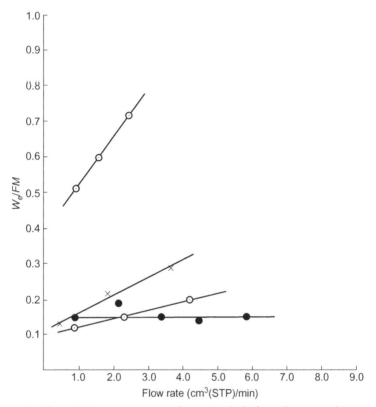

Figure 5.4 The *W/FM* parameter values needed for plasma polymerization to occur for various starting materials at various flow rates ⊙ *n*-Hexane, × *c*-Hexane, ⊡ *c*-Hexene, ● Benzene. *Reprinted from H. Yasuda, T. Hirotsu, Critical evaluation of conditions of plasma polymerization, J. Polym. Sci. Part A: Polym. Chem. 16 (1978) 743–759, Copyright (2003), with permission from John Wiley and Sons.*

of operation. At higher frequencies the electron energy distribution function flattens out and, hence, the number of electrons in the entire energy spectrum is appreciable [32]. This indicates that the starting material is completely dissociated in high frequency plasma and, hence, the probability of product formation also increases.

5.4.3 Flow Rate

The amount of the reactants in the system is represented using the mass flow rate or the volume flow rate at standard temperature and pressure. Instead of flow rate, the flow rate per unit pressure (F/p), which indicates the velocity of the flow, is often used to represent the amount of the reactants in plasma-based polymerization systems [32].

5.4.4 System Pressure

Polymerization in plasma is essentially a conversion from gaseous phase to solid phase. Hence the pressure inside the reactor drops when the process is initiated. An efficient process results in a heavy pressure drop inside the reactor. Hence the initial pressure (p_0) will be significantly different from the process pressure (p_p). The production of H_2 during the process also contributes to the process pressure. This process pressure depends on the flow rate and not on the initial pressure, as depicted in Fig. 5.5 which represents polymerization of Ethylene [32]. However, with increasing process pressure, Kim et al. have observed an increase in the deposition rates when using SiH_4/NH_3 RF discharges in the pressure range between 80 and 160 Pa. [33]. This can be accounted to the fact that with increasing pressure, the plasma sheath shrinks, and the width of the active plasma zone (d_{act}) increases which, in turn, causes an increase in the energy input for plasma polymerization.

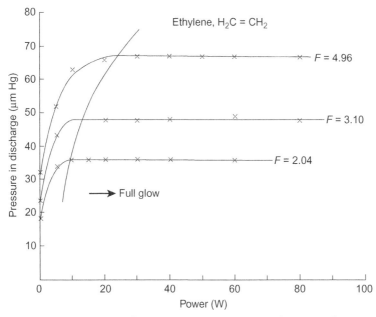

Figure 5.5 The dependence of the process pressure on the flow rate of the reactant (Ethylene) for identical initial chamber pressure. *Reprinted from H. Yasuda, T. Hirotsu, Critical evaluation of conditions of plasma polymerization, J. Polym. Sci. Part A: Polym. Chem. 16 (1978) 743–759, Copyright (2003), with permission from John Wiley and Sons.*

5.5 RATE OF POLYMERIZATION

As discussed in the previous section, the polymerization process and thereby the rate of polymerization is directly influenced by the parameters discussed. However, other parameters and processes, like the bypass ratio and the etching of the product polymer, have a marked influence on the polymerization rate, but cannot be easily controlled. Hence an analytical expression describing the polymerization rate to the process parameters is difficult to conceive.

Discharge power is one of the major factors that affect the rate of polymerization. Fig. 5.6 shows the general trend of variation in deposition rate with discharge power for a constant monomer flow rate [24]. There exists a linear relationship between the deposition rate and the power to certain extent, and then the curve flattens out, indicating saturation of the deposition rate. This can be due to the increase in competing phenomena like heating and etching due to bombardment of energetic charged particles, etc. Flow rate, on the other hand, increases the

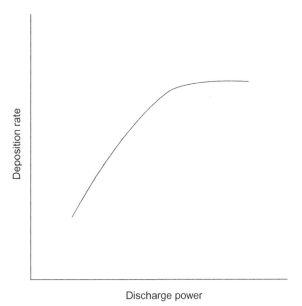

Discharge power

Figure 5.6 Graph depicting the generic variation of polymer deposition rate with the discharge power in a capacitive discharge system. *Reprinted from H. Yasuda, Glow discharge polymerization, in: J.L. Vossen, W. Kern (Eds.), Thin Film Processes, Academic Press, New York, 1978, pp. 361–398, Copyright (1978), with permission from Elsevier.*

deposition rate initially, but soon drops at high flow, as shown in Fig. 5.7 [24]. A similar observation has been made by several authors [34–38]. This often happens due to the low residence time of the activated species in the plasma at high monomer flow rates, so that the polymerization process will not proceed efficiently. These two parameters have been combined into a single factor W/FM. As long as this factor remains above a critical value, the input energy is sufficient for polymerization and increasing the flow rate serves to increase the feed-in rate and, hence, the deposition rate also increases. However, if W/FM drops below a certain value as F increases at constant W, then the discharge power is not sufficient to polymerize all the starting materials and consequently, the polymerization mechanism changes, accompanied by a drop in deposition rate. So the discharge power has to be increased along with increasing flow rate to keep the W/FM factor constant and, thereby, ensuring constant deposition rate [24].

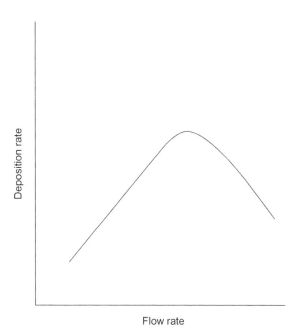

Figure 5.7 Graph depicting the generic variation of polymer deposition rate with the monomer flow rate in a capacitive discharge system. *Reprinted from H. Yasuda, Glow discharge polymerization, in: J.L. Vossen, W. Kern (Eds.), Thin Film Processes, Academic Press, New York, 1978, pp. 361–398, Copyright (1978), with permission from Elsevier.*

5.6 MATERIALS FOR POLYMERIZATION

Almost all organic compounds serve as starting material for glow discharge polymerization; however, their influence on polymerization can be different as mentioned below.

5.6.1 Hydrocarbons

Hydrocarbons with triple bonds and aromatic compounds polymerize in the plasma state by opening the triple bonds or aromatic chains. Hence, the amount of hydrogen evolved during this process is minimal. However, they require low energy input to break the bonds and this energy required barely changes with the flow rate of the compound. Compounds with double bonds and cyclic structures polymerize through a combination of breaking the double bonds as well as through hydrogen abstraction. Hence the hydrogen yield involving these compounds is higher compared to the triple-bonded materials. The energy required and its dependence on the flow rate is slightly higher compared to triple-bonded materials. With these two compounds, plasma-induced polymerization wins over plasma polymerization and hence the product polymer will be a long chain of the starting material, similar to conventional polymerization processes. Single-bonded hydrocarbons polymerize primarily through plasma polymerization reactions involving hydrogen abstraction and, hence, their hydrogen yield will be quite high. These compounds require the highest input energy and the energy dependence on the flow rate is high. The hydrogen yield per molecule for several hydrocarbons is given in Fig. 5.8 [39].

5.6.2 Nitrogen-Containing Compounds

N_2 can easily incorporate into the compounds and hence the product always includes amines and nitrile groups [40]. Additional N_2 can be added to the plasma polymerization process for better yield of polymer.

5.6.3 Fluorine/Oxygen-Containing Compounds

Atomic fluorine/oxygen evolved during the plasma process can etch out a significant amount of the polymer formed and it is not uncommon to see zero product formation with fluorine-containing compounds. By keeping the discharge power as low as possible, the fluorine evolution can be minimized and the process proceeds to produce the require polymer [41]. Hence, fluorine-containing triple-bonded and double-bonded

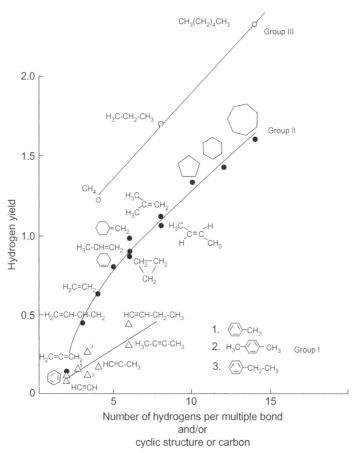

Figure 5.8 Hydrogen yield per molecule for different hydrocarbons. *Reprinted from H. Yasuda, Modification of polymers by plasma treatment and by plasma polymerization, Radiat. Phys. Chem. 9 (1977) 805–817, Copyright (1977), with permission from Elsevier.*

compounds are favored for these reactions [42]. Apart from these, hydrogen can be intentionally added to the process to neutralize the reactive fluorine atoms as HF and pumped out.

5.6.4 Silicon-Containing Compounds

The silanes and siloxanes yield a significant amount of silicon-containing compounds because of silicon's high tendency to remain in the solid phase. The rate of deposition is also very high because of the high molecular weight and high vapor pressure of Si-containing compounds [43–45].

5.7 POLYMER-BASED NANOCOMPOSITE COATINGS

Plasma-enhanced MOCVD and physical or chemical sputtering of a solid metal target along with the plasma polymerization are the two strategies of obtaining the nanoparticle or nanocomposites embedded polymer films. These plasma polymers can be utilized for many applications such as optical coatings, protective films, and biomedical materials, etc. A new class of materials based on plasma polymers has emerged where an inorganic component is admixed to the plasma polymer matrix during deposition. Thus, nanomaterial/plasma polymer films have been studied in detail with metals such as Ag, Au, Ni, Mo, Cu, and Ti [46−50]; and metal oxides such as SiO_2 and TiO_2 [51−54] used as inorganic fillers dispersed in a plasma polymer matrices from a hydrocarbon precursor. Metal oxide (SiO_2 and TiO_2) plasma polymers are nanocomposites with nanometer scale inclusions.

Plasma-based methods of material synthesis have certain advantages over wet chemistry methods that are commonly used for synthesizing a wide variety of polymer materials. Plasma techniques are particularly suitable for production of thin, smooth, and homogeneous coatings or films on a wide range of substrate materials, whereas wet chemistry that uses toxic and environmentally dangers substances, can be avoided. Also, it offers the possibility to finely tune the properties of deposited films by deposition parameters (e.g., pressure, applied power, working gas mixture, or monomer used, etc.) that are highly suitable for the implementation into the current technological processes [55, 56].

Plasma technique is a versatile method to prepare polymers and nanocomposite-embedded polymer films. Plasma polymer thin films can be prepared by using plasma polymerization processes. These films have only few similarities with conventional polymers where a repeated structure is observed. Plasmas used in materials processing are partially ionized gases and most commonly used plasma sources are RF or DC and microwave [57,58,59]. Plasma polymerization has emerged as an integral part of wide verities of polymer coatings.

5.7.1 Nanomaterial/Polymer Film

Plasma polymers films are usually amorphous with branched and cross-linked chains (as shown in Fig. 5.9A). The properties of polymer films can be significantly extended by admixing other components such as metal and metal oxide nanocomposites into the plasma polymers

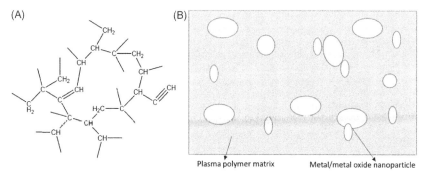

Figure 5.9 A schematic structure of a hydrocarbon polymer (A) and metal/plasma polymer composite film (B).

(Fig. 5.9B). These composite films have exhibited interesting physical (mechanical, electrical, and optical), chemical, and biological properties that are different from individual components. Most of the investigations in this area were concentrated on metal/plasma polymers and metal oxide/plasma polymer films. These polymer films can be used for applications such as optics and electronics as filters, optical memories, sensors of humidity and lithography, etc. Nanocomposite/plasma polymers consist of two phases: plasma polymer matrix and metal or metal oxide nanostructure inclusions. The size of the inclusion is usually about 1−100 nm. When a metal nanostructure is incorporated into the polymer matrices, then the dielectric component is influenced by both the presence of the metal itself and by the modified plasma processes during the deposition, thus changing the optical and electrical properties of the polymer film.

5.7.2 Processing Technique for Polymer Nanocomposite Matrices

The size of the nanoparticle depends on the volume fraction and can be controlled by various deposition parameters. Deposition rate and substrate temperature have a strong influence in the nanoparticle grain size [60−62].

The metal content in the polymer matrix can be described using filling factor or volume fraction ratio f, defined as:

$$f = V_{metal} / V_{composite} \tag{5.1}$$

By measuring thickness of the polymer composite film, the filling factor can be calculated by:

$$f = \left(\partial_{comp} - \partial_{p}\right) / \left(\partial_{met} - \partial_{p}\right) \tag{5.2}$$

where ∂_{comp}, ∂_{met}, and ∂_{p} are the density of composite, metal, and plasma polymer, respectively.

Several deposition techniques can be used for the deposition of nanomaterial/plasma polymer composites. Simultaneous plasma polymerization with sputter deposition from a metal target with RF or DC power are commonly used for obtaining the polymer composite film [63,64]. Various other methods of deposition are:

1. Simultaneous plasma polymerization of an organic gas and evaporation of a metal.
2. Simultaneous plasma polymerization and sputter-etching using a RF or DC plasma from a metal target.
3. Plasma polymerization of metal organic compounds using RF or microwave plasma
4. Cosputtering from composite metal/polymer target.
5. Reactive plasma deposition of metal target (for metal oxide) and simultaneous plasma polymerization.

5.7.3 Metal/Metal Oxide Embedded Plasma Polymer

In order to deposit nanocomposite metal/plasma polymer films, simultaneous plasma polymerization with RF/DC sputtering of a metal or metal coevaporation can be used. These kinds of films were also prepared by RF sputtering of a composite target or sputtering from two targets of polymer and metal. Harding and Craig sputtered metals like Fe and Cu in a mixture of argon and acetylene using DC cylindrical magnetron sputtering [65−67]. Laurent and Ka sputtered gold and cobalt in an argon/propane mixture with a simple RF diode system [68]. Using the same method, hard carbon (C = H) or even amorphous hydrogenated carbon a-C:H films containing Ag, Ti, Ta, and W metals were also synthesized [69,70]. In addition, Mo and Ni were added into C:H films and investigated [47]. These types of coatings (especially with Ti, Ta and W) are now in commercial use for tribological applications.

Combining the antibacterial and catalytic properties of silver nanoparticles (Ag-NPs) with polymer films have attracted some specialized applications in biomedical and health-care devices. Poly ethylene terephthalate coated with poly acrylic acid and embedded with silver nanoparticles are used for such applications [71]. Plasma polymerization of fluorocarbons and sputter etching of Au was investigated by Biederman et al. [61]. Nanocomposites of Pt/plasma polymer attracts the catalytic activities for fuel cell applications. For this, plasma polymerization of C_2H_4 combined with Pt sputtering has been done [72].

Metal oxide nanostructures such as silicon dioxide (SiO_2) and titanium dioxide (TiO_2) embedded in a hydrocarbon or fluorocarbon plasma polymer matrix [73] were used for specialized applications in the fields of solar batteries for spacecraft, barrier coatings for food packages, and water-repellent coatings, etc. SiO_x/fluorocarbon plasma polymer films were deposited through various methods such as codeposition from a composite SiO_2/PTFE (polytetrafluoroethylene) target in an argon ion beam sputtering configuration, microwave plasma polymerization from a mixture of TMS, FAS-17, and Ar [74], RF cosputtering from the targets of PTFE and silica (SiO_2) [52], etc. TiO_x/hydrocarbon plasma polymer films are being used in the applications of photo-catalysis, photo-electrochemical solar energy conversion [75], self-cleaning, and for high refractive index material in optical filter applications [76, 77].

5.8 POLYMER SUPER-HARD COATINGS

Generally for hard coatings, materials based on nitrides and carbides of transition metals are used. The hardness of these compounds is in the range of 20−40 GPa [78,79]. The hardness and stiffness (elastic modulus) are completely governed by the nature of chemical bonding and distance between the two atoms, etc., in monolithic coatings such as TiN and WC, etc., In the case of super-hard coatings, where the hardness is in the range of 40−60 GPa [80], it is achieved using nanosize and composite effect methods and these coatings are widely used insurface engineering industries. Polymer-based coatings are also gaining momentum to replace some inorganic hard coatings, by combining the additional functionalities. In general, polymer coatings exhibit lower hardness compared with inorganic coatings and the hardness range of polymer coatings is of the order of 0−3 GPa [81]. That means, if one could increase the hardness of polymer coatings by 60%−70% by proper engineering, this could be called super-hard polymer coatings. Plasma polymer coatings are formed by allowing a monomer into the reaction chamber and polymerizing it using plasma. These coatings can be produced on any three-dimensional object at low temperature. This promotes the usage of polymer coatings for various applications. For example, polymeric coatings are widely used in making hydrophobic and hydrophilic surfaces, where surfaces will provide the tendency of being repellent or attractive to water molecules. But, polymer coatings will not be able to sustain the contact-induced stress. In case of such coatings, mechanical properties like hardness is the key limitation to use them in surface wear resistance or contact stress induced applications [82].

Hardness is defined as the resistance offered by a given material to external mechanical action. It depends on many factors, for example, in the case of inorganic covalent and ionic solids, it is governed by the creation and motion of dislocations [83]. In a covalent crystal, bonding is highly stereospecific and dislocation energy depends strongly on its position. Regardless of details, the basic fact remains that, in order to plastically shear such a crystal, electron-pair bonds must be broken first and then remade, resulting in two unpaired electrons when an atomic shear process is half completed. For pure covalent crystal, the resistant force of the bond can be evaluated by energy gap E_g and the hardness (H) and are related by the following [83]:

$$H(\text{GPa}) = A N_a E_g \tag{5.3}$$

where N_a is the covalent bond number per unit area and A is a proportional coefficient. In the case of ionic crystals, the ionicity plays an important role in governing the hardness of the materials. The hardness of the ionic crystals is correlated using ionicity factor (f_i) and bonding length in crystals (d). Hardness measurement methods for coatings are extensively reported in scientific literature and one has to follow the standards while performing the experiments. For visco elastic polymers, dynamic nanoindentation techniques are widely used [81]. Variation of hardness with ionicity is shown in Fig. 5.10.

$$H_v(\text{GPa}) = 556 \frac{N_a e^{-1.191 f_i}}{d^{2.5}} \tag{5.4}$$

The above hardness equation can be extended somewhat even to the polymeric network. That means, in the case for polymers also, increased level of ionicity by modifying bonding length and increasing the covalence (large number of sigma bonds) nature, one can achieve the desired high-hardness values. The basic mechanical properties of materials are governed by many factors such as molecular weight, crosslinking and

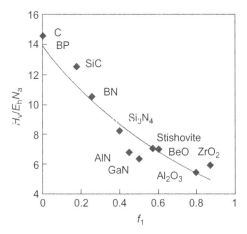

Figure 5.10 Variation of hardness with as function of hardness [83].

branching, crystallinity and crystal morphology, copolymerization (random, block, and graft), plasticization, molecular orientation, fillers, blending, phase separation, and orientation in blocks, grafts, and blends [84]. This has been tried and achieved to a certain extent in bulk polymer materials. Polymers are generally in the glassy state and the critical stress required to plastically deform the amorphous polymer involves displacement of bundles of chain segments against the local restraints of cohesive secondary bond forces and internal rotations [85]. The intrinsic stiffness of these glassy polymers below the glass transition temperature, T_g, leads to microhardness values which may even be two to three times larger than those obtained for typical flexible semi-crystalline polymers [86]. The glass transition temperature (T_g) is another measure of hardness, as T_g generally increases with increasing cohesive energy. The available correlation between the microhardness and T_g for a number of amorphous glassy polymers has been given in literature [87−89]. One obtains a fairly good relationship between H values measured at room temperature (T_{room}) and T_g, in the interval between 300K and 500K and the relation between T_g and H is shown in Fig. 5.11, which can be expressed as [87]:

$$H = kT_g + C \qquad (5.5)$$

Generally, during real-time applications, most of the wear resistance-based functional coatings experience larger contact stresses. This will result in the delamination of coatings or wear at a faster rate. That means, most of the polymer materials will not work very effectively compared to

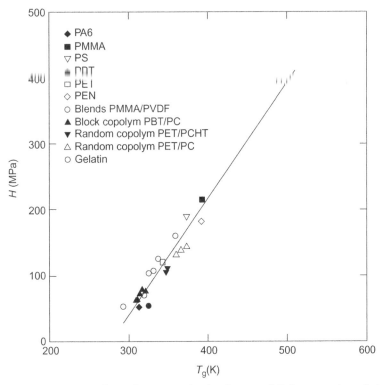

Figure 5.11 Linear correlation between microhardness and T_g for a number of glassy polymers with dominating single bonds in the main chain [85].

other inorganic coatings. This shows that polymer materials alone cannot support such resistance to high contact stress values during the deformation process. Enhancing the hardness of the polymer could be achieved not only by strengthening the bonds, but also could be influenced by extrinsic factors such as impurities, precipitates, and secondary phases in the polymer matrix [84]. The method of incorporating a secondary phase during the polymerization process needs to be understood and feasible experimental methods need to be established. There is little scientific and technical literature available on the incorporation of secondary materials or phases during the polymerization process. In principle, introducing covalent and hard metallic materials into the polymer matrix will increase the hardness.

The incorporation of secondary materials or phases into the system could be achieved by two processes. The first process is based on the organometallic compound combined plasma polymerization process. This will enhance the stiffness of the polymer network. For example, incorporating organosilicon in a polymer network would increase the matrix modulus and hardness significantly [90,91], and Fig. 5.12 shows the variation of hardness with plasma power. There are similar organometallic

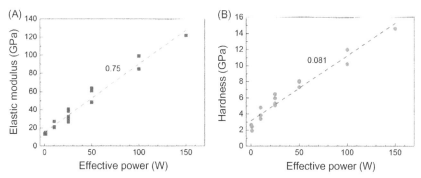

Figure 5.12 (A) Elastic modulus and (B) hardness for pp-TVS films deposited at different effective powers. The linear character of this dependence is marked by dashed line with a slope value [90].

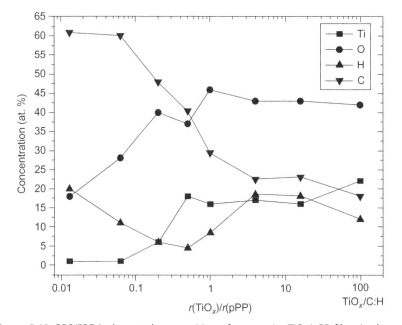

Figure 5.13 RBS/ERDA elemental composition of composite TiO_x/pPP films in dependence on the $r(TiO_x)/r(pPP)$ deposition rate ratio. A separately prepared TiO_x/C:H film is taken as a limiting case with its $r(TiO_x)/r(pPP)$ ratio equal to 100/1 [92].

precursors that could be used during the polymerization process and pre-cipitating the secondary phase is also a way of achieving higher hardness [92]. At present, this kind of research is least explored, because of restrictions with reactor designs for usage for such complicated chemical precursors. This type of incorporation in the gaseous phase will enable the coating process on complex 3D objects.

The other way of incorporating the secondary phases into the polymer is by using a hybrid of physical vapor deposition and plasma polymerization process. Sputter deposition of metals and oxides, etc., during the polymerization process will enable the incorporation of a controlled quantity of secondary material into the polymer [92], and Fig. 5.13 shows the variation in TiO_2 percentage in composite coating. In a nutshell, achieving super-hard coating using plasma-processed polymers is a less-travelled path and there is much research that needs to be adopted in making composite coatings and understanding them under higher contact stress.

5.9 CONCLUSIONS

Plasma-induced polymeric coatings is an interesting area of research due to the controllable process. Added with additional processes into this, one can incorporate a secondary material into the polymer matrix and make a composite. The polymers and their composites have a variety of applications ranging from hydrophobic/hydrophilic coatings, optical coatings, and tribological coatings. Although what happens inside the plasma is complex, the parameters are easily controllable to achieve the required behavior from these coatings. We tried to address the issues involved in the deposition, characterization, and functional behavior of these coatings. Several unanswered questions exist and, thus, makes research on polymeric coatings a fertile field for future research.

REFERENCES

[1] F. Khelifa, S. Ershov, Y. Habibi, R. Snyders, P. Dubois, Free-radical-induced grafting from plasma polymer surfaces, Chem. Rev. 116 (2016) 3975−4005.
[2] N. De Vietro, L. Belforte, V.G. Lambertini, F. Fracassi, Low pressure plasma modified polycarbonate: a transparent, low reflective and scratch resistant material for automotive applications, Appl. Surf. Sci. 307 (2014) 698−703.
[3] D.A. Dragatogiannis, E. Koumoulos, K. Ellinas, A. Tserepi, E. Gogolides, C.A. Charitidis, Nanoscale mechanical and tribological properties of plasma nanotextured cop surfaces with hydrophobic coatings, Plasma Process. Polym. 12 (2015) 1271−1283.

[4] M.E. Alf, A. Asatekin, M.C. Barr, S.H. Baxamusa, H. Chelawat, G. Ozaydin-Ince, et al., Chemical vapor deposition of conformal, functional, and responsive polymer films, Adv. Mater. 22 (2010) 1993−2027.

[5] S. Bhatt, J. Pulpytel, M. Mirshahi, F. Arefi-Khonsari, Plasma co-polymerized nano coatings—as a biodegradable solid carrier for tunable drug delivery applications, Polymer (Guildf) 54 (2013) 4820−4829.

[6] B. Wang, W. Liang, Z. Guo, W. Liu, Biomimetic super-lyophobic and super-lyophilic materials applied for oil/water separation: a new strategy beyond nature, Chem. Soc. Rev. 44 (2015) 336−361.

[7] M. Psarski, D. Pawlak, J. Grobelny, G. Celichowski, Hydrophobic and superhydrophobic surfaces fabricated by plasma polymerization of perfluorohexane, perfluoro (2-methylpent-2-ene), and perfluoro (4-methylpent-2-ene), J. Adhesion Sci. Technol. 29 (2015) 2035−2048.

[8] S. Ershov, F. Khelifa, M.E. Druart, Y. Habibi, M.G. Olivier, R. Snyders, et al., Free radical-induced grafting from plasma polymers for the synthesis of thin barrier coatings, RSC Adv. 5 (2015) 14256−14265.

[9] L. Ejenstam, M. Tuominen, J. Haapanen, J.M. Mäkelä, J. Pan, A. Swerin, et al., Long-term corrosion protection by a thin nano-composite coating, Appl. Surf. Sci. 357 (2015) 2333−2342.

[10] K. Vasilev, N. Poulter, P. Martinek, H.J. Griesser, Controlled release of levofloxacin sandwiched between two plasma polymerized layers on a solid carrier, ACS Appl. Mater. Interfaces 3 (2011) 4831−4836.

[11] T.D. Michl, B.R. Coad, M. Doran, M. Osiecki, M.H. Kafshgari, N.H. Voelcker, et al., Nitric oxide releasing plasma polymer coating with bacteriostatic properties and no cytotoxic side effects, Chem. Commun. 51 (2015) 7058−7060.

[12] J. Schneider, K.M. Baumgärtner, J. Feichtinger, J. Krüger, P. Muranyi, A. Schulz, et al., Investigation of the practicability of low-pressure microwave plasmas in the sterilisation of food packaging materials at industrial level, Surf. Coat. Technol. 200 (2005) 962−966.

[13] S. Plog, J. Schneider, M. Walker, A. Schulz, U. Stroth, Investigations of plasma polymerized SiO_x barrier films for polymer food packaging, Surf. Coat. Technol. 205 (2011) 165−170.

[14] E. Makhneva, A. Manakhov, P. Skládal, L. Zajíčková, Development of effective QCM biosensors by cyclopropylamine plasma polymerization and antibody immobilization using cross-linking reactions, Surf. Coat. Technol. 290 (2016) 116−123.

[15] V. Anand, S. Ghosh, M. Ghosh, G.M. Rao, R. Railkar, R.R. Dighe, Surface modification of PDMS using atmospheric glow discharge polymerization of tetrafluoroethane for immobilization of biomolecules, Appl. Surf. Sci. 257 (2011) 8378−8384.

[16] D.T. Clark, W.J. Feast, W.K. Musgrave, I. Ritchie, Applications of ESCA to polymer chemistry. Part VI. Surface fluorination of polyethylene. Application of ESCA to the examination of structure as a function of depth, J. Polym. Sci. Part A: Polym. Chem. 13 (1975) 857−890.

[17] D.T. Clark, A. Dilks, ESCA applied to polymers. XV. RF Glow-discharge modification of polymers, studied by means of ESCA in terms of a direct and radiative energy-transfer model, J. Polym. Sci. Part A: Polym. Chem. 15 (1977) 2321−2345.

[18] R.F. Baddour, R.S. Timmins, Application of Plasmas to Chemical Processing, MIT Press, Cambridge, 1967.

[19] F.K. McTaggart, Plasma Chemistry in Electrical Discharges, Elsevier, New York, 1967.

[20] F. Cabannes, J. Chapelle, M. Venugopalan, Reactions under plasma conditions, Spectroscopic Plasma Diagnostics, vol. 1, Wiley-Interscience, New York, 1971, p. 367.

[21] J.R. Hollahan, A.T. Bell, Techniques and Applications of Plasma Chemistry, John Wiley & Sons, New Jersey, 1974.

[22] M.C. Shen, Plasma Chemistry of Polymers, Marcel Dekker, New York, 1976.

[23] H. Yasuda, T. Hsu, Plasma polymerization investigated by the comparison of hydrocarbons and perfluorocarbons, Surf. Sci. 76 (1978) 232–241.

[24] H. Yasuda, Glow discharge polymerization, in: J.L. Vossen, W. Kern (Eds.), Thin Film Processes, Academic Press, New York, 1978, pp. 361–398.

[25] D.K. Lam, R.F. Baddour, A.F. Stancell, Fundamentals of plasma polymerization, in: M. Shen (Ed.), Plasma Chemistry of Polymers, Dekker, New York, 1976.

[26] H.U. Poll, M. Arzt, K.H. Wickleder, Reaction kinetics in the polymerization of thin films on the electrodes of a glow-discharge gap, Eur. Polym. J. 12 (1976) 505–512.

[27] H. Yasuda, Plasma Polymerization, Academic Press, New York, 1985.

[28] E. Kay, Invited paper at International Round Table on Plasma Polymerization and Treatment, in: IUPAC Symposium on Plasma Chemistry, Limoges, France, 1977.

[29] H. Yasuda, H.C. Marsh, E.S. Brandt, C.N. Reilley, ESCA study of polymer surfaces treated by plasma, J. Polym. Sci. Part A: Polym. Chem. 15 (1977) 991–1019.

[30] H.K. Yasuda, Competitive ablation and polymerization (CAP) mechanisms of glow discharge polymerization, in: M. Shen, A.T. Bell (Eds.), Plasma Polymerization. ACS Symposium Series, vol. 108, American Chemical Society, Washington, DC, 1979, pp. 37–52.

[31] D. Hegemann, M.M. Hossain, E. Körner, D.J. Balazs, Macroscopic description of plasma polymerization, Plasma Process. Polym. 4 (2007) 229–238.

[32] H. Yasuda, T. Hirotsu, Critical evaluation of conditions of plasma polymerization, J. Polym. Sci. Part A: Polym. Chem. 16 (1978) 743–759.

[33] B. Kim, K. Park, D. Lee, Use of neural network to model the deposition rate of PECVD-silicon nitride films, Plasma Sources Sci. Technol. 14 (2005) 83–88.

[34] F.F. Chen, J.P. Chang, Lecture Notes on Principles of Plasma Processing, Springer Science & Business Media, Berlin, 2012.

[35] A.R. Westwood, Glow discharge polymerization—I. Rates and mechanisms of polymer formation, Eur. Polym. J. 7 (1971) 363–375.

[36] H. Kobayashi, A.T. Bell, M. Shen, Plasma polymerization of saturated and unsaturated hydrocarbons, Macromolecules 7 (1974) 277–283.

[37] K.C. Brown, Polymerization in radio frequency glow discharges—I, Eur. Polym. J. 8 (1972) 117–127.

[38] K.C. Brown, M.J. Copsey, Polymerization in radio frequency glow discharges–II, Eur. Polym. J. 8 (1972) 129–135.

[39] H. Yasuda, Modification of polymers by plasma treatment and by plasma polymerization, Radiat. Phys. Chem. 9 (1977) 805–817.

[40] H. Yasuda, M.O. Bumgarner, J.J. Hillman, Polymerization of organic compounds in an electrodeless glow discharge. V. Amines and nitriles, J. Appl. Polym. Sci. 19 (1975) 1403–1408.

[41] H. Yasuda, T.S. Hsu, E.S. Brandt, C.N. Reilley, Some aspects of plasma polymerization of fluorine-containing organic compounds. II. Comparison of ethylene and tetrafluoroethylene, J. Polym. Sci. Part A: Polym. Chem. 16 (1978) 415–425.

[42] H. Yasuda, T.S. Hsu, Some aspects of plasma polymerization of fluorine-containing organic compounds, J. Polym. Sci. Part A: Polym. Chem. 15 (1977) 2411–2425.

[43] M.J. Vasile, G. Smolinsky, Organosilicon films formed by an RF plasma polymerization process, J. Electrochem. Soc. 119 (1972) 451–455.

[44] L.F. Thompson, K.G. Mayhan, The plasma polymerization of vinyl monomers. I. The design, construction, and operation of an inductively coupled plasma generator and preliminary studies with nine monomers, J. Appl. Polym. Sci. 16 (1972) 2291−2315.

[45] L.F. Thompson, K.G. Mayhan, The plasma polymerization of vinyl monomers. II. A detailed study of the plasma polymerization of styrene, J. Appl. Polym. Sci. 16 (1972) 2317−2341.

[46] V. Stundžia, P. Bílková, H. Biederman, D. Slavínská, P. Hlídek, Electrical properties of plasma-polymerized C: H films, Vacuum 50 (1998) 23   25.

[47] H. Biederman, R.P. Howson, D. Slavínská, V. Stundžia, J. Zemek, Composite metal/ C: H films prepared in an unbalanced magnetron, Vacuum 48 (1997) 883−886.

[48] H. Biederman, P. Hlidek, J. Pešička, D. Slavinska, V. Stundžia, Deposition of composite metal/C:H films—the basic properties of Ag/C:H, Vacuum 47 (1996) 1385−1389.

[49] H. Biederman, P. Hlidek, J. Pešička, D. Slavinska, V. Stundžia, J. Zemek, et al., Composite metal/C:H films prepared by unbalanced magnetron sputtering: Ni/C: H, Vacuum 47 (1996) 1453−1463.

[50] H. Biederman, D. Slavinska, Plasma polymer films and their future prospects, Surf. Coat. Technol. 125 (2000) 371−376.

[51] A. Choukourov, Y. Pihosh, V. Stelmashuk, H. Biederman, D. Slavinska, M. Kormunda, et al., Rf sputtering of composite SiO_x/plasma polymer films and their basic properties, Surf. Coat. Technol. 151−152 (2002) 214−217.

[52] Y. Pihosh, H. Biederman, D. Slavínská, J. Kousal, A. Choukourov, M. Trchova, et al., Composite SiO_x/hydrocarbon plasma polymer films prepared by RF magnetron sputtering of SiO_2 and polyethylene or polypropylene, Vacuum 81 (2006) 32−37.

[53] Y. Pihosh, H. Biederman, D. Slavínská, J. Kousal, A. Choukourov, M. Trchova, et al., Composite SiO_x/fluorocarbon plasma polymer films prepared by r.f. magnetron sputtering of SiO_2 and PTFE, Vacuum 81 (2006) 38−44.

[54] M. Drabik, J. Kousal, Y. Pihosh, A. Choukourov, H. Biederman, D. Slavínská, et al., Composite SiO_x/hydrocarbon plasma polymer films prepared by RF magnetron sputtering of SiO_2 and polyimide, Vacuum 81 (2007) 920−927.

[55] H. Biederman, Y. Osada, Plasma Polymerization Processes, Elsevier, Amsterdam, 1992.

[56] H. Yasuda, Plasma Polymerization, Academic Press, New York, 1985.

[57] M.A. Lieberman, A.J. Lichtenberg, Principles of Plasma Discharges and Materials Processing, John Wiley & Sons, Inc, New York, 1994.

[58] Y.P. Raizer, Gas Discharge Physics, Springer, Berlin, 1991.

[59] M.I. Boulos, P. Fauchais, E. Pfender, Thermal Plasmas: Fundamentals and Applications, Springer Science & Business Media, Berlin, 2013.

[60] J. Perrin, B. Despax, V. Hanchett, E. Kay, Microstructure and electrical conductivity of plasma deposited gold/fluorocarbon composite films, J. Vac. Sci. Technol. A 46 (1986) 46−51.

[61] H. Biederman, K. Kohoutek, Z. Chmel, V. Stary, R.P. Howson, Hard carbon and composite metal/hard carbon films prepared by a dc unbalanced planar magnetron, Vacuum 40 (1990) 251−255.

[62] R. d'Agostino, Plasma Deposition, Treatment and Etching of Polymers, Academic Press, New York, 1990.

[63] H. Biederman, Plasma Polymer Films, Imperial College Press, London, 2004.

[64] S. Craig, G.L. Harding, Composition, optical properties and degradation modes of Cu/(graded metal-carbon) solar selective surfaces, Thin Solid Films 101 (1983) 97−113.

[65] G.L. Harding, S. Craig, Magnetron-sputtered metal carbide solar selective absorbing surfaces, J. Vac. Sci. Technol. 16 (1979) 857—862.

[66] S. Craig, G.L. Harding, Structure, optical properties and decomposition kinetics of sputtered hydrogenated carbon, Thin Solid Films 97 (1982) 345—361.

[67] C. Laurent, E. Kay, Properties of metal clusters in polymerized hydrocarbon versus fluorocarbon matrices, J. Appl. Phys. 63 (1989) 1717—1720.

[68] C.P. Klages, R. Memming, Microstructure and physical properties of metal-containing hydrogenated carbon films, Mater. Sci. Forum 52 (1990) 609—644.

[69] M. Grischke, K. Bewilogua, H. Dimigen, Preparation, properties and structure of metal containing amorphous hydrogenated carbon films, Mater. Manuf. Process. 8 (1993) 407—417.

[70] V. Kumar, C. Jolivalt, J. Pulpytel, R. Jafari, F. Arefi-Khonsari, Development of silver nanoparticle loaded antibacterial polymer mesh using plasma polymerization process, J. Biomed. Mater. Res. A 101 (2013) 1121—1132.

[71] H. Biederman, L. Holland, Metal doped fluorocarbon polymer films prepared by plasma polymerization using an RF planar magnetron target, Nucl. Instrum. Methods Phys. Res. 212 (1983) 497—503.

[72] E. Dilonardo, A. Milella, F. Palumbo, G. Capitani, R. d'Agostino, F. Fracassi, One-step plasma deposition of platinum containing nanocomposite coatings, Plasma Process. Polym. 7 (2010) 51—58.

[73] A. Hozumi, H. Sekoguchi, N. Kakinoki, O. Takai, Preparation of transparent water-repellent films by radio-frequency plasma-enhanced chemical vapour deposition, J. Mater. Sci. 32 (1997) 4253—4259.

[74] A. Hozumi, O. Takai, Preparation of ultra water-repellent films by microwave plasma-enhanced CVD, Thin Solid Films 303 (1997) 222—225.

[75] A. Fujishima, T.N. Rao, D.A. Tryk, Titanium dioxide photocatalysis, J. Photochem. Photobiol. C: Photochem. Rev. 1 (2000) 1—21.

[76] C.A. Linkous, G.J. Carter, D.B. Locuson, A.J. Ouellette, D.K. Slattery, L.A. Smitha, Photocatalytic inhibition of algae growth using TiO_2, WO_3, and cocatalyst modifications, Environ. Sci. Technol. 34 (2000) 4754—4758.

[77] H. Hoppe, N.S. Sariciftci, Organic solar cells: an overview, J. Mater. Res. 19 (2004) 1924—1945.

[78] M.A. Lieberman, A.J. Lichtenberg, Principles of Plasma Discharges and Materials Processing, John Wiley & Sons, New York, 1994.

[79] R. Riedel (Ed.), Handbook of Ceramic Hard Materials, Wiley-VCH, Weinheim, 2008.

[80] A. Cavaleiro, J.T. de Hosson (Eds.), Nanostructured Coatings, Springer, New York, 2006.

[81] A. Flores, F. Ania, F.J. Baltá-Calleja, From the glassy state to ordered polymer structures: a microhardness study, Polymer (Guildf) 50 (2009) 729—746.

[82] B. Bhushan, Micro/nanotribology and materials characterization studies using scanning probe microscopy, Nanotribology and Nanomechanics, Springer, Berlin, Heidelberg, 2005, pp. 315—387.

[83] F. Gao, J. He, E. Wu, S. Liu, D. Yu, D. Li, et al., Hardness of covalent crystals, Phys. Rev. Lett. 9 (2003) 1015502.

[84] R.F. Landel, L.E. Nielsen, Mechanical Properties of Polymers and Composites, CRC Press, Boca Raton, 1993.

[85] F.J. Baltá-Calleja, A. Flores, F. Ania, in: G.H. Michler, F.J. Baltá-Calleja (Eds.), Mechanical Properties of Polymers Based on Nanostructure and Morphology, CRC press, Taylor and Francis, Boca Raton, 2005 (Chapter 8).

[86] B. Crist, Plastic deformation of polymers, in: E.L. Thomas (Ed.), Materials Science and Technology, vol. 12, Wiley Verlag, Weinheim, 1993.

[87] S. Fakirov, F.J. Baltá -Calleja, M. Krumova, On the relationship between microhardness and glass transition temperature of some amorphous polymers, J. Polym. Sci. Part B: Polym. Phys. 37 (1999) 1413−1419.

[88] F.B. Calleja, E.G. Privalko, A.M. Fainleib, T.A. Shantalii, V.P. Privalko, Structure-microhardness relationship in semi-interpenetrating polymer networks, J. Macromol. Sci. B 39 (2000) 131−141.

[89] T. Scrivani, R. Benavente, E. Pérez, J.M. Pereña, Stress-strain behaviour, microhardness, and dynamic mechanical properties of a series of ethylene-norbornene copolymers, Macromol. Chem Phys. 202 (2001) 2547 2553.

[90] V. Cech, J. Lukes, E. Palesch, T. Lasota, Elastic modulus and hardness of plasma-polymerized organosilicones evaluated by nano indentation techniques, Plasma Process. Polym. 12 (2015) 864−881.

[91] I.S. Bae, S.J. Cho, S.H. Jeong, H.J. Cho, B. Hong, J.H. Boo, Organic and inorganic hybrid-polymer thin films by PECVD method and characterization of their electrical and optical properties, Plasma Process. Polym. 4 (2007) S812−S816.

[92] M. Drabik, J. Hanus, J. Kousal, A. Choukourov, H. Biederman, D. Slavinska, et al., Composite TiO_x/hydrocarbon plasma polymer films prepared by magnetron sputtering of TiO_2 and poly (propylene), Plasma Process. Polym. 4 (2007) 654−663.

CHAPTER 6

Application of Plasma in Printed Surfaces and Print Quality

Joanna Izdebska-Podsiadły

Department of Printing Technology, Institute of Mechanics and Printing, Faculty of Production Engineering, Warsaw University of Technology, Warszawa, Poland

6.1 APPLICATION OF PLASMA IN PRINTED SURFACES

The basic printing substrate is paper, but printing can be done on various other types of materials: plastics, metal, glass, wood, ceramics, and textiles, etc. Beside paper substrates, plastics are the second-most commonly used group of printing substrates. The largest use of printed plastics is in the packaging industry (Fig. 6.1). The packaging materials are the largest group of commercial materials (counting flexible and rigid plastic together) and their market share is expected to increase even further in the next few years [1]. Therefore, attention will be paid in this chapter to the plasma modification of plastics rather than glass or metal.

Unlike paper, which is usually absorbent, plastics are nonabsorbent substrates. This means that during the printing process the ink does not penetrate into the printing substrate, but is only fixed on its surface. Therefore, problems with wettability of the substrate and adhesion of inks may occur during printing. Before the activation process, most polymers are characterized by the value of surface free energy that does not allow good wetting of their surface by printing ink and does not guarantee satisfactory adhesion. Wettability is one of the basic characteristics of printing substrates, which depends on the chemical composition and morphology of the material which, in turn, affects hydrophilicity. The value of the adhesion force is affected by: impurities on the surface of the material, surface typography, surface free energy, and the value of the difference between the surface free energy of the substrate and the surface tension of the printing ink. Therefore, before the printing process or before lamination, coating, metallization, extrusion, or other special processes on the plastic surface, it is necessary to modify it [2−4].

Non-Thermal Plasma Technology for Polymeric Materials
DOI: https://doi.org/10.1016/B978-0-12-813152-7.00006-8

Figure 6.1 An example of printed plastic packaging.

Modification of the surface layer of the polymer is carried out in order to change the hydrophilicity of the material, increase its surface free energy, create new functional groups, change the surface morphology, its roughness and crosslinking, as well as remove impurities or for sterilization or improving biocompatibility in the case of medical polymers [3,5–8]. Activation can be achieved by physical or chemical methods. The former are considered to be more environmentally friendly as they do not require the use of chemical solutions [9–12]. Physical activation methods include corona, plasma, flame, laser, ultraviolet, X-ray, and high-energy electron beam activation [3,13,14]. Corona activation is a common industrial method due to low cost and high process efficiency, as well as the possibility to apply it directly on the production line. However, the plasma methods allow for more uniformity of the modification of the surface layer of the material [15], moreover, it does not affect the bulk properties [16]. Additionally, plasma activation is more effective than corona activation; it enables a greater increase in surface free energy [17].

For several years now, printing facilities have been able to equip their printing machines with industrial atmospheric plasma systems. Currently, the number of these devices in relation to the number of corona systems is negligible, but due to the unique surface characteristics that can be achieved with the right gas and performance parameters, it complements the needs of the market by eliminating the need for primer coating of areas for which corona activation was not sufficient [17].

Printed plastic materials are primarily printed packaging and labels, but also all kinds of plastic products, finished and decorated by printing used in such industries as: construction, transport, automotive, electrical and electronic industry, and agriculture, etc. Printing not only gives or enhances the aesthetic value of the material, but also plays an important role in providing information. Printing on the product is also an important marketing tool, for example, in the packaging industry. Therefore, the quality of printing is very important. Printing quality is influenced by many factors [3,18]. The most important ones are: printing substrate and its possible preparation (modification of nonabsorbent media), printing technique, and type of printing inks used.

The most widely used plastics in the packaging industry, which are commonly printed or laminated are polyethylene (LDPE and HDPE), polypropylene (PP), and poly(ethylene terephthalate) (PET). Other printed plastics are poly (vinyl chloride) (PVC), polyamide (PA), polystyrene (PS), and polyamide (PC) [3,19]. In addition to plastics, interest in biodegradable plastics has increased in recent years. The most widely used are polylactide (PLA), cellulose, and starch-based plastics [20,21]. These biodegradable materials may be an alternative to traditional plastics and replace them, among others, in the packaging segment, where the packaging life is short and the amount of postconsumer waste is large and generates significant problems with its utilization.

Plastics can be printed in the form of film, rigid and semi-rigid sheets, or molded components. Regardless of the form in which they are used and printed, plastic materials are rather nonpolar hydrophobic materials that must have hydrophilic properties prior to printing.

Films are the largest group of plastic products. This name is used for materials of nominal thickness not exceeding 250 μm [22]. They are used mainly on the packaging market, which usually requires multicolor high quality printing. They are used in the form of monofilms and multilayer films, and laminates. The monofoils are homogeneous films made of one plastic. Laminates are multilayer films which are built of two or more layers of one plastic, different layers of several plastics, or combine the plastic or several layers with paper or aluminum foil [3]. The films are mainly printed with flexographic and rotogravure printing techniques. They can be supplied by the manufacturer as either corona or chemically activated materials, where the activation is (or is not) to be refreshed directly on the printing machine (due to the loss of activation during storage), or less frequently, can only be activated on the printing machine.

Semi-rigid and rigid plastic sheets are mainly used in the form of unorientated sheets with a thickness of 150–1600 μm. They are used primarily for thermoforming of products, which are later labeled or printed. After activation, the extruded products can be decorated using printing techniques such as dry offset, screen printing, or digital printing [23].

Molded components are, on the other hand, injection-manufactured products, which can have very different shapes and sizes. They are manufactured by injection molding a high-pressure plasticized polymer into the mold. They are often decorated directly in the manufacturing process (in-mold labeling technique), but also by direct printing on finished products (especially for low-volume production). Printing on the injection mold can be carried out using the following techniques: screen printing and pad printing, dry offset or digital thermal transfer. Before printing, the plastic molded components are generally subject to corona or flame activation [23,24].

Printing of plastics, regardless of the substrate form (film, sheet, molded components) and the printing technique used, requires that the problem of insufficient ink wettability and unsatisfactory adhesion to the substrate is addressed before starting the process. Printability depends primarily on: the properties of the substrate determined by its porosity and surface properties determined by the value of surface free energy [25]. The porosity of the material (i.e., the size and quantity of pores on the surface) influences its permeability, the ability to absorb liquid and greasy substances, that is, among others ink. Plastics are nonporous or very slightly microporous materials and are, therefore, nonabsorbent [18,26].

As plastics are nonabsorbent materials, the key determinant of printability is their surface free energy, which determines the wettability of the material and the adhesion of ink to the substrate. In the case of finished printing, it translates into the thickness of the fixed ink layer, that is, the optical density of the print (see Fig. 6.2). In addition to ensuring that the surface free energy (SFE) value for the substrate is adequate, it is also important to choose a suitable ink with proper surface tension. It is always worth remembering that the surface tension of the ink should be lower than the surface free energy of the substrate. Some practitioners and theoreticians claim that the difference between these values of at least 10 mN/m guarantees good printability for a given set of ink-base

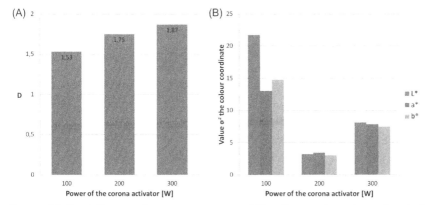

Figure 6.2 The influence of corona treatment of PLA film on the values of optical density and colour coordinates of prints.

[27,28]. However, on the basis of my own work, this is not always a sufficient determinant [29].

6.2 NONTHERMAL PLASMA FOR PRINTED PLASTICS

Plasma is a totally or partially ionized gas which contains electrons, ions, high-energy neutrals, and radicals. It is considered to be the fourth state of matter, which is the greatest part of the matter of the universe. The basic method for plasma classifying is the classification based on relative energy levels of electron and heavy species and the particle distribution function. According to this classification, it is distinguished by thermal and nonthermal plasma. Only low-temperature plasma is suitable for modifying plastics and biodegradable plastics due to their thermal sensitivity [5,10,20,30−32].

Nonthermal plasma is otherwise referred to as cold, low temperature or nonequilibrium plasma. Cold plasma is formed by electrical discharges in gas, during which UV radiation is usually emitted. Plasma can be produced when gases are excited into energetic states by radio frequency (RF), microwave (MW), or electrons from a hot filament discharge. Plasma contains reactive species such as fast-moving electrons, ions, free radicals, metastables, and photons in the short-wave ultraviolet range, which initiate physical and chemical reactions on the polymer surface. The overall plasma gas temperature is lower than 473 K, so it can be used for the modification of plastics [20,33].

Low-temperature plasma activation can be performed as a low pressure plasma or plasma under atmospheric conditions, depending on the device. Low-pressure plasma requires the use of vacuum chambers, and the process is carried out under pressure from 1 Pa to 1 kPa [10], with the most common values used being those in the range of 10−75 Pa [9,19,34−37]. Low pressure values make it easier to control the plasma reaction, as discharges are more stable [20]. Undoubtedly, the use of chambers is a significant limitation of the application of this process on an industrial scale due to the reduction of sample size resulting from reactor size and its inability to be used in continuous production in the production line (batch or semi-batch processes). Nonetheless, the use of low-pressure plasma has many advantages. The main ones are: high process repeatability associated with its easy control, high homogeneity of the modified surface, high concentration of reactive species, the best chemical selectivity allowing to obtain the desired surface properties ([10,30]; Bryak et al., 2009).

An alternative to a low-pressure plasma that solves the problem of vacuum chambers may be atmospheric plasma. Atmospheric plasma systems operate at standard atmospheric pressure and can work with high-speed printing or laminating machines, without restricting their speed. The disadvantage of such solutions may be the costs resulting from the high consumption of process gases. In order to get the right plasma atmosphere and to obtain a drift of the reactants to the surface, it is necessary to ensure a high flow rate of process gas.

Among the types of atmospheric plasma, we can distinguish are dielectric barrier discharge (DBD), atmospheric-pressure glow discharge (APGD), and atmospheric-pressure plasma jet (APPJ). Although in some research corona discharges are considered as a kind of atmospheric plasma [10,30], in this study corona activation is treated as another type of surface modification and is not included in plasma activation. The subject of corona activation of polymeric substrates before the printing process was described in another publication [38]. DBD plasma has a higher electron density compared to corona activation, but, like corona activation, can lead to uneven substrate activation. This is caused by micro discharges that are not completely uniform and have a short discharge time. The APGD plasma is more uniform and homogeneous than the DBD plasma and gives rise to a better uniformity of the activated plastic film. APPJ plasma is the latest type of glow discharge at atmospheric pressure used to modify the surface of plastics. It produces a more homogeneous gas than DBD plasma. In addition, it can be used for molded components and not just flat materials such as DBD [10,30].

Choosing a plasma type is strongly dependent on the effects of modifications we want to obtain, the costs we can incur, the productivity of the process we require. Each type of plasma, whether a low-pressure plasma or a kind of atmospheric plasma, has some advantages and disadvantages.

6.3 MECHANISM OF PLASMA TREATMENT

As a result of plasma activation, as well as corona activation, physical and chemical digestion, and chemical surface modification occur. Plasma activation affects only the surface of the material, without affecting its bulk properties. It is widely believed that changes are taking place in the top layer limited to 5−10 nm [39,40], although some have shown that the changes may be more extensive and even change the bulk properties [41].

Polymer ablation and grafting of new species on polymer surfaces are the main two mechanisms of plasma treatment. Both processes run simultaneously, but to a different degree depending on the type of plasma, the material, and the conditions of the process. In case of the plasma activation of polymers, more attention is paid to grafting polar species on the surfaces, considering it to be the main mechanism of activation [42].

The etching causes ablation and surface cleaning, that is, removal of organic impurities from the surface. It improves adhesion due to the removal of impurities and increased surface roughness. Indirectly, however, it can also influence the change of wetting, since at higher roughness values ($R_a \geq 100$ nm) the increase in roughness affects the change of the contact angle [43].

The changes in chemical properties are responsible for achieving the desired hydrophilic or hydrophobic properties, which are important for the proper printing process due to good or bad printing-based surface wetting. In order to prepare the material for the printing process or other decorative processes, it is important to increase its hydrophilicity. Changes in the chemical properties of the surface polymer layer are visible in the form of reduction of C−C and C−H bonds and increase in the number of free radicals. New functional groups are created on the surface of the material [30,32,44,45].

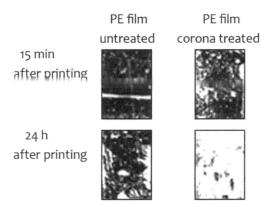

Figure 6.3 The influence of corona treatment of PE film on ink adhesion to the film; images of adhesive tape fragments after adhesion tests performed 15 min after printing and 24 h after printing.

6.3.1 Polymer Ablation

Surface ablation (surface destruction) of the polymer is a result of complex processes occurring during the exposure of the material to the plasma. The plasma components react with the polymer chain during the activation process, changing surface properties, and leading to polymer degradation. The top layer of the polymers (e.g., PLA, PP) forms low molecular mass and gaseous products and crosslinking processes may occur. The ablation rate can be determined by gravimetric analysis [46]. Ablation is important in eliminating weak domains and increasing surface roughness [30], which has a significant impact on the adhesion of ink to the substrate (see Fig. 6.3) or the strength of laminate bond, etc.

6.3.2 Grafting of New Species on Polymer Surfaces

The polymer-plasma activation is carried out primarily in order to achieve the desired properties of the material, enabling further processing. From the industrial polymers application point of view, this will be improved adhesion and increased surface free energy value due to grafting of new polar species on the surface [47]. These features are combined with functions such as hydrophilization, hydrophobization, adhesibility, printability, and paintability, etc. However, functionalization can be categorized not only by the function, but also by the chemistry of the process, such as oxidation, nitration, and fluorination [30].

During chemical surface modification, the bonds $C-C$ and $C-H$ are bursting and free radicals are formed on the surface. In connection with the emergence of free radicals, the following processes may take place: recombination, unsaturation, branching, and crosslinking [30]. Surface functionalization and the type of functional groups generated on the polymer surface depend on the gas used for plasma activation [9,20,48].

6.4 EFFECTS OF PLASMA ON POLYMER SURFACE AND THEIR IMPACT ON PRINTABILITY

Plasma can be used to modify the surface before printing or to improve the print properties afterwards. In the first case, just like the traditional corona activation, it has to improve the printing properties of the substrate and enable high-quality printing. In the latter case, however, the aim is to obtain or improve the usability of the printed image. Both plasma applications are relatively new and are still evolving.

Plasma activation used before the printing process is primarily intended to improve the hydrophilicity of the substrate and can be performed using such gases as He, Ar, N_2, NH_3, and O_2, etc. In addition, some types of plasma (e.g., Ar) can significantly change the roughness of the substrate [20]. The effects of the individual gases used during plasma activation on the properties of the modified surface are discussed in detail in Chapter 5, Plasma-Induced Polymeric Coatings.

The basic test parameters to determine the influence of activation on the polymer surface layer are contact angle measurements with water, or extended with other measuring liquids. Knowing the contact angles of at least two fluids allows one to determine the value of surface free energy [49]. The assessment of chemical changes in the top layer and the identification of emerging functional group types is possible with Fourier transform infrared spectroscopy and X-ray photoelectron spectroscopy, energy dispersive X-ray spectrometers, and Time-of-Flight Secondary Ion Mass Spectrometry (TOF-SIMS). Changes occurring in surface morphology are analyzed using atomic force microscopy (AFM), optical microscopy, and scanning electron microscopy (SEM). AFM measurements also allow one to determine the change in surface roughness values, which affects adhesion and wettability of the material. Sometimes changes in material barrier properties, polymer crystallinity, elongation and tensile strength, and dynamic mechanical thermal properties, etc., are also evaluated. [14,45]. Such tests are important if the modified plastic is a packaging material as changes in the properties of the material may affect its suitability for protecting the product.

The application of plasma activation on already printed material is mostly in the functionalization of printed electronics. For example, oxygen plasma treatment of temperature sensors printed with screen printing on PEN and PET has made it possible to increase sensitivity [50]. Low temperature hydrogen plasma, on the other hand, has been successfully used for sintering of prints done with copper complex ion ink with inkjet on film substrates. The plasma treated pattern shows better microstructure and electrical properties than the pattern after conventional hydrogen thermal treatment [51].

6.4.1 Changes of Contact Angle

The contact angle with water is the basic parameter used to assess the hydrophilicity or hydrophobicity of a material. In addition, its measurements are commonly used to evaluate the plasma activation of plastics (see Fig. 6.4). Materials for which the contact angle is less than 90 degrees with water, are considered to be hydrophilic and for values above, the material is said to be hydrophobic. The smaller the contact angle, the more easily the surface is wetted with liquid, and at an angle close to 0 degree the liquid properly dissolves over the surface [52–54]. In order to obtain high quality prints, the surface of the substrate must be properly wetted by the ink that is available for the hydrophilic surface.

The plasma activation of plastics and biodegradable plastics is performed mainly in order to improve the hydrophilicity of the material, assessed by lowering the contact angle value with water. The change of contact angle due to plasma activation will depend on the type of process

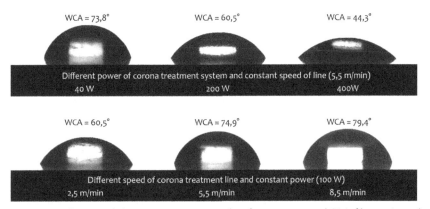

Figure 6.4 Change of the water contact angle of corona-treated PLA film activated with different power of corona treatment system and with different line speed.

gas used, the type of activated material and its structure, as well as the type of plasma activation and process parameters. Determination of the effect of a given type of plasma activation on the contact angle of a modified material requires tests to be carried out for this contouring material under specified process conditions. A change of the gas used, or an introduction of a gas mixture instead of a pure gas, a slight change of the gas pressure for a low-pressure plasma, change of the device's operating power, prolongation or shortening of the modification time, and even the omission of crystallinity of the material, will have an impact on the reduction of the contact angle.

As an example, Chen et al. [55] noted as a result of activation of PS with RF plasma with Ar that the contact angle of water was reduced from 88.9 to 14.3 degrees. The efficiency of the cold atmospheric pressure (CAP) atmospheric pressure plasma with Ar on polyethylene film (LDPE) was significantly lower and the contact angle decreased from 85.2 only to 83 degrees [40]). In other studies, using medium pressure DBD plasma with Ar on LDPE, very large changes (from c. 104 to c. 51 degrees) were observed [56]). Wang et al. [57] observed a maximum change in the contact angle from 69 to 34 degrees using plasma jet with Ar on PMMA, with the initial contact angle decreased with the time of modification, and then, with an increase in the activation time to more than 120 s, it no longer decreased, remaining at a minimum value.

Vesel et al. [34], using low pressure plasma on PET with oxygen and nitrogen, observed a decrease in contact angle as a result of activation, from an initial value of 75 degrees for the inactivated sample of to 20 and 25 degrees respectively for samples modified for 3 s. As a result of an increase in the exposure time regardless of the type of gas, a further decrease in contact angle has been observed, with a minimum contact angle of approximately 10 degrees achieved for nitrogen plasma after 1 min, meanwhile with oxygen plasma a further decrease in contact angles has been observed for samples with longer treatment time. The oxygen plasma-based activation was also used by Vishnuvarthanan and Rajeswari [19] but PP was used as a substrate. Even for the most aggressive activation conditions, contact angle changes were not as high. They were only able to reduce the contact angle from 74.56 to 43.85 degrees.

Yeo et al. [58] investigated the influence of plasma activation with O_2/Ar plasma on the change in hydrophilicity of FOTS-coated PET film. The contact angle of pure PET was 74.2 degrees, after coating it increased to 114.1 degrees, and after activation it decreased to

29.5 degrees. Activation was effective even for very hydrophobic surface, and what is more, thanks to the application of an electroformed nickel stencil mask it was possible to selectively modify the material and obtain hydrophobic/hydrophilic surfaces. PET substrate at different stages of modification (pristine, with FOTS silane coating, with coating and plasma treatment) was printed using the ink-jet technique. It was found that selective activation of the hydrophilic film allows one to obtain hydrophilic, well-wetted areas (activated areas) and hydrophobic, independent barriers, which enables controlled ink expansion.

In recent years, there has been a growing interest in biodegradable plastics, for which quite a number of studies have been conducted on the impact of activation on changes in hydrophilic properties. One of the most commonly used biodegradable polymers, both in medical and industrial applications, is PLA. Just like traditional plastics, PLA can be used as a printing substrate and is an alternative to traditional plastics in packaging and, similarly, it also needs to be modified to lower the contact angle with water. The contact angle of unmodified polymer with water is usually within 70–80 degrees, which means that the material can only be printed with selected printing inks [29].

The research on the improvement of PLA hydrophilicity was conducted by, among others Jordá-Vilaplana et al. [59], Moraczewski et al. [46], Pankaj et al. [31], Kim and Masuoka [60], De Geyter [5,61,62], Boselli et al. [63], Demina et al. [64], Garcia-Garcia et al. [65], Jacobs et al. [66], and Kudryavtseva et al. [67]. Jordá-Vilaplana et al. [59] used air atmospheric plasma, and depending on the conditions of the process, they stated that the contact angle was reduced from 73.4 degrees to less than 60 and 30 degrees respectively for the least and most aggressive conditions. Reducing the contact angle with water as a result of low pressure plasma with oxygen activation for maximum activation time (30 min) was even greater, from 78 to 17 degrees [46]. The use of CO_2, on the other hand, caused a significant degradation of the material (pinholes visible on the surface) and a smaller change in the contact angle, namely from ~ 80 to ~ 45 degrees [60]. Changes in the contact angle at a similar level were also recorded by De Geyter [61] using medium pressure DBD plasma with air and helium mixture (reduction from 75 to 40 degrees). The plasma, using only air alone, reduced the contact angle to a lower level of wetting, and with helium to the highest level, a reduction to 59 and 36 degrees was observed respectively. The contact angle values of 37–44 degrees for low-pressure plasma with O_2 and Ar were also observed by [5,62], with lower values obtained for an oxygen plasma.

6.4.2 Changes of Surface Free Energy Values

In addition to the water contact angle, contact angle measurements of diiodomethane are commonly performed in wettability studies. Unlike water, which is a very polar liquid, diiodomethane is a nonpolar liquid. The knowledge of contact angles with these fluids is commonly used to determine the values of surface free energy using the Owens-Wendt method [49,68]). However, surface free energy can also be determined by a number of other methods. In addition to the methods developed by Owens and Wendt, methods developed by Berthelot, Antonow, Girifalco and Good, Neumann et al., Fowkes, Zettlemoyer, Owens and Wendt, Wu, van Ossa-Chauhury-Good, and Zisman are available. Therefore, apart from water and diiodomethane, such liquids as formamide, ethylene glycol, and 1-bromo naphthalene, etc., can be used for contact angle measurement [52,53].

The surface free energy in the printing process determines the wettability of the material (see Fig. 6.5) and the ink adhesion to the substrate. In the case of finished printing, it influences the thickness of the fixed layer of ink, and thus densitometric parameters of prints, that is, its optical

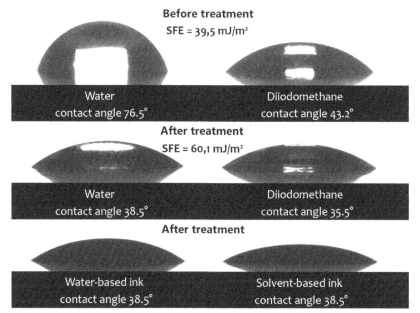

Figure 6.5 The contact angle measurements of different liquids on untreated and low-pressure argon plasma-treated PLA film and its surface free energy values.

density. One of the ways to improve nonabsorbent substrates' printability is to increase the surface free energy of the material by activating it. In the case of printing plastics with liquid inks, it is particularly important to increase the polar SFE component of the substrate [69]. If the modification is not sufficient or cannot be carried out, an attempt may be made to change the ink composition in order to reduce its surface tension or to use another ink with a lower surface tension. In the literature [27,28] there is information on the difference between the SFE value of the substrate and the surface tension of the ink at the level of 10 mN/m or higher guarantees high quality printing. In case of biodegradable plastics, as shown by previous research [29] such a value, or even higher, does not always guarantee high quality flexographic prints, however, in most cases an increase in this value has a positive effect on the obtained results.

The surface tension of the ink depends on its type and formula. The largest range of surface tension values, depending on their type and purpose, is characterized by ink-jet inks (22−45 mN/m), slightly smaller flexographic inks (28−38 mN/m, with lower values of up to 32 mN/m being typical for solvent-based inks, and higher for water-based inks), while offset and lithographic inks have surface tension of approx. 37 mN/m [70,71]. The values of surface free energy of selected unmodified polymers are as follows: LDPE 32−33 mJ/m^2, PP 29−30 mJ/m^2, PET 35−47 mJ/m^2, PMMA 33−49 mJ/m^2, PS 29−44 mJ/m^2, PVC 36−39 mJ/m^2, PLA 40−50 mJ/m^2, and starch film $\sim$60 mJ/m^2 [4,72]. Following the principle that the surface tension of the ink should be at least 10 mN/m below the surface free energy value, it can be concluded that some substrates for some inks do not require activation, which may (or may not) be appropriate. Activation does not have to be used to increase the SFE value, but can be used to clean the surface and improve adhesion. Generally speaking, it should be the rule that for new printing substrates and new inks tests are always carried out to assess whether a given level of SFE allows good wettability of the substrate by the ink, and whether adhesion of the fixed ink layer is satisfactory. If a satisfactory result is obtained without substrate activation, no activation is required.

It is generally agreed that, for example, PP for printing requires an SFE of at least 37 mJ/m^2 and for lamination of 42 mJ/m^2, while PE should have an SFE value of 46−52 mJ/m^2 before printing. For PBAT and PBAT-PLA Künkel et al. [73] stated that if the SFE value is less than 38 mJ/m^2 then activation of the substrate is necessary. After ensuring appropriate wettability, printing can be carried out with water-based and alcohol-based inks.

Plasma activation allows to increase the surface free energy of polymers by several dozen percent (up to 85%). For example, as a result of plasma activation of PLA using air atmospheric plasma, the value of SFE increased to 57 or 59 mJ/m^2 [74,75], using low pressure plasma with oxygen to over 60 mJ/m^2 [76] and for oxygen plasma at exposure for 30 min up to 75 mJ/m^2 [46]. For PET activated with SDBD plasma with oxygen a maximum value of SFE 82.6 mJ/m^2 was observed, with a slightly lower value of 80.2 mJ/m^2 for the same material activated with nitrogen plasma [77]. By using an oxygen plasma activation for PP it was possible to increase the SFE of the activated material to 69.4−94.0 mJ/m^2 depending on activation time and power [19]. On the other hand, the value of SFE increased to over 60 mJ/m^2 due to the APP plasma activation of PE [78].

6.4.3 Chemical Changes of Top Polymer Layer

Chemical changes of the surface layer of plasma-treated plastic are intended to introduce various functional groups in order to increase hydrophilicity and achieve specific surface functionalities [30,79]. During discharges, free radicals are formed that can react with reactive molecules present in the atmosphere (oxygen, ozone, water molecules, and hydroxyl groups, etc.) or with each other, modifying the surface chemically. This creates polar compounds in the top layer of the material. Creating strong polar function groups with a simultaneous decrease of nonpolar groups has a very positive effect on wettability. The introduction of oxygen-containing polar functional groups improves adhesional surface properties [30].

Plasma modification reduces the amount of carbon atoms and increases the amount of oxygen atoms. The changes may be significant [20,57,75] or only minimal ones [62]. The main groups identified in the top layer of plasma-modified plastics are: ketone, aldehyde, hydroxyl, carboxylic, and carbonyl groups [30]. For example, Wang et al. [57] reported an increase in the number of oxygen groups (C−O, C−OH, O=C−O) on the PMMA surface as a result of plasma action of Ar and He. For argon plasma, the share of C−C groups decreased from 58.5% to 53.93%, while C−O groups increased from 24.23% to 27.49%, and O=C−O groups from 17.27% to 18.58%. Van Deynse et al. [56] noted an increase in the surface area of oxygen groups as a result of plasma activation of

LDPE with Ar plasma, with the highest increase observed for C−O and O−C=O groups. As a result of PLA activation with argon plasma, similar observations were made (concentration of C−O and O−C=O groups is increased, while the number of C−C and C−H groups is decreased) [62]. The activation of PLA plasma from CO_2 caused an increase in the intensity of −C=O groups [60].

The use of nitrogen or helium plasma during modification leads to nitrogen groups appearing on the surface of the material [57,61]. Their presence in the top layer of the material has a beneficial effect on the wettability of the substrate.

6.4.4 Changes of Surface Topography

As a result of plasma activation, the surface topography changes and surface roughness is usually observed to increase significantly [40,57,75,80], although sometimes there may also be a slight decrease [81]. In addition, it has been observed that the increase in activation time and power increase leads to further increases in roughness [10,14,19,46,59,78]. Increased roughness will result in the polymer surface development, which translates into better adhesion of ink to the substrate.

The influence of roughness on wettability depends on roughness values and water contact angle of the material. Busscher et al. [43] found that there are certain correlations between the roughness of the substrate and its contact angle. The contact angle is affected by roughness with R_a of 100 nm or more, but the impact type depends on the contact angle. For relatively hydrophobic substrates, for which the contact angle is greater than 86 degrees, the increase in surface roughness leads to an increase in the contact angle. With contact angle values of 60−86 degrees, the change in roughness does not affect the contact angle. If the contact angle is less than 60 degrees, the increase in roughness results in a reduced contact angle.

6.4.5 Changes in the Surface Crosslinking

The polymer surface can be crosslinked by plasma activation with inert gas [79]. The effect is called CASING (crosslinking by activated species of inert gases). Furthermore, it is believed that energy ions contribute to the crosslinking of polymer chains. Vacuum ultraviolet energy (VUV) photons can also act on the polymer surface, leading to crosslinking [9].

The surface crosslinking effect is achieved by using argon and helium plasma. The cleanliness of the gas used plays an equally important role in the crosslinking process. If it contains contaminants, the activation effect may be completely the opposite [30].

The crosslinking of surfaces may improve the cohesive strength of the surfaces [30]. Furthermore, according to some studies, surface crosslinking may mitigate the effects of reorientation. It may impact the mobility or the rate of diffusion of low molecular-weight oxidized material into the bulk [9,10]. In addition, the crosslinking leads to increased bond strength, reducing the risk of material breakage. This has a positive effect on the printing process on high-speed web machines, where the film can sometimes be ripped off.

6.4.6 Surface Cleaning

Activation is not always carried out in order to improve wettability and increase the surface free energy of polymers. Sometimes the SFE value is high enough, but the substrate requires the removal of impurities and contaminants such as oils, greases, and oxides from the substrate surface to improve adhesion [10]. The low-molecular-weight oligomers on the plastic surface may also be a reason for the reduced adhesion of ink or adhesive to the substrate. The reduced binding strength will result from the weak binding of the oligomer layer with the substrate [75]. In addition, the contact angle will be lowered as a result of cleaning by removing organic impurities.

Plasma surface cleaning is a very effective method, working in two ways. Firstly, the removal of contaminants from the surface is carried out using VUV. Thanks to VUV energy most of the organic bonds (i.e., $C-H$, $C-C$, $C=C$, $C-O$, and $C-N$) are broken. The second cleaning process involves reactions between oxygen species created in the plasma (O_2^+, O_2^-, O_3, O, O^+, O^-, ionised ozone, metastably-excited oxygen, and free electrons) and organic contaminants. The reaction produces H_2O, CO, CO_2, and lower molecular weight hydrocarbons. The surface cleaned in this way is ultra-pure (sterilized) [31]. Cleaning of low-pressure surfaces in plasma equipment is carried out under pressure in the range of $10-200$ Pa [33]. The cleaning process is environmentally friendly, as small amounts of carbon dioxide and water vapor are produced as by-products with traces of carbon monoxide and other hydrocarbons [14,82].

6.5 INFLUENCE OF PLASMA PARAMETERS ON WETTABILITY AND PRINTABILITY

The substrate printability and printing quality depend on the characteristics of the treated printing substrate, while the various modification effects are achieved by changing plasma parameters. The commonly controlled parameters of the plasma process are: the chemical composition of the gas used, gas pressure, activation time, and parameters of the feeder system [5,10]. The influence of gas, pressure, and activation time on wettability, adhesion, and print quality will be discussed based on available studies.

6.5.1 Gas Composition

Plasma activation can be performed using various gases (Ar, O_2, He, CO_2, N_2, NH_3, H_2, CF_4, SF_4, C_4F_8, or air) and their choice depends on the expected effects of surface modification [20,42,75,83]. For example, most of these gases will be used to increase the energy of the surface material. However, if the wettability improvement is to occur directly as a result of oxygen functionalities during activation, an oxygen plasma should be used. Using other gases (with the exception of those containing fluorine) the wettability will be improved by postplasma reactions [84]. When nitrogen is applied, groups with atoms of nitrogen are formed on the surface. If the purpose of activation is to reduce chemical activity or to induce release properties, fluoride plasma should be used [10,84,85]. Good hydrophobic effects were achieved by, for example, Chaiwong et al. [86] using SF_6 plasma.

Gases such as CO_2, O_2, N_2, NH_3, H_2, or their blends (e.g., $N_2 + O_2$) should be used when reactive groups such as hydroxyl ($-OH$), carboxylic ($-COOH$), and amine ($-NH_2$) [87,88] are to be produced on the surface. Ar or He gases, on the other hand, can be used to produce radicals that will react with oxygen or water in the air to form peroxides, which will be able to initiate further the grafted polymerization of monomers (acrylic acid with hydrophilic surface preparation or surface crosslinking) [89]. However, if plasma activation is used in etching the polymer layer, gases such as argon, helium, nitrogen, and oxygen can be used [10].

A common problem in printing plastics, which makes it impossible to make a positive assessment of quality, is the unsatisfactory adhesion of ink to the printing base. Gas selection is important due to the type of functional groups that will be created on the surface. Some functional groups may not be relevant to improve adhesion and some may even have a negative impact. The desired function groups can be controlled by changing the gas and switching the activation process parameters [9].

The type of gas affects not only the type of chemical changes occurring on the surface, but also its roughness. Argon plasma activation causes a significant increase in roughness and formation of free electron pairs in the carbon atoms. Significantly smaller changes are observed during activation with oxygen and air [85].

Air and oxygen are commonly used gases for reasons of availability and cost. Argon is the most commonly used noble gas. Its high use results from high efficiency of ablation, chemical indifference to the surface of the modified material and relatively low application costs [10]. Apart from argon, helium is also very important [57]. This is due to the fact that the plasma stream induced by both gases is very stable.

Due to the significant influence of process parameters on the activation effects achieved for a given material, it is difficult to compare the available test results. However, publications are available where plasma activation with different gases has been used for the same material under the same conditions, which makes it possible to determine their impact on the material and compare process efficiency. And so, for example, De Geyter et al. [90] stated that, although plasma activation leads to a slight increase in surface roughness of the PLA surface regardless of the type of gas used (air, nitrogen, nitrogen, helium, or argon) the biggest changes occurred at argon activation, while the smallest changes occurred with helium plasma. The type of gas also determined the chemical composition of the PLA surface. While air and Ar generated oxygen groups, N_2 caused the formation of nitrogen groups. Helium also caused a small number of nitrogen groups to develop.

By contrast, De Geyter et al. [61], when comparing the effects of helium plasma, air and a two-gas mixture, found that helium plasma is the most effective in increasing hydrophilicity of PLA. Even better results from De Geyter et al. [90] were obtained by using nitrogen as a process gas during plasma activation of PLA.

The high effectiveness of the helium plasma was also confirmed by Wang et al. [57]. They used APPJ plasma with He and Ar on PMMA. As a result of the study, they found that helium plasma gave better results, and in addition, the aging process of the activated material was slower.

Zhang et al. [91] in turn used atmospheric plasma in the air and gas composed of a mixture of air and helium on polyester. The air helium addition used enabled the activation process to be more effective.

Moreover, the activation significantly improved wettability and roughness of the material, which had a positive impact on the amount of ink fixed on the surface and the quality of ink-jet printing.

The low-pressure plasma with oxygen and argon used for PLA modification was compared by Izdebska-Podsiadły [35]. A better wettability improvement (lower contact angle) was achieved using an oxygen plasma than argon plasma. However, the repeatability of results and homogeneity of changes occurring on the surface were greater for argon plasma. Differences observed in the improvement of wettability have an impact on wetting of modified substrates with flexographic solvent-based inks, however, no such effect was observed in the case of water-based inks. However, both types of activation had a positive effect on the wettability of the material with both flexographic inks.

Kang and Oh [92] studied the effect of added argon and oxygen to C_4F_8 used for substrate activation prior to ink-jet printing. The increase of O_2 or Ar gases resulted in a decrease of the density of the fluorine functional group in the plasma that have hydrophobic properties. This led to improved hydrophilicity of the substrate, which translated into an increase in the diameter of dots printed on the substrate. In another study, Kang et al. [93] used a gas mixture of C_4F_8 and O_2, which similarly increased the addition of oxygen, increased wettability, and increased the diameter of the printed dots.

Fleischman et al. [94], on the other hand, used an atmospheric plasma with two different gas mixtures (helium and oxygen and helium and water vapor) to modify the substrate with polyethylene naphthalate (PEN). Both plasma types have improved wettability and increased surface free energy, allowing good quality, uniform lines to be printed using ink-jet. Precise analysis by SEM of individual droplets has shown, however, that the He/H_2O plasma has enabled better results. Drops printed with PEDOT:PSS dispersion on PEN modified with this plasma were characterized by greater homogeneity and less agglomeration of PEDOT: PSS.

Park et al. [95], prior to ink-jet printing, used a hydrophobic O_2 and Ar plasma coating to change the wettability of PI substrate. With the increase in activation time, regardless of plasma type, they observed a greater removal of hydrophobic film due to surface etching, with argon plasma being larger than for an oxygen plasma. Also, the print dots were larger in size range, $38-70\,\mu m$ and $38-92\,\mu m$ for O_2 plasma and Ar plasma respectively.

6.5.2 Pressure

Low-temperature plasma activation is performed at atmospheric pressure or low pressure. The pressure applied during low-pressure plasma activation has an impact on the efficiency of the process and the changes in the top layer. The number of publications testing the pressure effect is limited, usually the process is carried out with a fixed value, however, in some publications this parameter is used as a variable.

Izdebska-Podsiadły [76] has shown that the gas pressure during plasma activation affects the contact angle. A greater effect was observed for oxygen activation than for PLA argon activation, however, regardless of the type of gas, lower gas pressure allowed to obtain lower contact angle values with water. The influence of gas pressure on wettability was also reported by Kang and Oh [96]. They stated that plasma power and gas pressure have a significant influence on the size of droplets printed on PI substrate modified with C_4F_8 plasma.

Kuvaldina et al. [97] have shown that the change in pressure determines the chemical properties of the surfaces and the type of functional groups to be formed. They noted that increased pressure leads to higher concentrations of ester groups, aldehyde and ketone carbonyl groups, and lactone groups, while reducing the concentration of double bonds. However, the concentrations of various OH and carboxylic groups weakly depend on the pressure. Grace and Gerenser [9] studied the influence of the parameters of the feed and pressure system on the creation of specific nitrogen functionalities. They did not find, for the range studied, that there was a significant influence of pressure on plasma density, but noted that the increase in pressure had an effect on increasing the concentration of nitrogen atoms.

In addition, the pressure may also influence the crystallinity of the polymer. Pressure increase during PP argon plasma activation caused a decrease of its crystallinity [98]. Crystallinity change, in turn, affects the aging of the modified material.

6.5.3 Exposure Time

The influence of exposure time on wettability and, thus, the improvement of plastic printing capacity depends, among other things, on the type of activated material and plasma type used. Řezníčková et al. [81] observed a greater effect of activation time on the change of water contact angle with PTFE than HDPE, PS, PET, and PP. Regardless of the activation time and type of material, however, they found that plasma activation with Ar had a beneficial effect on the wettability of all materials, significantly lowering the contact angle even when using the shortest duration of activation.

In the studies conducted by Izdebska-Podsiadły [35] it has been observed that PLA activation time has a greater effect on wettability with oxygen plasma, while for argon plasma it is only minor. The smallest contact angle was obtained for the shortest activation time with oxygen plasma activation. Extending the activation time adversely affected wettability, but the very long activation time extension has resulted in the contact angle decreasing again. In the case of argon plasma activation, the smallest contact angle was obtained using the longest activation time (10 min), however, differences between values were very small.

The fact that the increase in activation time does not always have a positive effect on the improvement of wettability was also observed by Moraczewski et al. [75]. Using air atmospheric plasma, they found that the lowest contact angle value was obtained by using medium exposure times (5 and 10 min) and that the increase in treatment time led to an increase in contact angle value. Goo et al. [99] also obtained good results for shorter, and not the longest, times. They used a variable activation time with PI oxygen plasma to prepare the substrate for ink-jet printing. Satisfactory results were achieved for a relatively short period of time, namely 3 s, with an activation time range of $1-120$ s. As a result of the substrate treatment, it was possible to make continuous lines with a well-sintered microstructure with nickel conductive paint.

However, numerous studies have reported that the increase in activation time has further improved wettability [19,34,46,77,81]. Vesel et al. [34], using low pressure oxygen and nitrogen plasma on PET, observed a beneficial effect of the activation time extension on the wettability of the material (further reduction of contact angle). However, for nitrogen after 60 s, the further increase of the activation time did not change the contact angle. Novák et al. [77] used the same gases (nitrogen and oxygen) to activate the PET film, using SDBD plasma in atmospheric pressure. They also stated that on the basis of an increase in surface free energy, the increased activation time favorably influences the wettability of the surface. Van Deynse et al. [100] also noted a positive effect of increasing the activation time on a decrease in contact angle for LDPE-modified Ar.

However, the activation time should be optimized, not only due to the improved wettability of the substrate, but also due to possible degradation of the material. The short activation of the plastic material leads to external chain breaking, which can lead to improved surface properties, but more intense activation also causes internal chain breaking, also supported by VUV radiation [84].

In industrial applications, longer activation times will reduce process efficiency and increase process costs. It is important to find the right balance between results and process efficiency. Lee et al. [101] used different activation times and different gases to improve both hydrophilic and hydrophobic properties of the substrate. To increase hydrophilicity, they used plasma in the mixture of Ar and O_2 gases and they did not notice a significant influence of time on the change of wettability. The increase in the activation time resulted in a slight increase in the contact angle with a later decline with a further increase in the exposure, but the changes were very small compared to the changes observed for plasma used to obtain hydrophobicity (plasma with He and C_4F_8 gases). Extension of the He/C_4F_8 plasma activation time resulted in a significant increase of the contact angle. The Ar/O_2 plasma led to an increase in SFE, making the water and nanopaste used for printing easy to cover the surface. As a result, the width of the printed lines was greater than that obtained on modified substrates in order to increase hydrophobicity.

Morsy et al. [28] used DBD atmospheric plasma activation on PP to improve the printing properties of the substrate before gravure printing. The increase in activation time significantly influenced the change of PP SFE. Apart from improved wettability, the activation caused an increase in surface roughness. A positive effect of activation on substrate adhesion and optical density of printing has been noted, however, the increase of activation time did not have a significant influence on the further increase of optical density on prints. In addition, the activation improved the shape of the printed dots, which was visible in the increased regularity of their shapes. However, due to the increased thickness of the ink layer as a result of activation, the gloss of the print was reduced.

6.6 THE AGING PROCESS OF PLASMA TREATMENT

Modified surfaces are subject to changes during storage and the basic change observed is a decrease in their hydrophilicity, observed by a decrease in the value of surface free energy and an increase in the contact angle with water [42,77,78]. This phenomenon is called hydrophobic recovery. It may be related to the formation of weak boundary layers or mobility of functional groups on the surface of the material [55]. The main causes of deterioration of the effects of modification during material storage are: migration of low molecular weight additives from polymer mass to the surface layer and reorientation and diffusion of polar groups formed during the modification [46]. Changes in the top layer of the treated polymer during storage may adversely affect the wettability of the substrate by the ink and the adhesion of the ink to the substrate. However, the rate of changes caused by storage depends on a number of factors and may affect the printing of nonabsorbent substrates in different ways. Therefore, the activation of nonabsorbent substrates is usually performed in-line on the printing machine immediately before the printing process.

The rate of change in the modified material depends on the storage conditions [31,102] and, as shown by the research on material type and activation [16,55,103]. A higher temperature is expected to increase the rate of decay of changes resulting from activation [104]. It will accelerate the surface rearrangement. A higher humidity, on the other hand, will have a positive effect on the course of changes—it will slow them down. The water molecules will be adsorbed on the hydrophilic surface, which will disrupt the rotation or diffusion of polar groups [105]. However, with a longer storage time (30 days for UHMPE fibers) the temperature and humidity may no longer affect the contact angle [57]. In order to slow the aging process and preserve the properties of the modified sample, these samples may be preserved in water rather than air. The water environment causes polar groups to remain on the surface of the material, due to the similar, polar environment. On the other hand, the nonpolar environment causes polar groups to move towards the inside of the material [56,104]. The ageing process can also be reduced by storing modified films at significantly reduced temperatures [16,56].

The properties of the modified material, its crosslinking, and crystallinity influence the course of changes [78]. Smaller hydrophobic lesions will occur in the case of higher crystallinity polymers [56,105]. Banik et al. [6] carried out a study of crystalline changes in the material and its crosslinking on hydrophobic recovery. They stated that the crosslinking may change due to the appropriate selection of the gas mixture used for polymer treatment. Increasing the argon's share in argon-oxygen mixture led to an increase in the degree of crosslinking and, thus, to a decrease in the aging effect. Equally, increased crystallinity of HDPE resulted in reduced hydrophobic recovery.

Moreover, the course of changes depends on both the parameters of the activator and the type of gas used. Poncin-Epaillard et al. [106] examined the aging process with different PP gases and concluded that the slowest changes occur for activation with nitrogen, then helium, air, and the largest changes occur with oxygen. The changes, regardless of the speed of their course, result from branch scissions and formation of low molecular weight organic molecules. Canal et al. [107] concluded that wettability lost during storage can be recovered by submerging the modified PA6 in water, with both the hydrophobicity restoration and hydrophilicity recovery rate depending on the type of gas used during activation. The smallest decrease and fastest increase of contact angle and then the most stable behavior of the sample was observed for a sample activated with nitrogen. Also, it was the least sensitive to the restoration of hydrophilicity in water. In both cases, the biggest gradually progressing changes were observed in PA6 activated in H_2O plasma. For this, the contact angle directly after activation was also smaller than for plasma-activated samples with nitrogen or oxygen.

Knowledge of the rate of material decay is very important in further processing, as it determines their useful life without significant changes in properties. However, it should be emphasized that it is very variable and can have a very diverse course. This means that for some materials, for example, within the first few hours of activation, rapid hydrophobic recovery can occur and then the material will show good stability, or alternatively, continuous deepening of hydrophilic decay will be observed over the next few weeks. Extensive amount of research has been carried out in this area (Refs. [45,46,55,56,62,84,100,107−111]).

For example, Ba et al. [108] observed the biggest changes in modified PS during the first 2 weeks of storage. After this time, continuous, slow oxidation of the surface was observed for N_2 plasma with extended storage time. For PS modified with oxygen plasma, on the other hand, the surface did not undergo any further chemical changes and was stable when extending the storage time. Hegemann et al. in turn stated that Ar and O_2 plasma on PC is less stable in time than N_2 plasma. Van Deynse et al. [56] noted that for LDPE activated with Ar plasma after 14 days of storage there was hydrofobic recovery observed, but the material still had a much better wetting effect than before modification (40% reduced contact angle). Other studies by Van Deynse et al. [103] observed that during the first hours of storage after plasma treatment, modified LDPE showed a large increase in the contact angle, after which changes became slower and after 4 days a constant contact angle value was observed, still much lower than the for an unmodified material. Li and Liao [112] used atmospheric plasma activation to change the properties of the surface of PDMS (polydimethylsiloxane), commonly used as a printing substrate for printing electronics, and determined the effect of storage time on adhesion of ink to the substrate. The oxidation of the material resulted in a hydrophilic silica surface layer, which enabled the printing of high-quality conductor patterns using screen printing. Prints made on unmodified substrates were characterized by an unsatisfactory adhesion of ink to the substrate. As a result of activation of both oxygen plasma and nitrogen plasma, printed conductive tracks showed great adhesion on PDMS. However, oxygen activation was only effective if the printing was made immediately after the substrate was activated. Storage of the material for up to 2 h caused significant changes in the wettability of the substrate and negatively affected the adhesion of conductive ink to PDMS. On the other hand, the aging of the nitrogen-treated sample compared to the oxygen-treated sample was very slow and the changes observed after 2 h did not affect the adhesion of the printed layer to the substrate.

6.7 CONCLUDING REMARKS

The change in material properties achieved by plasma activation depends on a number of factors, which are closely related to each other. Thanks to the use of a different type of device, pressure, power, distance between the plasma source and the material, the speed of material movement (if it is transported during modification), the gas used, or various environmental conditions, it is possible to obtain different properties of the surface layer of the treated material. Extremely varied modification effects are possible due to easy control of plasma activation parameters.

In addition, the improvement in wettability achieved by plasma activation is impressive and exceeds the changes that can be achieved by corona treatment. Moreover, compared to chemical modification, it is much more environmentally friendly. Plasma makes it possible to reduce the contact angle with water up to several degrees and increase the surface free energy to over 80 mJ/m^2, which makes the substrate very hydrophilic, well-wetted and very useful for printing or other processes of refining. High SFE values will also translate into good adhesion of inks to the substrate.

Nowadays, corona treatment is commonly used in the modification of film in industrial conditions. However, in order to obtain the special properties of the material, it can be increasingly replaced by atmospheric plasma [17], which makes it possible to achieve a more even and efficient modification of the substrate. Low-pressure plasma and atmospheric plasma can effectively be used for the treatment of parts before painting in the automotive industry [113]. Moreover, plasma is important for electronic printing, where selective plasma is widely used [94,112,114]. Due to the substrate's hydrophilic and hydrophobic properties, it is possible to obtain the right point size and line width. The use of atmospheric plasma can also grow in the area of printed textiles, where it gives good results [115—118], although it is a separate topic from the one discussed in this paper.

The use of plasma for modifying materials before printing, coating, and other refining is a still evolving and relatively new area of knowledge. Further development of this area can be expected in the near future.

REFERENCES

[1] Pira, Packaging material outlooks—towards a $1 trillion milestone in 2020 (2017). Available from: < http://www.smitherspira.com/resources/2016/february/global-packaging-material-outlooks > . [Accessed: August, 2017].
[2] M. Żenkiewicz, Adhezja i modyfikowanie warstwy wierzchniej tworzyw wielkoczą-steczkowych. WNT, 2000.
[3] J. Izdebska, S. Thomas (Eds.), Printing on Polymers. Fundamentals and Applications, Elsevier, Oxford, 2016.
[4] T. Karbowiak, F. Debeaufort, A. Voilley, Importance of surface tension characterization for food, pharmaceutical and packaging products: a review, Crit. Rev. Food. Sci. Nutr. 46 (5) (2006) 391—407.
[5] J. Izdebska-Podsiadły, E. Dörsam, Article sent to the publisher, 2017b.
[6] I. Banik, et al., Inhibition of aging in plasma-treated high-density polyethylene, J. Adhesion Sci. Technol. 16 (9) (2002) 1155—1169.
[7] A. Kahouli, et al., Effect of O2, Ar/H2 and CF4 plasma treatments on the structural and dielectric properties of parylene-C thin films, J. Phys. D. Appl. Phys. 45 (2012) 215306 (7 pp).
[8] F. Poncin-Epaillard, G. Legeay, Surface engineering of biomaterials with plasma techniques, J. Biomater. Sci. Polym. Ed. 14 (10) (2003) 1005—1028.
[9] J.M. Grace, L.J. Gerenser, Plasma treatment of polymers, J. Disper. Sci. Technol. 24 (3—4) (2003) 305—341.
[10] R.A. Jelil, A review of low-temperature plasma treatment of textile materials, J Mater Sci. 50 (2015) 5913—5943.
[11] J. Lv, et al., Environmentally friendly surface modification of polyethylene terephthalate (PET) fabric by low-temperature oxygen plasma and carboxymethyl chitosan, J. Clean. Prod. 118 (2016) 187—196.
[12] R. Morent, et al., Plasma surface modification of biodegradable polymers: a review, Plasma. Process. Polym. 8 (2011) 171—190.

[13] W. Decker, et al., Long lasting surface activation of polymer webs, in: 43rd Annual Technical Conference Proceedings ISSN 0737-5921, 2000.

[14] S.K. Pankaj, et al., Applications of cold plasma technology in food packaging, Trends Food Sci. Technol. 35 (2014) 5–17.

[15] M. Niaounakis, Surface treatment. Ch. 8, Biopolymers: Processing and Products, Elsevier Inc., William Andrew, 2015.

[16] J.R. Rocca-Smith, et al., Impact of corona treatment on PLA film properties, Polym. Degrad. Stab. 132 (2016) 109–116.

[17] Vetaphon, 2017. Available from: http://www.vetaphone.com/wp-content/uploads/2017/09/Vetaphone_iPlasma_2017_ENG-low-resolution.pdf [Accessed: September, 2017].

[18] H. Kipphan (Ed.), Handbook of Print Media, Springer, Berlin, 2001.

[19] M. Vishnuvarthanan, N. Rajeswari, Effect of mechanical, barrier and adhesion properties on oxygen plasma surface modified PP, Innovat. Food Sci. Emerg. Technol. 30 (2015) 119–126.

[20] N. De Geyter, R. Morent, Cold plasma surface modification of biodegradable polymer biomaterials. Ch.7, in: P. Dubruel, S. Van Vlierberghe (Eds.), Biomaterials for Bone Regeneration, Elsevier, Oxford, 2014.

[21] E. Rudnik, Compostable Polymer Materials, Elsevier, Oxford, 2007.

[22] D.V. Rosato, et al. (Eds.), Concise Encyclopedia of Plastics, Springer Science + Business Media, New York, 2000.

[23] A. Emblem, H. Emblem, Packaging Technology. Fundamentals, Materials and Processes, Woodhead Publishing Ltd, Cambridge, 2012.

[24] R. Coles, D. McDowell, M.J. Kirwan (Eds.), Food Packaging Technology, Blackwell Publishing Ltd, Oxford, 2003.

[25] J. López-García, et al., Enhanced printability of polyethylene through air plasma treatment, Vacuum 95 (2013) 43–49.

[26] R.H. Leach, R.J. Pierce (Eds.), The Printing Ink Manual, Springer, Dordrecht, 2007.

[27] R.W. Bassemir, R. Krishnan, Surface phenomena in water based flexo inks for printing on polyethylene films, in: F.J. Micale, M.K. Sharma (Eds.), Surface Phenomena and Fine Particles in Water-Based Coatings and Printing Technology, Springer, 1991.

[28] F.A. Morsy, et al., Surface properties and printability of polypropylene film treated by an air dielectric barrier discharge plasma, Surf. Coat. Int. Part B: Coat. Trans. 89 (B1) (2006) 49–55.

[29] J. Izdebska, Evaluation of quality of flexographic print on selected biodegradable films. PhD Thesis. OWPW, 2011.

[30] Y. Kusano, Atmospheric pressure plasma processing for polymer adhesion: a review, J. Adhesion 90 (9) (2014) 755–777.

[31] S.K. Pankaj, et al., Characterization of polylactic acid films for food packaging as affected by dielectric barrier discharge atmospheric plasma, Innovat. Food Sci. Emerg. Technol. 21 (2014) 107–113.

[32] M.J. Shenton, G.C. Stevens, Surface modification of polymer surfaces: atmospheric plasma versus vacuum plasma treatments, J. Phys. D. Appl. Phys. 34 (2001) 2761–2768.

[33] K. Kryża, G. Szczepanik, Zastosowanie techniki zimnej plazmy jako nowoczesna technologia zabezpieczania surowców żywnościowych, 2017. Available from: http://www.food.rsi.org.pl/dane/Artyku__._Plasma._Kry__a__Szczepanik.pdf [Accessed: September, 2017].

[34] A. Vesel, et al., Surface modification of polyester by oxygen and nitrogen-plasma treatment, Surf. Interf. Anal. 40 (2008) 1444–1453.

[35] J. Izdebska-Podsiadły, Impact of low-temperature plasma treatment parameters on wettability and printability of PLA film. Proceedings of 49th IC Conference, China, 2017a.

[36] E.V. Kuvaldina, V.V. Rybkin, Interaction of active particles of oxygen plasma with polypropylene, High Energy Chem. 42 (1) (2008) 59−63.

[37] M. Stepczyńska, Surface modification by low temperature plasma: sterilization of biodegradable materials, Plasma. Process. Polym. 13 (2016) 1080−1088.

[38] J. Izdebska, Corona treatment, in: Izdebska, Thomas (Eds.), Printing on Polymers. Fundamentals and Applications, Elsevier, 2016.

[39] I. Junkar, et al., Influence of oxygen and nitrogen plasma treatment on polyethylene terephthalate (PET) polymers, Vacuum. 84 (2010) 83−85.

[40] K.N. Pandiyaraj, et al., Effect of cold atmospheric pressure plasma gas composition on the surface and cyto-compatible properties of low density polyethylene (LDPE) films, Curr. Appl. Phys. 16 (2016) 784−792.

[41] A.R. Calchera, A.D. Curtis, J.E. Patterson, Plasma treatment of polystyrene thin films affects more than the surface, ACS Appl. Mater. Interfaces. 4 (2012) 3493−3499.

[42] N. Vandencasteele, F. Reniers, Plasma-modified polymer surfaces: characterization using XPS, J. Electron Spectrosc. Relat. Phenom. 178−179 (2010) 394−408.

[43] H.J. Busscher, et al., The effect of surface roughening of polymers on measured contact angles of liquids, Colloids Surf. 9 (4) (1984) 319−331.

[44] S. Ebnesajjad, C. Ebnesajjad, Plasma treatment of polymeric materials. Chapter 9, Surface Treatment of Materials for Adhesive Bonding, William Andrew, 2014.

[45] A.Y. Song, et al., Cold oxygen plasma treatments for the improvement of the physicochemical and biodegradable properties of polylactic acid films for food packaging, J. Food. Sci. 81 (1) (2016) E86−E96.

[46] K. Moraczewski, et al., Stability studies of plasma modification effects of polylactide andpolycaprolactone surface layers, Appl. Surf. Sci. 377 (2016) 228−237.

[47] L. Sabbatini (Ed.), Polymer Surface Characterization, Walter de Gruyter GmbH & Co KG, 2014.

[48] K.S. Siow, et al., Plasma methods for the generation of chemically reactive surfaces for biomolecule immobilization and cell colonization—a review, Plasma Process. Polym. 3 (2006) 392−418.

[49] M. Żenkiewicz, Methods for the calculation of surface free energy of solids, J. Achi. Mater. Manuf. Eng. 24 (1) (2007) 137−145.

[50] A. Aliane, et al., Enhanced printedtemperaturesensorson flexiblesubstrate, Microelectr. J. 45 (2014) (2014) 1621−1626.

[51] Y.-T. Kwon, et al., Full densification of inkjet-printed copper conductive tracks on a flexible substrate utilizing a hydrogen plasma sintering, Appl. Surf. Sci. 396 (2017) 1239−1244.

[52] Y. Yuan, T.R. Lee, Contact angle and wetting properties, in: G. Bracco, B. Holst (Eds.), Surface Science Techniques. Springer Series in Surface Sciences 51, Springer-Verlag Berlin, Heidelberg, 2013.

[53] M. Zielecka, Methods of contact angle measurement as a tool for characterization of wettability of polymers, Polimery 49 (5) (2004) 327−332.

[54] D. Tian, Y. Song, L. Jiang, Patterning of controllable surface wettability for printing techniques, Chem. Soc. Rev. 42 (2013) 5184−5209.

[55] Y. Chen, et al., Surface modification and biocompatible improvement of polystyrene film by Ar, O_2 and Ar + O_2 plasma, Appl. Surf. Sci. 265 (2013) 452−457.

[56] A. Van Deynse, et al., Influence of ambient conditions on the aging behavior of plasma-treated polyethylene surfaces, Surf. Coat. Technol. 258 (2014) 359−367.

[57] R. Wang, et al., Comparison between helium and argon plasma jets on improving thehydrophilic property of PMMA surface, Appl. Surf. Sci. 367 (2016) 401—406.

[58] L.P. Yeo, et al., Selective surface modification of PETsubstrate for inkjet printing, Int. J. Adv. Manuf. Technol. 71 (2014) 1749—1755.

[59] A. Jordá-Vilaplana, et al., Surface modification of polylactic acid (PLA) by air atmospheric plasma treatment, Eur. Polym J. 58 (2014) 23—33.

[60] M.C. Kim, T. Masuoka, Degradation properties of PLA and PHBV films treated with CO_2-plasma, React. Funct. Polym. 69 (2009) 287—292.

[61] N. De Geyter, Influence of dielectric barrier discharge atmosphere on polylactic acid (PLA) surface modification, Surf. Coat. Technol. 214 (2013) 69—76.

[62] J. Izdebska-Podsiadły, E. Dörsam, Effects of argon low temperature plasma on PLA film surface and aging behaviors, Vacuum. 145 (2017) 278—284.

[63] M. Boselli, et al., Comparing the effect of different atmospheric pressure nonequilibrium plasma sources on PLA oxygen permeability, J. Phys. Conf. Ser. 406 (2012) 1—8. Available from: https://doi.org/10.1088/1742-6596/406/1/012038.

[64] T.S. Demina, A.B. Gilman, A.N. Zelenetskii, Application of high-energy chemistry methods to the modification of the structure and properties of polylactide (a review), High Energy Chem. 51 (4) (2017) 302—314.

[65] D. Garcia-Garcia, et al., Surface modification of polylactic acid (PLA) by air atmospheric plasma treatment, Eur. Polym J. 58 (2014) 23—33.

[66] T. Jacobs, et al., Plasma surface modification of polylactic acid to promote interaction with fibroblasts, J. Mater. Sci. Mater. Med. 24 (2) (2013) 469—478.

[67] V.L. Kudryavtseva, M.V. Zhuravlev, S.I. Tverdokhlebov, Surface modification of polylactic acid films by atmospheric pressure plasma treatment, AIP Conf. Proc. 1882 (2017). Available from: https://doi.org/10.1063/1.5001616. 020037-1—020037-4.

[68] M. Tryznowski, J. Izdebska, Z. Żołek-Tryznowska, Wettability and surface free energy of NIPU coatings based on bis(2,3-dihydroxypropyl)ether dicarbonate, Prog. Org. Coat. 109 (2017) 55—60.

[69] J. Izdebska, H. Podsiadło, L. Harri, Influence of surface free energy of biodegradable films on optical density of ink coated fields of prints, 39th International Reaserch Conference of iarigai, ISSN 978-3-9812704-5-7, 2012, pp. 245—251.

[70] FFTA, Flexography: Principles and Practices, fifth ed, Foundation of Flexographic Technical Association, Inc, USA, 1999.

[71] R.H. Leach, R.J. Pierce (Eds.), The Printing Ink Manual, Springer, Dordrecht, 2007.

[72] KBA, Substrates for printing and packaging, 2008. Available from: http://www2.kba.com/fileadmin/user_upload/KBA_Report/r33_beilage_en.pdf. [Accessed: September, 2017].

[73] A. Künkel, et al., Polymers, Biodegradable, Wiley-VCH Verlag GmbH & Co. KGaA, Weinheim, 2016.

[74] A. Jordá-Vilaplana, et al., Effects of aging on the adhesive properties of poly(lactic acid) by atmospheric air plasma treatment, J. Appl. Polym. Sci. (2016). Available from: https://doi.org/10.1002/APP.43040.

[75] K. Moraczewski, et al., Comparison of some effects of modification of a polylactide surface layer by chemical, plasma, and laser methods, Appl. Surf. Sci. 346 (2015) 11—17.

[76] J. Izdebska-Podsiadły, The influence of the operating parameters of the plasma activator on improvement of PLA film wettability, Opakowanie. 10 (2017) 67—71.

[77] I. Novák, et al., Adhesive properties of polyester treated by cold plasma in oxygen and nitrogen atmospheres, Surf. Coat. Technol. 235 (2013) 407—416.

[78] V. Fombuena, et al., Optimization of atmospheric plasma treatment of LDPE films: influence on adhesive properties and ageing behavior, J. Adhesion Sci. Technol. 28 (1) (2014) 97−113.

[79] M. Bryjak, et al., Plazmowa modyfikacja membran polimerowych, Membrany. Teoria i praktyka. Zeszyt III 3 (2009) 64−79.

[80] K. Gotoh, et al., Surface hydrophilization of two polyester films by atmospheric-pressure plasma and ultraviolet excimer light exposures, J. Adhesion Sci. Technol. 29 (6) (2015) 473−486.

[81] A. Řezníčková, et al., Comparison of glow argon plasma-induced surface changes of thermoplastic polymers, Nucl. Instrum. Methods Phys. Res. Sect. B. 269 (2011) 83−88.

[82] W. Petasch, et al., Low-pressure plasma cleaning: a process for precision cleaning applications, Surf. Coat. Technol. 97 (1997) 176−181.

[83] H.-H. Chien, et al., Effects of plasma power and reaction gases on the surface properties of ePTFE materials during a plasma modification process, Surf. Coat. Technol. 228 (2013) S477−S481.

[84] D. Hegemann, H. Brunner, C. Oehr, Plasma treatment of polymers for surface and adhesion improvement, Nucl. Instrum. Methods Phys. Res. Sect. B. 208 (2003) 281−286.

[85] A.E. Wiącek, et al., Effect of low-temperature plasma on chitosan-coated PEEK polymer characteristics, Eur. Polym J. 78 (2016) 1−13.

[86] B. Chaiwong, et al., Effect of plasma treatment on hydrophobicity and barrier property of polylactic acid, Surf. Coat. Technol. 204 (2010) 2933−2939.

[87] D. Daranarong, et al., Effect of surface modification of poly(L-lactide-co-ε-caprolactone) membranes by low-pressure plasma on support cell biocompatibility, Surf. Coat. Technol. 306 (2016) 328−335.

[88] J. Peyroux, et al., Enhancement of surface properties on commercial polymer packaging films using various surface treatment processes (fluorination and plasma), Appl. Surf. Sci. 315 (2014) 426−431.

[89] G.-Q. Ma, et al., Plasma modification of polypropylene surfaces and its alloying with styrene in situ, J. Sheng, Appl. Surf. Sci. 258 (2012) 2424−2432.

[90] N. De Geyter, et al., Plasma modification of polylactic acid in a medium pressure DBD, Surf. Coat. Technol. 204 (2010) 3272−3279.

[91] C. Zhang, et al., Effect of atmospheric-pressure air/He plasma on the surface properties related to ink-jet printing polyester fabric, Vacuum. 137 (2014) 42−48.

[92] B.J. Kang, J.H. Oh, Geometrical characterization of inkjet-printed conductive lines of nanosilver suspensions on a polymer substrate, Thin. Solid. Films. 518 (2010) 2890−2896.

[93] B.J. Kang, et al., Effects of plasma surface treatments on inkjet-printed feature sizes and surface characteristics, Microelectr. Eng. 88 (2011) 2355−2358.

[94] M.S. Fleischman, et al., Hybrid method involving atmospheric plasma treatment and inkjet deposition for the development of conductive patterns on flexible polymers, Surf. Coat. Technol. 206 (2012) 3923−3930.

[95] H.Y. Park, et al., Control of surface wettability for inkjet printing by combining hydrophobic coating and plasma treatment, Thin. Solid. Films. 546 (2013) 162−166.

[96] B.J. Kang, J.H. Oh, Influence of C_4F_8 plasma treatment on size control of inkjet-printed dots on a flexible substrate, Surf. Coat. Technol. 205 (2010) S158−S163.

[97] E.V. Kuvaldina, et al., Oxidation and degradation of polypropylene in an oxygen plasma, High Energy Chem. 38 (6) (2004) 411−414.

[98] D. Akbar, Surface modification of polypropylene (PP) using single and dual highradio frequency capacitive coupled argon plasma discharge, Appl. Surf. Sci. 362 (2016) 63−69.

[99] Y.-S. Goo, et al., Ink-jet printing of Cu conductive ink on flexible substrate modified by oxygen plasma treatment, Surf. Coat. Technol. 205 (2010) S369−S372.

[100] A. Van Deynse, et al., Surface activation of polyethylene with an argon atmospheric pressure plasma jet: influence of applied power and flow rate, Appl. Surf. Sci. 328 (2015) 269−278.

[101] Y.-C. Lee, et al., Effects of atmospheric pressure plasma surface treatments on the patternability and electrical property of screen-printed Ag nanopaste, Met. Mater. Int. 19 (4) (2013) 829−834.

[102] A. Vesel, M. Mozetič, Surface modification and ageing of PMMA polymer by oxygen plasma treatment, Vacuum. 86 (2012) 634−637.

[103] A. Van Deynse, et al., Surface modification of polyethylene in an argon atmospheric pressure plasma jet, Surf. Coat. Technol. 276 (2015) 384−390.

[104] J. Nakamatsu, et al., Ageing of plasma-treated poly(tetrafluoroethylene) surfaces, J. Adhesion Sci. Technol. 13 (7) (1999) 753−761.

[105] Y.I. Yun, et al., Aging behavior of oxygen plasma-treated polypropylene with different crystallinities, J. Adhesion Sci. Technol. 18 (11) (2004) 1279−1291.

[106] F. Poncin-Epaillard, J.-C. Brosse, T. Falher, Reactivity of surface groups formed onto a plasma treated poly(propylene) film, Macromol. Chem. Phys. 200 (1999) 989−996.

[107] C. Canal, et al., Wettability, ageing and recovery process of plasma-treated polyamide 6, J. Adhesion Sci. Technol. 18 (9) (2004) 1077−1089.

[108] O.M. Ba, et al., Surface composition XPS analysis of a plasma treated polystyrene: Evolution over long storage periods, Colloids Surf., B. 145 (2016) 1−7.

[109] D. Bodas, C. Khan-Malek, Hydrophilization and hydrophobic recovery of PDMS by oxygen plasma and chemical treatment—An SEM investigation, Sensors Actuators B 123 (2007) (2007) 368−373.

[110] C. Borcia, I.L. Punga, G. Borcia, Surface properties and hydrophobic recovery of polymers treated by atmospheric-pressure plasma, Appl. Surf. Sci. 317 (2014) 103−110.

[111] R. Morent, et al., Influence of Discharge Atmosphere on the Ageing Behaviour of Plasma-Treated Polylactic Acid, Plasma Chem. Plasma Process. 30 (2010) 525−536.

[112] C.-Y. Li, Y.C. Liao, Adhesive stretchable printed conductive thin film patterns on PDMS surface with an atmospheric plasma treatment, ACS Appl. Mater. Interfaces. 8 (2016) 11868−11874.

[113] Y. Tsuchiya, K. Akutu, A. Iwata, Surface modification of polymeric materials by atmospheric plasma treatment, Prog. Org. Coat. 34 (1998) 100−107.

[114] K.N. Kim, et al., Atmospheric pressure plasmas for surface modification of flexible and printed electronic devices: a review, Thin. Solid. Films. 598 (2016) 315−334.

[115] A. Zille, F.R. Oliveira, A.P. Souto, Plasma treatment in textile industry, Plasma. Process. Polym. 12 (2015) 98−131.

[116] C.W. Kan, C. Yuen, W. Tsoi, Using atmospheric pressure plasma for enhancing the deposition of printing paste on cotton fabric for digital ink-jet printing, Cellulose. 18 (2011) 827−839.

[117] M. Radetic, et al., The effect of lowtemperature plasma pre-treatment on wool printing, Text Chem. Color Am. Dyest Report. 32 (2000) 55−60.

[118] U.M. Rashed, et al., Surface characteristics and printing properties of PET fabric treated by atmospheric dielectric barrier discharge plasma, Eur. Phys. J. Appl. Phys. 45 (2009) 1−7.

FURTHER READING

K. Gotoh, A. Yasukawa, K. Taniguchi, Water contact angles on poly(ethylene terephthalate) film exposed to atmospheric pressure plasma, J. Adhesion Sci. Technol. 25 (1—3) (2011) 307—322.

L. Nguyen, Minimizing Hydrophobic Recovery of Polydimethylsiloxane after Oxygen Plasma Treatment. Student Thesis Collection, Mount Holyoke College, 2014.

R.M. Rasal, A.V. Janorkar, D.E. Hirt, Poly(lactic acid) modifications, Prog. Polym. Sci. 35 (2010) 338—356.

Y. Ren, et al., Effect of dielectric barrier discharge treatment on surface nanostructure and wettability of polylactic acid (PLA) nonwoven fabrics, Appl. Surf. Sci. 426 (2017) (2017) 612—621.

CHAPTER 7

Plasma Treatment of Powders and Fibers

Taťana Vacková, Petr Špatenka and Syam Balakrishna
Department of Material Science, Faculty of Mechanical Engineering, Czech Technical University of
Prague, Prague, Czech Republic

7.1 INTRODUCTION

Granules or powders are applied in many different industry branches, such as basic construction materials, composite materials, synthetic polymer fillers, agriculture, pharmaceutical, cosmetic, paintings, and food. According to Duran [1], the worldwide annual production of grains and aggregates of various kinds reached before the year 2000 was around ten billion metric tons. And with extensive development of the application of nanotechnology, this number is increasing annually.

Plasma treatment, in general, represents a process in which energetic particles and plasma-generated photons interact with surfaces in a way that generates free radicals. During plasma treatment of substrate surfaces, effects such as modification of chemical surface structure, crosslinking and scission of molecules near the surface, etching and surface cleaning can be observed. The occurrence of the effects depends on the substrate surface nature, the working gas, the construction of the apparatus, and the process parameters.

Plasma processing can be efficiently used to produce high-quality powders [2,3]. Thermal plasmas have several advantages in high-temperature material processing compared with conventional processes. These plasmas are characterized by high temperature, good thermal and electrical conductivity, enhanced chemical activity, intense fluxes of charged particles and radiation, electromagnetic response, and other exotic properties. Such characteristics make thermal plasmas unique and useful for processing materials like spheroidization, melting, densification extractive metallurgy, synthesis, and chemical vapor deposition (CVD) [4—7].

Cold plasma surface modification has been established as an inexpensive and highly effective technology for influencing surface properties of

Non-Thermal Plasma Technology for Polymeric Materials
DOI: https://doi.org/10.1016/B978-0-12-813152-7.00007-X
193

polymers without altering the bulk material. Nevertheless, the plasma modification of polymer powders has not found many applications as a modification of flat bulk polymers. This is due to problems connected with the (1) three-dimensional geometry of powders, (2) large surface area of powders, and (3) the aggregation phenomenon and the necessity of solid mixing of treated powders. Therefore, the application of plasma treatment of powders and nanopowders is dealt within this chapter. Near the end of the chapter plasma treatment of fibers is also briefly discussed.

7.2 PARTICLES FORMATION IN PLASMA

Nanometer or micrometer-sized particles can be formed and grown inside the so-called dusty plasma. Dusty, or complex, plasma is normal electron-ion plasma with an additional highly charged component of massive particulates and/or dust grains. A significant amount of dust particles was also frequently encountered in industrial reactors for semiconductor manufacturing, in processing discharges for etching, sputtering, and ion implantation, etc. [8–14], and in the chambers of fusion devices [15]. Powder formation has also been a critical concern for the microelectronics industry because dust contamination can severely reduce the yield and performance of fabricated devices. Submicron particles deposited on the surface of process wafers can obscure device regions, cause voids and dislocations, and reduce the adhesion of thin films [16,17]. These dust particles, which may have sizes ranging from tens of nanometers to hundreds of microns, are typically billions of times more massive than protons, and can have between one thousand and several hundred thousand elementary charges. Therefore, the presence of dust grains can significantly alter the properties and behavior of plasma in which they are immersed [18,19].

Powder formation in plasma requires sufficient precursor concentration and sufficient energy input in localized space in plasma reactors. These conditions result in high radicals concentration promoting volume reaction in plasma. The detailed description of plasma formation is described in [20] and references cited there.

Nowadays, dusty plasmas have grown into a vast field and new applications of plasma-processed dust particles are emerging. The industrial demand for particles with special properties and for particle-seeded composite materials has been increased. Low-pressure plasmas offer a unique possibility of confinement, control and fine tailoring of particle properties. The increased interest in the field of plasma–particle interaction in respect to dusty plasmas is due to applied research related to material science and surface processing technology [19,21−24]. Powders produced using plasma technology have very interesting and potentially useful properties, for example, very small sizes (nanometer to micrometer range), uniform size distribution, and chemical activity. Size, structure, and composition can be tailored to specific requirements, dependent on the desired application [22,24,26,27].

Nanopowders or nanoparticles (NPs) have a great interest in the field of applied research related to material science and surface processing technology. It is known that nanopowders are referred to as ultra-fine solid particles having a diameter between about 1 and 100 nm (10^{-9} m to 10^{-7} m). Nanopowders are replacing conventional powders in many applications because of their unique properties, such as higher surface area and easier formability, and because of improved performance of end products. Some current applications of nanopowders are catalysts, fillers, lubricants, abrasives, explosives, magnetically sensitive materials, decorative materials, taggants, and reflective materials. However, before use, powders need adequate surface treatment in order to acquire new surface properties required in the different applications.

Nanosized SiC was synthesized by the solid-state method followed by plasma treatment. Solid state-synthesized SiC particles were 1−5 μm in size whereas the plasma-synthesized particles were below 30 nm. There was no change in structure or phase of SiC before and after plasma treatment except for a few free silicon peaks observed after plasma treatment. DC current in terms of length and width of plasma arc greatly affected the particle size which was reduced with increasing DC current. Nanosized SiC was formed due to the dissociation of grains from their grain boundary by a strong plasma gas stream. It was found that a higher percentage of carbon compared to silicon was necessary for the complete conversion of silicon into SiC. Furthermore, the synthesis of nanosized SiC, by thermal plasma, using presynthesized SiC was better than the use of a mixture of Si and C powders [28].

Patel et al. [29] synthesized colloidal and electrostatically stabilized surfactant-free gold NPs (AuNPs) by the direct interaction of atmospheric pressure plasma at the surface of a gold (III)-chloride-trihydrate ($HAuCl_4.3H_2O$) aqueous solution. The interactions of the gas-phase plasma with the solution initiate liquid-based reactions that determine the nucleation and growth of AuNPs from the reduced gold precursor ($HAuCl_4$). The plasma−liquid interface is possibly represented by a gas/

water vapor plasma environment where electrons are believed to be responsible for initiating reactions that then cascade into the liquid solution [29]. More specifically, the plasma was generated across a gap between the liquid surface and a stainless-steel capillary. Inside the capillary, helium was blown so that the plasma was largely formed in helium gas.

This synthesis technique only requires a water-based solution with a metal precursor. Plasma processing parameters affect the particle shape, size, and the rate of the AuNP synthesis process. Particles of different shapes (e.g., spherical, triangular, hexagonal, pentagonal, etc.) can be synthesized and the size was tuned from 5 nm to several hundred nanometers by varying the initial gold precursor ($HAuCl_4$) concentration from 2.5 μM to 1 mM. At low molar concentrations (2.5 μM) all the particles were spherical. For medium molar concentrations (0.05 and 0.1 mM) the particle shape distribution included various shapes of AuNPs such as spherical, hexagonal, pentagonal, triangular, etc., with some gold nanorods. At higher molar concentrations (0.2 and 1 mM) more than 80% of the AuNPs were found to be spherical or hexagonal. The predominance of given shapes at different concentrations may indicate that reduction of the gold precursor is activated differently under various conditions.

Wang et al. [31] reported a novel approach for producing Magnetite, Fe_3O_4, NPs in a liquid by nonthermal argon plasma. The Fe_3O_4 NPs synthesis was performed in a U-shaped electrochemical cell, which consisted of a plasma cathode and a platinum foil anode. Well-dispersed Fe_3O_4 NPs with an average size of 12.5 $\pm$ 2.4 nm were produced in the solution under the right pH and argon gas buffer. They demonstrated good aqueous-phase stability and excellent superparamagnetic properties with a saturation magnetization of 60.1 emu/g and a small coercivity (<10 Oe). The simplicity of the apparatus and room temperature operation makes this plasma−liquid interaction-based approach a promising method for the production of metal oxide NPs in liquids for various nano-technological applications.

7.3 PLASMA TREATMENT OF POWDERS

7.3.1 Theoretical Background

Assuming free packing of the powder particles in a container, the particles do not fill the whole container volume. A quantity of the powder can be characterized by a filling factor F and typically the total sum of particles volume occupies between 0.45% and 0.64% of the total volume of the container [32]. According to Samarian et al. [11], the total surface area S_{total} of the particles in a container of a volume V_{box} can be expressed as:

$$S_{total} = N.S_{particle} = \frac{3V_{box}F}{R}$$

For example, the total surface area of particles of 1 mm in diameter filled in a one liter container with the filling factor of 0.5 is 1.5 m^2 but in case of particles with a diameter of 1 µm the total area increases to 1500 m^2, thus, one problem connected with powder treatment is the large total area which has to be treated.

Randomly filled powder in a plasma reactor does not form a compact body but resembles a porous material. Reactive radicals can, under certain conditions, penetrate into porous materials [33−35] resulting in that even particles placed in deeper layers can be treated under some conditions without any fluidization. As has been shown by Špatenka et al. [36], a certain degree of powder hydrophilization was observed even 10 mm deep under the surface layer for particles when the average diameter of particles was 200 µm. This penetration significantly contributes to the efficiency of powder treatment in large-scale applications.

Lowering the diameter brings additional problems connected to the enhancement of the treated area and lowering the radicals' penetration. Generally, for powder with particles sized under 20 µm, the penetration practically vanishes [37] and the particles tend to form agglomerates [38]. To ensure homogenous treatment of the whole surface of all particles, fluidization of the particle bed during plasma treatment is necessary.

The problems of particles fluidization have been addressed by various papers in connection with increasing application of microparticles and NPs in different products. The dispersion of particles is of great importance for many industrial areas, such as the pharmaceutical and cosmetics industry, and materials for printer toners, etc., to deagglomerate particles.

The agglomeration is based on cohesive forces between particles. A comprehensive description of particle deagglomeration theory can be found in review articles [39,40] and literature cited there. Dry deagglomeration of particles is based on an application of relative "hard forces" to overcome the cohesive forces. According to Calvert et al. [40], dry methods of particle dispersion are based either on inter-particle collisions and/or collisions of particles with walls, or on the production of significant aerodynamic forces acting on the particles.

Among various aerodynamic methods listed in Calvert et al. [40], for CVD and PE/CVD the fluidized bed is the only practically applied method. In this method, the particle bed is placed over a gas-distributing plate. Gas flow through this plate allows the particles to levitate due to the aerodynamic forces in the gas stream. The method is described in more detail, for example, in Sefcik et al. [39]. For the enhancement of the deagglomeration process, several methods combined the fluidized bed with an additional segregation mechanism have been proposed [38,41−43].

An important step in the deagglomeration process is preventing reagglomeration. Electrostatic charging could significantly contribute to this process. Charging particles in plasma media was investigated predominately in connection with particle levitation [43,44]. It has been shown that the potential of the dust particles can reach nearly 20 V [43]. Bartoš et al. [45] calculated the electrostatic potential in the particulate vicinity and showed that this potential significantly increases by lowering the distance from the particulate. Electrostatic forces between charged particles, thus, can significantly contribute to the stabilization of the system in active plasmas and prevent particles' reagglomeration. Several investigations of particle flow and similar crystal formations [46,47] in low-pressure plasmas support these hypotheses.

7.3.2 Plasma Grafting on Particles

Plasma grafting is based on the incorporation of functional groups on particle surfaces. Incorporation of a polar group like $-OH$ or $-NH_2$ causes enhancement of surface energy resulting in better powder dispersion. Reactive groups also contribute to bond formation between matrix and filler in composite materials. The principle of plasma grafting and polymer modification is a standard process that has been described in many review papers and books (e.g., 48). The treatment of powders is connected to the necessity of the application for any mixing method to ensure homogeneous end efficient powder treatment.

Quitzau et al. [49] modified PE powders by means of a DC hollow cathode glow discharge (HCGD) as the plasma source placed in a spiral conveyor. This experimental set-up allowed a continuous treatment and homogeneous surface modification of the powder at low gas flow rates and variable treatment times. Due to the vibrations of the reactor, the powder was agitated from the bottom of the chamber along the spiral path up to the top. Consequently, it fell down through the HCGD plasma.

In the HCGD, the excitation, dissociation, and ionization are much more efficient compared to other glow discharges. Fast, secondary electrons which are released from the cathode surface by ion bombardment are electrostatically confined inside the hollow cathode. They are repelled by the opposite cathode and can oscillate between both cathode surfaces in the negative glow. These "pendular" electrons dissipate most of their energy acquired in the cathode fall by collisions in the plasma. The pendular effect causes increased ionization and dissociation rates as well as a strongly increased current density at a constant operating voltage (hollow cathode effect) compared to a linear glow discharge [50–53].

Bretagnol et al. [54] modified the surface properties of low-density polyethylene (LDPE) powders by nitrogen and ammonia RF plasma treatments in a low-pressure fluidized bed reactor. For low gas flow rates (<15 sccm), in the used reactor, powders remain immobile. The flow rate, corresponding to the minimum of fluidization state, is determined through the relationship of the pressure drop versus the gas flow rate. The minimum fluidization flow rate corresponds to the maximum pressure drop [55]. A flow rate of 30 sccm was found as the threshold for particle fluidization at the working pressure of 50 Pa with both nitrogen and ammonia gases.

After plasma treatment in the low-pressure fluidized bed reactor, LDPE powders show a considerable increase in hydrophilic character and surface functionality. It has been found that the hydrophilicity and the nitrogen incorporation on the surface of LDPE powders increase both with treatment time and RF power. A comparison between nitrogen and ammonia plasmas showed that nitrogen plasma was more efficient than ammonia plasma in order to incorporate new nitrogen moieties in this particular configuration.

Shidong et al. used an inductively coupled RF argon plasma reactor with a stirrer to modify the surfaces of polypropylene (PP) powders [56]. With an increase in plasma treatment time, the carbon atom concentration decreased while the oxygen atom concentration increased on the surface of the plasma-treated PP powders. The XPS and ATR-FTIR results showed that the oxygen-containing polar functional groups, such as $C-O$, $C=O$, and $O-C=O$, were introduced onto the surfaces and caused both the decrease in water contact angle and the improvement of the hydrophilic properties of the plasma modified PP powders. The inductively coupled plasma reactor with a stirrer is an interesting and environmentally efficient plasma treatment device to modify the surface of low surface-energy polymers.

Atmospheric plasma is a promising alternative for plasma hydrophilization of hydrophobic polymer particles. Gilliam et al. [57] used low-temperature atmospheric plasma for the chemical surface modification of polymethylmethacrylate (PMMA) and PP particles. The main components of the apparatus include the atmospheric pressure plasma generator, injection of water or an additional chemical into the plasma, and a feeder of particles to the plasma. A gravity-driven setup was used to deliver the particles which were held in a hopper with a gate which was open during the treatment process enabling the particles to fall through a tube with an outlet placed directly in the plasma discharge stream. The PMMA and PP powder consisted of free-flowing particles, which allowed a continuous stream with the assistance of a vibration mechanism attached to the delivery tube. A water spray was placed downstream of the particles to enable collection of the treated particles into water.

Microsized polymer particles of PMMA and PP were successfully surface modified using a fast, low-temperature, atmospheric plasma process. The treatment enhanced surface hydrophilicity of the particles that was correlated to an increase in polar functional groups. Contact angle analysis of treated PMMA revealed effects of treatment conditions on the surface hydrophilicity. Air plasma produced higher contact angles than pure nitrogen, which is attributed to the primary surface degradation from oxygen. Higher energy and flow rate of water resulted in decreased contact angles. X-ray photoelectron spectroscopy increased $C-O$ and $C=O$ for treated samples.

Plasma modification on an industrial scale has been introduced to the market by the SurfaceTreat Company. A reactor equipped with two 1000 W microwave power sources is able to treat more than 30 kg/h of PE powder. Mechanical mixing is used to ensure homogeneous treatment of all the particles in a batch [58].

Besides polymer treatment of inorganic particles, the treatment of carbon nanotubes can also be found in the literature. Lapčík et al. [59] demonstrated the enhancement of wettability and flow properties of kaolinite powder after air plasma treatment in a low-pressure drum-type reactor. Inorganic Laponic clay was also treated by Fatyeyeva and Poncin-Epaillard. Low-pressure RF plasma and sulfur dioxide (SO_2) as a process gas were used to graft sulfonic acid groups onto the inert clay surface [60].

The introduction of chemically reactive sites, either by a surface site or a lattice site modification, is a general method to improve the performance of carbon materials such as graphite, graphene, and carbon black. Mueller et al. [61] modified two different grades of graphite powders by nitrogen plasma treatment at different excitation energies. Other experiments were performed in a specially customized inductively coupled RF plasma CVD reactor [28] on a pilot plant scale for powder treatment in controlled gas atmospheres. The reactor setup allowed high flexibility regarding reaction time and energy input as well as a statistical treatment of the individual powder particles within a sample. Graphites were directly functionalized with nitrile groups by such a gaseous phase process without the need for any additional wet-chemical treatment and, thus, could directly introduce polar nitrogen-containing groups. This increases the hydrophilic character of the materials and their surface interactions with polar solvents or any composite materials, thus, improving their dispersibility. The reactor loading depends on the density of the material and typically 40.0 g for spherical particle graphite and 20.0 g for flake graphite was used. An N_2 gas flow rate of 300 sccm, suitable for satisfactory fluidization of the materials, was used.

Oh et al. [62] treated carbon NPs derived from polypyrrole with low-temperature O_2/NH_3 radio frequency plasma in order to modify the surface of the carbon NPs. The plasma reactor used (Korea Vacuum Co.) was the parallel-electrode type. The carbon NPs were inserted into the plasma chamber, the chamber was evacuated, and the carrier gas O_2 and/or NH_3 was introduced into the chamber.

In the case of O_2 plasma treatment, the hydroxyl group appeared on the surface of the carbon NPs, while the amino group was introduced on the surface by NH_3 plasma treatment. The morphology of plasma-treated carbon NPs was retained without NP aggregation. The plasma-treated carbon NPs exhibited enhanced dispersibility in an aqueous solution, compared to the pristine carbon NPs.

Amino and carboxyl groups were incorporated in inductively coupled discharge on carbon nanotubes. Better electrochemical response of treated nanotubes could contribute to the development of sensitive and simple detectors of harmful environmental pollutants [63]. Amino-groups were incorporated on the surface of carbon nanotubes also by other researchers (e.g., [64,65]). In all the described experiments the treatment was performed on carbon nanotubes placed on a solid substrate without any mixing during the plasma treatment.

7.3.3 Encapsulation

Coating of particles was investigated by several authors as a tool for enhancement of particles flowability, a decrease of particles agglomeration, and as a protective coating for functional particles. Plasma enhanced chemical vapor deposition (PECVD) or plasma polymerization was predominately used for these processes. A few papers also applied physical vapor deposition for particles metallization, but with a very low modification yield [66,67].

Brueser et al. demonstrated in their early works the encapsulation of NaCl by SiO_x containing thin film by using a dielectric barrier discharge from HMDSO and TEOS as precursors. They demonstrated that the clear shell remained compact even after the NaCl particle was washed out [68,69].

Systematically the particle coating by SiO_x containing layer was investigated by Roth et al. [70]. Lactose powder was coated in a low-pressure plasma downer reactor with oxygen, argon, and the organosilicon monomer HMDSO as the working gas. The powder was fed from the storage container over a metering screw with 2 kg/h to the downer pipe and mixed with the process gases through a nozzle. Below the plasma zone, the particles were separated from the gas stream by a downcomer, cyclone, and filter unit and settled in the solid collection vessels. The flowability of all investigated powder fractions was considerably increased after the plasma treatment. Roth et al. concluded that the coated layer significantly reduced the van der Waals interactions between the fine particles.

Reduction of the gas flow in the fluidized bed and enhancement of the deposition rate was motivation for reactor construction proposed by Quitzau et al. [71]. They proposed a low-pressure plasma reactor equipped with a hollow cathode filled with HMDS/Ar mixture. Due to the vibrating spiral conveyor, the particles entered the active plasma zone several times. The film formed on polyethylene powder was the SiO_2 thin film with various polar groups on its surface.

An arc discharge has been used to fabricate graphite-encapsulated magnetic NPs. Hu et al. [72] described the fabrication of encapsulating graphite shell over iron composite NPs with tunable surface morphologies and qualities through an one-step direct current arc discharge method. The synthesis of graphite-encapsulated iron composite nanoparticles (GEICNPs) was performed in a DC arc discharge device. Typically, the anode was a sintered graphite rod containing the catalyst. The cathode was a pure sharpened graphite rod.

By carefully tuning the parameters of the arc plasma, tunable surface morphologies and qualities of encapsulating graphite shell have been systemically investigated. Through detailed microstructural characterization, it was found that: (1) the outermost layer of coating graphite shell tends to be smoother and cleaner as the concentration of CH_4 increases from

0% to 50%; (2) the surface morphology with higher graphitization degree was achieved by varying collecting distances from 10 to 1 cm to the arc center; and (3) carbon nano-loops on the outmost shell expanded volumetrically to form a continuous amorphous cover as the working gas pressure decreases from 100 to 25 Torr.

Particle encapsulation by plasma polymer has been described in Shahravan et al. [73] and Dierkes et al. [74] Both research teams used a tubular reactor with 13.53 MHz RF discharge. The polymer film contributed to the reduction of silica filler agglomeration in the rubber matrix and resulted in better vulcanization properties [73]. It was shown in Shahravan et al. [72] that the film thickness controlled by permeation of small molecules enabling regulated release of the core material which can be used, for example, for controlled drug release.

Coating particles is more complicated than particle grafting. Demand for homogeneous encapsulation of whole particles increases the demands on the homogeneous distribution of particles in the reaction zone. Additionally, the process time is determined by the film deposition which is significantly longer than the time which is necessary for grafting of functional groups on the particles surface. This leads to a very low efficiency of the plasma-based particles encapsulation. Shahravan et al. [73] were able to modify 100 g of silica powder in 90 min. In Roth et al. [70] the feeding of particles into the reactor was even 2 kg/h, but the coverage of the particles was estimated to be between 5% and 15%. The particles metallization by PVD method was even less effective. The process described in Yu et al. [67] coated 4 g of particles in one batch and the deposition took 90 min. Thus, the process efficiency of plasma encapsulation has to be significantly increased for a reasonable industrial application.

7.4 PLASMA TREATMENT OF FIBERS

Similarly to powders, plasma treatment of fibers is based on incorporation of functional groups (e.g., $-COOH$, $-OH$, $C-O$, $C=O$, $-NH_2$) on their surface. Fibers are made from almost all kinds of materials, such as cellulose, glass, carbon, and polymeric and even metallic fibers have been treated. Numerous papers and books deal with plasma treatment of fibers, yarns, or fabrics [75,76]. This technology has been established as a standard method for surface treatment before painting/dyeing or for the reinforcement of composites. Various types of atmospheric pressure plasma [77−79] are most frequently used for this purpose. A detailed review devoted to plasma treatment of fibers is the subject of Chapter 13, Plasma

Treatment on High Performance Fibrous. Thus, we here focus on some problems connected with fiber coatings only.

Similarly to the way powder stacks, also fabric or fiber bundles form a porous complex shape body. But because of the impossibility of mixing the coating of whole fibers in the bundle depends on the penetration of active species between particular fibers. Nevertheless, it was proved both experimentally and by computer modeling that both the polymeric and also ceramic TiO_2 thin films prepared using PECVD process can grow between the particular fibers [80,81]. Thus, the PECVD seems to be a proper technology for fiber coating.

Fiber coating is necessary, for example, for carbon fiber protection during metal matrix composite manufacturing [82]. Various ceramic coatings (zirconia or aluminum oxide) have been investigated as barrier protection of the carbon fibers prior oxidation and/or chemical reaction with the molted ceramic matrix [83].

RF-PECVD of SiO_2 as an interface layer on carbon fiber tow was investigated by Pillai et al. [84]. The coated fibers evidenced better thermal oxidative resistance than an uncoated one, indicating uniform fiber tow coating. The film structure was amorphous with a needle-like nanostructure.

Pillai et al. [85] also reported low-temperature deposition of SiC film on carbon fiber. They used a low-pressure bipolar pulsed RF PECVD system and the substrate was again the carbon fiber tow. The film has a crystalline structure and the scanning electron micrographs indicated homogeneous covering of the fibers.

The application of fiber coating was also investigated as a tool for adhesion promotion between fibers and epoxy or thermoplastic matrix. Friedric et al. [86] investigated plasma–polymerized thin film on carbon fibers. The single fiber pull-out test proved significant adhesion enhancement between the fibers coated by plasma deposited poly(acrylic acid) and/or poly(hydroxyethylmethacrylate) to the epoxy resin.

Cech et al. [87] demonstrated interfacial strength enhancement between glass fiber coated by plasma–polymerized tetravinylsilane to polyester matrix. They found that the interfacial shear strength for the optimized plasma coating was 26% higher than that for the industrially sized glass fibers.

Low-temperature, low-pressure plasma treatment is one of the most promising environmentally friendly and rapid methods for surface modification. But in the case of fiber and fabric surface treatment, the method requires continuous winding under vacuum. Hence very low yield leads to small doses and/or small series. Moreover, the homogeneity of this treatment has to be still optimized in order to assure modification of the whole surface of fibers. In general, the plasma processes are suitable for tailoring physicochemical properties and have potential applications in the subsequent improvement of composite interphases through the construction of gradient interlayers required for the novel conception of composites, but the low deposition speed and butch-process results in relatively high costs, thus, nowadays this technique could have only some specialized applications.

7.5 CONCLUSION

Plasma has been used for the production, surface treatment, and surface encapsulation of powders and fibers. Various technologies, including plasma arc, low-pressure plasma, and plasma-liquid interaction have been described in the literature. Plasma arc was used predominately for the production of various ceramics and metallic NPs. Low-temperature plasma excited both at atmospheric and low-pressure have been most frequently applied for plasma grafting and plasma encapsulation of powder. Whereas industrial-scale plasma-based hydrophilization of hydrophobic polyolefins has already been described, the plasma encapsulation is still at laboratory scale with very low efficiency. Thus, efficiency enhancement of plasma-based coating of particles is a challenge for further development.

Plasma treatment of fibers has been investigated predominately aimed at the improvement of bonding strength between fibers and matrix in composite materials. Plasma treatment of fibers enables surface functionalization without altering the structure and, thus, the mechanical properties of the fiber. Low-pressure atmospheric plasma seems to be the most prospective technology for the treatment of fibers. In connection with the rapid growth of 3D-print technologies and their application in the production of thermoplastics composite parts, new challenges in the surface treatment of fibers will arise. Here plasma treatment may play a vital role.

ACKNOWLEDGMENT

This work was supported by the Ministry of Education, Youth and Sport of the Czech Republic, program NPU1, project No. LO1207 and project No. CZ.02.1.01./0.0./0.0./ 16_ 019/0000826. The authors are grateful to Zdeňka Jeníková and Vladimír Mára for their assistance.

REFERENCES

[1] J. Duran, Sands, Powders, and Grains, Springer, New York, 2000.

[2] G. Bonizzoni, E. Vassallo, Plasma physics and technology; industrial applications, Vacuum 64 (2002) 327–336.

[3] Z. Karoly, J. Szepvolgyi, Plasma spheroidization of ceramic particles, Chem. Eng. Process. 44 (2005) 221–224.

[4] G. Soucy, M. Rahmane, X. Fan, T. Ishigaki, Heat and mass transfer during in-flight nitridation of molybdenum disilicide powder in an induction plasma reactor, Mater. Sci. Eng. A 300 (2001) 226–234.

[5] X.-L. Jiang, M. Boulos, Induction plasma spheroidization of tungsten and molybdenum powders, Trans. Nonferrous Metals Soc. China 16 (2006) 13–17.

[6] N. Kobayashi, Y. Kawakami, K. Kamadal, J.G. Li, R. Ye, T. Watanabe, et al., Spherical submicron-size copper powders coagulated from a vapor phase in RF induction thermal plasma, Thin Solid Films 516 (2008) 4402–4406.

[7] H. Kersten, H. Deutsch, E. Stoffels, W.W. Stoffels, G.M.W. Kroesen, Plasma-powder interaction: trends in application and diagnostics, Int. J. Mass. Spectrom. 313 (2003) 223–224.

[8] M.A. Lieberman, A.J. Lichtenberg, Principle of Plasma Discharges and Material Processing, Wiley, New York, 1994.

[9] J.E. Daugherty, D.B. Graves, Derivation and experimental verification of a particulate transport model for a glow discharge, J. Appl. Phys. 78 (1995) 2279–2287.

[10] M. Takai, T. Nishimoto, M. Kondo, A. Matsuda, Effect of higher-silane formation on electron temperature in a silane glow-discharge plasma, Appl. Phys. Lett. 77 (2000) 2828–2830.

[11] A.A. Samarian, B.W. James, S.V. Vladimirov, N.F. Cramer, Self-excited vertical oscillations in an rf-discharge dusty plasma, Phys. Rev. E 64 (2001) 025402–025404.

[12] A.A. Samarian, S.V. Vladimirov, Charge of a macroscopic particle in a plasma sheath, Phys. Rev. E 67 (2003) 066404–066405.

[13] A.A. Samarian, S.V. Vladimirov, B.W. James, Dust particle alignments and confinement in a radio frequency sheath, Phys. Plasmas 12 (2005) 022103–022106.

[14] N.C. Adhikary, H. Bailung, A.R. Pal, J. Chutia, Y. Nakamura, Observation of sheath modification in laboratory dusty plasma, Phys. Plasmas 14 (2007) 103705–103707.

[15] J. Winter, Dust in fusion devices-experimental evidence, possible sources and consequence, Plasma Phys. Controlled Fusion 40 (1998) 1201–1210.

[16] G.S. Selwyn, J. Singh, R.S. Benett, In situ laser diagnostic studies of plasma-generated particulate contamination, J. Vacuum Sci. Technol. A: Vacuum Surf. Films 7 (1989) 2758–2765.

[17] A. Bouchoule, Dusty plasmas, Phys. World 6 (1993) 47–52.

[18] R.L. Merlino, A. Barkan, C. Thomson, N. D'Angelo, Laboratory studies of waves and instabilities in dusty plasmas, Phys. Plasmas 5 (1998) 1607–1614.

[19] P.K. Shukla, A survey of dusty plasma physics, Phys. Plasmas 8 (2001) 1791–1803.

[20] H. Yasuda, Plasma Polymerization, Academic Press Inc., 1985.

[21] G.S. Selwyn, J.S. McKillop, K.L. Haller, J.J. Wu, In situ plasma contamination measurements by HeNe laser light scattering: a case study, J. Vacuum Sci. Technol. A: Vacuum Surf. Films 8 (1990) 1726–1731.

[22] A. Bouchoule, Dusty Plasmas: Physics, Chemistry and Technological Impacts in Plasma Processing, Wiley, New York, 1999.

[23] E. Stoffels, W.W. Stoffels, H. Kersten, G.H.P.M. Swinkels, G.M.W. Kroesen, Surface processes of dust particles in low pressure plasmas, Phys. Scripta T89 (2001) 168–172.

[24] E. Stoffels, W.W. Stoffels, G.M.W. Kroesen, F.J. Hoog, Dust in plasmas: fiend or fried, Electron Technol. 31 (1998) 255–274.

[25] C. Hollenstein, The physics and chemistry of dusty plasmas, Plasma Phys. Control. Fusion 42 (2000) R93–R104.

[26] H. Hofmeister, J. Duta, H. Hofmann, Atomic structure of amorphous nanosized silicon powders upon thermal treatment, Phys. Rev. B 54 (1996) 2856–2862.

[27] H. Kersten, P. Schmetz, G.M.W. Kroesen, Surface modification of powder particles by plasma deposition of thin metallic films, Surf. Coat. Technol. 108 (1998) 507–512.

[28] H. Sachdev, P. Scheid, Formation of silicon carbide and silicon carbonitride by RF-plasma CVD, Diamond Related Mater. 10 (2001) 1160–1164.

[29] J. Patel, L. Nemcova, P. Maguire, W.H. Graham, D. Mariotti, Synthesis of surfactant-free electrostatically stabilized gold nanoparticles by plasma-induced liquid chemistry, Nanotechnology 24 (2013) 245604.

[30] D. Mariotti, J. Patel, V. Svrcek, P. Maguire, Plasma–liquid interactions at atmospheric pressure for nanomaterials synthesis and surface engineering, Plasma Process. Polym. 9 (2012) 1074–1085.

[31] R. Wang, S. Zuo, W. Zhu, J. Zhang, J. Fang, Rapid synthesis of aqueous-phase magnetite nanoparticles by atmospheric pressure non-thermal microplasma and their application in magnetic resonance imaging, Plasma Process. Polym. 11 (2014) 448–454.

[32] L.N. Smith, P.S. Midha, A computer model for relating powder density to composition, employing simulations of dense random packings of monosized and bimodal spherical particles, J. Mater. Process. Technol. 72 (2) (1997) 277–282.

[33] E. Krentsel, S. Fusselman, H. Yasuda, T. Yasuda, M. Miyama, Penetration of plasma surface modification. II. CF 4 and C 2 F 4 low temperature cascade arc torch, J. Polym. Sci. Part A: Plasma Chem. 32 (10) (1994) 1839–1845.

[34] J.F. Friedrich, W.E.S. Unger, A. Lippitz, I. Koprinarov, G. Kuhn, S. Weidner, et al., Chemical reactions at polymer surfaces interacting with a gas plasma or with metal atoms—their relevance to adhesion, Surf. Coat. Technol. 119 (1999) 772–782.

[35] P. Bartoš, L. Volfová, P. Špatenka, Thin film deposition in limited volume of geometrically complicated substrates, Europ. Phys. J. D 54 (2) (2009) 173–177.

[36] P. Spatenka, J. Hladík, A. Kolouch, A. Pfitzmann, P. Knoth, Plasma treatment of polyethylene powder—process and application, in: 48th Annual Technical Conference Proceedings of the Society of Vacuum Coaters 48, 2005, pp. 95–98.

[37] P. Špatenka, T. Vacková, Z. Jeníková, Plasma and Fusion Science, From Fundamental Research to Technological Applications, Apple Academic Press, 2016 (Chapter Plasma Treatment of Powder and Granulates).

[38] J.M. Valverde, A. Castellanos, Fluidization, bubbling and jamming of nanoparticle agglomerates, Chem. Eng. Sci. 62 (2007) 6947–6956.

[39] J. Sefcik, M. Soos, A. Vaccaro, M. Morbidelli, Effects of mixing on aggregation and gelation of nanoparticles, Chem. Eng. Process. 45 (2006) 936–943.

[40] G. Calvert, M. Ghadiri, R. Tweedie, Aerodynamic dispersion of cohesive powders: a review of understanding, Adv. Powder Technol. 20 (2009) 4–16.

[41] C. Vahlas, B. Caussat, P. Serp, G.N. Angelopoulos, Principles and applications of CVD powder technology, Mater. Sci. Eng. R 53 (2006) 1–72.

[42] H.N. Lari, J. Chaouki, J.R. Tavares, De-agglomeration of nanoparticles in a jet impactor-assisted fluidized bed, Powder Technol. 316 (2017) 455–461.

[43] A. Akhavan, F. Rahman, S. Wang, M. Rhodes, Enhanced fluidization of nanoparticles with gas phase pulsation assistance, Powder Technol. 284 (2015) 521–529.

[44] C. Mogre, A.U. Thakurdesai, J.R. van Ommen, S. Salameh, Long-term fluidization of titania nanoparticle agglomerates, Powder Technol. 316 (2017) 441–445.

[45] P. Bartoš, J. Blažek, P. Jelínek, P. Špatenka, Hybrid computer simulations: electrical charging of dust particles in low-temperature plasma, Europ. Phys. J. D 54 (2009) 319−323.

[46] L. Couëdel, M. Mikikian, L. Boufendi, A.A. Samarian, Residual dust charges in discharge afterglow, Phys. Rev. E 74 (2006) 026403.

[47] A. Usachev, A. Zobnin, O. Petrov, V. Fortov, M. Thoma, M. Kretschmer et al., The project "Plasmakristall-4" − a dusty plasma experiment in a combined dc/rf discharge plasma under microgravity conditions, Czechoslovak J. Phys. 54 (2004) C649.

[48] R. d'Agostino, Plasma Deposition, Treatment and Etching of Polymers, Academic Press, Inc., San Diego, CA, 1990.

[49] M. Quitzau, M. Wolter, H. Kersten, Plasma treatment of polyethylene powder particles in a hollow cathode glow discharge, Plasma Process. Polym. 6 (2009) S392−S396.

[50] J.-H. Lee, K.-J. Jeong, Measurement of plasma characteristics of the hollow cathode discharge, J. Phys. Soc. Japan 72 (2003) 2530−2532.

[51] M.E. Pillow, A critical review of spectral and related physical properties of the hollow cathode discharge, Spectrochim. Acta Part. B. At. Spectrosc. 36 (1981) 821−843.

[52] H. Barankova, L. Bardos, Hollow cathode plasma sources for large area surface treatment, Surf. Coat. Technol. 146 (2001) 486−490.

[53] R.R. Arslanbekov, A.A. Kudryavtsev, R.C. Tobin, On the hollow-cathode effect: conventional and modified geometry, Plasma Sour. Sci. Technol. 7 (1998) 310−322.

[54] F. Bretagnol, M. Tatoulian, F. Arefi-Khonsari, G. Lorang, J. Amouroux, Surface modification of polyethylene powder by nitrogen and ammonia low pressure plasma in a fluidized bed reactor, React. Funct. Polym. 61 (2004) 221−232.

[55] D. Kuniii, O. Levenspiel, Fluidisation Engineering, Krieger, 1977.

[56] F. Shidong, M. Yuedong, S. Jie, C. Jie, Surface treatment of polypropylene powders using a plasma reactor with a stirrer, Plasma Sci. Technol. 13 (2011) 217−222.

[57] M. Gilliam, S. Farhat, A. Zand, B. Stubbs, M. Magyar, G. Garner, Atmospheric plasma surface modification of PMMA and PP micro-particles, Plasma Process. Polym. 11 (2014) 1037−1043.

[58] www.surface-treat.cz

[59] L. Lapčík, B. Lapčíková, I. Krásny, I. Kupská, R.W. Greenwood, K.E. Waters, Effect of low temperature air plasma treatment on wetting and flow properties of kaolinite powders, Plasma Chem. Plasma Process. 32 (2012) 845−858.

[60] K. Fatyeyeva, F. Poncin-Epaillard, Sulfur dioxide plasma treatment of the clay (laponite) particles, Plasma Chem. Plasma Process. 31 (2011) 449−464.

[61] A. Mueller, M.G. Schwab, N. Encinas, D. Vollmer, H. Sachdev, K. Mullen, Generation of nitrile groups on graphites in a nitrogen RF-plasma discharge, Carbon. N. Y. 84 (2015) 426−433.

[62] W.-K. Oh, W. Hoon, J. Jang, Characterization of surface modified carbon nanoparticles by low temperature, Diamond Related Mater. 18 (2009) 1316−1320.

[63] R. Yang, Y. Wei, Y. Yu, C. Gao, L. Wang, J.-H. Liu, et al., Make it different: the plasma treated multi-walled carbon nanotubes improve electrochemical performances toward nitroaromatic compounds, Electrochim. Acta 76 (2012) 354−362.

[64] J.Y. Yook, J. Jun, S. Kwak, Amino functionalization of carbon nanotube surfaces with NH_3 plasma treatment, Appl. Surf. Sci. 256 (2010) 6941−6944.

[65] C. Chen, B. Liang, D. Lu, A. Ogino, X. Wang, Masaaki Nagatsu: amino group introduction onto multiwall carbon nanotubes by NH_3 /Ar plasma treatment, Carbon. N. Y. 48 (2010) 939−948.

[66] H. Kersten, P. Schmetz, G.M.W. Kroesen, Surface modification of powder particles by plasma deposition of thin metallic films, Surf. Coat. Technol. 108−109 (1998) 507−512.

[67] X. Yu, Z. Shen, Metal copper films coated on microparticle substrates using an ultrasonic-assisted magnetron sputtering system, Powder Technol. 187 (2008) 239–243.

[68] V. Brueser, M. Haehnel, H. Kersten, Thin film deposition on powder surfaces using atmospheric pressure discharge, in: CAPPSA Proceedings, 2005, p. 261–264.

[69] V. Brueser, M. Haehnel, H. Kersten, Thin film deposition on powder surfaces using atmospheric pressure discharge, New Vistas Dusty Plasmas Book Series: AIP Conf. Proc. 799 (2005) 343–346.

[70] C. Roth, Z. Kunsch, A. Sonnenfeld, P.R. von Rohr, Plasma surface modification of powders for pharmaceutical applications, Surf. Coat. Technol. 205 (2011) S597–S600.

[71] M. Quitzau, M. Wolter, V. Zaporojtchenko, H. Kersten, F. Faupel, Modification of polyethylene powder with an organic precursor in a spiral conveyor by hollow cathode glow discharge, Europ. Phys. J. D 58 (2010) 305–310.

[72] R. Hu, T. Furukawa, X. Wang, M. Nagatsu, Morphological study of graphite-encapsulated iron composite nanoparticles fabricated by a one-step arc discharge method, Appl. Surf. Sci. 416 (2017) 731–741.

[73] A. Shahravan, T. Matsoukas, Encapsulation and controlled release from core–shell nanoparticles fabricated by plasma polymerization, J. Nanoparticle Res. 14 (2012) 668–679.

[74] W.K. Dierkes, M. Tiwari, R.N. Datta, A.G. Talma, J.W.M. Noordermeer, Plasma polymerization of monomers onto fillers, to tailor their surface properties in tire compounds, Rubber Chem. Technol. 83 (2010) 405–426.

[75] S.K. Nema, P.B. Jhala, Plasma Technologies for Textile and Apparel, CRC Press, New Delhi, 2015.

[76] P. Mallick, Fiber-Reinforced Composites: Materials, Manufacturing, and Design, third ed., CRC Press, Boca Raton, 2007, p. 638.

[77] N. Encinas, M. Lavat-Gil, R.G. Dillingham, J. Abenojar, M.A. Martinez, Cold plasma effect on short glass fibre reinforced composites adhesion properties, Int. J. Adhesion Adhesives 48 (2014) 85–91.

[78] Y. Kusano, S. Teodoru, C.M. Hansen, The physical and chemical properties of plasma treated ultra-high-molecular-weight polyethylene fibers, Surf. Coat. Technol. 205 (2011) 2793–2798.

[79] C. Zhang, C. Li, B. Wang, H. Cui, Effects of atmospheric air plasma treatment on interfacial properties of PBO fiber reinforced composites, Appl. Surf. Sci. 276 (2013) 190–197.

[80] P. Bartoš, L. Volfová, P. Špatenka, Limited volume of geometrically complicated substrates, Europ. Phys. J. D 54 (2) (2009) 1–5.

[81] P. Bartoš, P. Špatenka, L. Volfová, Deposition of TiO_2-based layer on textile substrate: theoretical and experimental study, Plasma Process. Polym. 6 (2009) 5897–5901.

[82] K.K. Chawla, Composite Materials Science and Engineering, third ed., Springer, 2013.

[83] C. Durkin, Low-Cost Continuous Production of Carbon-Fiber-Reinforced Aluminum Composites, Thesis of Georgia Institute of Technology, USA, December 2007.

[84] R. Pillai, C. Jariwala, K. Kumar, S. Kumar, Process optimisation of SiO2 interface coating on carbon fibre by RF PECVD for advanced composites, Surf. Interf. 9 (2017) 21–27.

[85] R. Pillai, N. Batra, L.M. Manocha, N. Machinewala, Deposition of silicon carbide interface coating on carbon fibre by PECVD for advanced composites, Surf. Interf. 7 (2017) 113–115.

[86] J. Friedrich, K. Altmann, S. Wettmarshausen, G. Hidde, Coating of carbon fibers with adhesion-promoting thin polymer layers using plasma polymerization or electrospray ionization technique—a comparison, Plasma Process. Polym. (2017) 1—14.
[87] V. Cech, A. Knob, H.-A. Hosein, A. Babik, P. Lepcio, F. Ondreas, et al., Enhanced interfacial adhesion of glass fibers by tetravinylsilane plasma modification, Compos. Part A: Appl. Sci. Manuf. 58 (2014) 84—89.

CHAPTER 8

Plasma Treatment of Polymeric Membranes

Ahmed Al-Jumaili[1], Surjith Alancherry[1], Daniel Grant[1], Avishek Kumar[1], Kateryna Bazaka[1,2] and Mohan V. Jacob[1]
[1]Electronics Materials Lab, College of Science and Engineering, James Cook University, Townsville, QLD, Australia
[2]School of Chemistry, Physics, Mechanical Engineering, Queensland University of Technology, Brisbane, QLD, Australia

8.1 INTRODUCTION

Engineered polymeric membranes are widely used to separate gases, and suspended and dissolved materials in various medical, industrial, food, and environmental applications. These membranes have become a competitive cost-effective technology compared to the conventional separation unit processes such as chemical and physical absorption, ion exchangers, and cryogenic distillation [1]. Membranes technology has revealed several advantages including works without the addition of any chemicals, relatively high efficiency, mild-working conditions, low-energy consumption, and the potential of being integrated with other membranes and production/separation procedures.

The performance of membrane separation can be described in terms of efficiency (e.g., flux, membrane permeability, permeate quality, and permeate recovery) and effectiveness (e.g., rejection of selectivity) that are typically determined by membrane characteristics and surface properties [2]. Although flux and rejection of a membrane largely depend on size exclusion, surface characteristics (e.g., surface hydrophilicity, roughness, and charge) of the polymeric membrane play a significant role in processes such as microfiltration or ultrafiltration, and reduce membrane fouling [3]. Thus, modifications in surface properties can increase the range of applications; which can be accomplished by applying a chosen type of surface treatment such as polymer blending approach, coating with inorganic additives, and plasma treatment [4,5]. In particular, non-equilibrium plasma treatment is a versatile technique that can be effectively applied for any polymeric material [6], where it creates a reactive

Non-Thermal Plasma Technology for Polymeric Materials
DOI: https://doi.org/10.1016/B978-0-12-813152-7.00008-1

211

environment consisting of electrons, radicals, ions, and excited molecules [7]. As soon as these interact with a solid polymeric surface, plasma particles transfer their energy through elastic and inelastic collisions that potentially induce significant chemical and/or physical changes in that surface. In general, the main changes occur in surface energy, roughness, hydrophilicity, crosslinking, and molecular weight of the topmost layer of the material [8,9]. Owing to the small depth of penetration of active-species, plasma − material interactions were observed to only change the top layer of an organic surface in a range of few nanometres [10]. Furthermore, the treatment can create reactive sites on the irradiated surface, which can be used for subsequent modification and decoration with biologically-active molecules [11].

Broadly, plasma treatment can be utilized directly or indirectly to adapt polymeric surfaces through the creation of functional groups (see Fig. 8.1). Direct treatment involves the use of reactive plasma gases (e.g., NH_3, O_2) that are known to produce desirable functionalities (e.g., amines, COOH, and free radicals) [12,13]. Indirect treatment often refers to the grafting of polymers, which results in the introducing of desired functionalities onto the surface. In both treatment types, the processing parameters, such as treatment time, pressure, power, and processing gas, etc., and the nature of the irradiated surface, will determine the resultant modification that potentially can be either ablation, substitution, or film deposition [14].

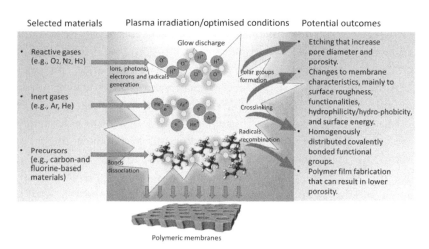

Figure 8.1 Diagram of plasma modification of polymeric membranes.

It is well-documented that fouling of polymer membranes is highly influenced by membrane chemistry and morphology. Polymers used in porous membrane manufacture have sufficient chemical and mechanical stability, but are mostly hydrophobic in nature, which makes it highly susceptible to adsorption of foulants [15]. Commercial approaches to reduce fouling mainly involve plasma/graft polymerization of hydrophilic monomers on the membrane surface. Plasma assisted membrane modification is often utilized due to the following advantageous characteristics [16]:

- Ability to induce uniform surface changes irrespective of the geometry of the exposed membranes.
- Easy and efficient way of modifying the membrane surface with a wide range of functionalities without affecting the bulk properties of the material.
- Devoid of toxic solvent treatments, which reduces the probability of surface contamination or membrane degradation.
- Safe for temperature-sensitive polymers.

This chapter will focus on the use of plasma to expand biological properties of polymeric membranes and mitigate their biofouling though surface functionalisation. The next section will discuss plasma-induced grafting on surfaces of membranes and surface modification using nonpolymerizable gas plasmas.

8.2 PLASMA TREATMENT OF BIOLOGICAL MEMBRANES

In the most general sense, membranes operate at the nano-, micro-, or macro-scale to physically separate two spaces. In this capacity, they find significant industrial utility in a variety of processes, including water desalination, drinking and waste water purification, air filtration, oil/water separation, and so on. In these and other roles, membrane performance is quantified with respect to specificity (i.e., the ability to isolate only those components of interest), and flux (i.e., the amount of fluid or gas processed per unit time). These performance parameters are often coupled with secondary interests centred on membrane longevity, environmental friendliness, and economic factors.

Within the biological domain, polymeric membranes have proved to be invaluable in haemodialysis [17], protein separation, cell culture work [18], tissue scaffolds, sensors [19,20], and drug release devices [21]. The utilisation of polymeric membranes in biological environments requires membranes capable of not only satisfying specificity and flux performance constraints, but also of eliciting specific responses at the biointerface between the polymer surface and biological medium. With respect to blood contacting or indwelling membranes (e.g., in haemodialysis) there is a need to maintain hemocompatibility and prevent thrombosis or deleterious immunogenic responses. Membranes in contact with aqueous environments (e.g., in water filtration and bioreactors) must prevent nonspecific protein adhesion and its concomitant biofouling effects [22]. Membranes used in the treatment of wounds or serving as tissue scaffolds must also foster endothelisation while simultaneously staving off infection [23,24]. Regardless of how well a membrane performs its primary function, failure to mitigate detrimental biointerfacial interactions will nonetheless render the device nonviable.

Attributes of both the biological medium and the polymer's surface exert considerable influence over interactions at the biointerface. With respect to the biological medium, numerous key factors have been identified, including chemical composition, nutrient availability, temperature, and flow conditions. These factors are particular to a given medium, and it is often times unfeasible (or even impossible) to induce changes in these parameters that would foster favourable biointerface responses. Fortunately, however, the polymeric device's surface attributes are readily accessible and can be tailored to promote beneficial interactions. In this respect surface topography, chemistry, energy, and wettability have been identified as dominant factors in biointerface interactions [14,25,26], regulating protein absorption, cellular adhesion, and the subsequent proliferation of cells. Strategies to circumvent deleterious interactions at biointerfaces have progressed beyond the use of chemically inert polymeric materials, to the point of engineering surfaces with specific chemical functionalities and topographies capable of inducing targeted responses [27]. Often this is accomplished by first selecting a cost-effective bulk polymeric material with the requisite mechanical properties and stability for the task. Common polymers for bio-applications include poly(methyl methacrylate) (PMMA), polyurethanes (PU), polyvinyl alcohol (PVA), polypropylene (PP), poly(ethylene terephthalate) (PET), polyethersulfone (PES), and polycarbonate (PC) [17,28]. The chemical stability exhibited by these materials is often due to their lack of functional groups, which commonly leads to the expression of surface hydrophobicity and an associated increase in the absorption of proteins and other foulants (which are typically hydrophobic) [29]. As such, the second phase of preparing polymeric devices for bio-applications involves using surface modification techniques to decouple surface properties from bulk properties. With respect to polymeric membranes in particular, the desired change in surface characteristics must occur without inducing detrimental alterations in physical properties—such as pore size, flexibility, and mechanical strength—which may influence primary performance traits (i.e., selectivity and flux).

The growing demand for low cost, rapid, and environmentally friendly surface processing technologies has served to enhance the attractiveness of nonthermal plasma-based techniques. Plasma processing has been applied to numerous polymeric materials requiring improved biointerfacial compatibility and may be categorised as either plasma treatment or plasma deposition. Plasma treatment entails subjecting the target material to a cold plasma environment supplied with a nonpolymerizable precursor or combination of precursors such as O_2, Ar, CF_4, or air [30]. This treatment induces hydrogen abstraction along with crosslinking of the surface layer and cyclization of macromolecules, both of which may serve to enhance the stability of the membrane. Radical formation in the top few nanometers of the polymer structure also transpires [31] and has been used in conjunction with nitrogen containing plasmas to introduce amine, imine, amide, nitrile, and other functionalities [32], with particular utility in the preparation of biocompatible surfaces [33]. Other precursor gases may also be chosen to introduce a variety of functional groups (e.g., carboxy, hydroxyl, aldehydes, etc.) or radicals suitable for further covalent immobilisation of biomolecules.

The capacity for plasma processing to control surface parameters that underlie biointerfacial interactions have been demonstrated on a variety of polymeric membranes. For example, inductively coupled H_2O vapor plasma has been used to introduce OH and O radicals into PC and PET track-etched membranes [34]. This increase in oxygen functionality served to increase wettability and produce hydrophilic membranes without the use of wetting agents (e.g., polyvinylpyrrolidone). Furthermore, the plasma treatment was able to enhance water flux through the polymeric membranes, without altering pore structure or damaging surface architecture. In instances where the polymer has shown susceptibility to plasma-induced degradation, it is also possible to mitigate such degradation by placing the target polymer outside of the glow discharge region. In one study, low temperature Ar plasma modification of PE hollow fibre membranes was carried out using downstream processing (i.e., placing sample downstream of the plasma volume). Here, the membrane surface contact angle was reduced from $\sim 120°$ to $\sim 60°$ due to the implantation of polar groups and the plasma-treated membranes showed reduced protein fouling, increased water flux, and hydrophilic stability for at least one month after treatment [35]

Age-related hydrophobic recovery is a common problem in surface-treated polymers, involving the migration of surface polar groups into the bulk polymer volume and an associated movement of untreated hydrophobic polymer chains to the surface. This hydrophobic recovery may reduce product shelf-life and yield a material unsuitable for certain bio-applications that require stable and well characterised surfaces (e.g., controlled studies of surface-protein or surface-cell interactions). Methods for circumventing hydrophobic recovery and creating stable hydrophilic polymeric membranes include blending with hydrophilic additives, or copolymerisation with hydrophilic polymers. Unfortunately, these approaches may produce undesirable variation in bulk mechanical properties, pore geometry, and surface morphology [34] which can, in turn, lead to further variation in membrane performance [36].

Regardless of the challenges posed by hydrophobic recovery, plasma treatment has been shown to produce polymeric surfaces with long-lived hydrophilicity. In one study, plasma supplied with an O_2/NH_3 mixture (in a 5:3 gas flow ratio) was shown to modify microporous PES membranes and create permanently hydrophilic surfaces. The incorporation of NH_x and OH species reduced the untreated PES membrane static contact angle measurement from $96°$ to $0°$, and most importantly, this post-treatment wettability was retained for up to one year. As with the previous plasma treatment examples, no membrane damage occurred and the treated PES membranes demonstrated enhanced water flux, reduced protein fouling, and greater flux recovery after gentle cleaning (relative to untreated membranes) [37].

Polymeric membranes intended for healing and tissue-growth applications must encourage coverage of anchorage-dependent cells (such as fibroblasts, osteoblasts, and endothelial cells) that need to adhere to the membrane surface prior to proliferating. Given its biocompatibility, biodegradability, and low toxicity, chitosan (an abundant natural polysaccharide derived from the exoskeletons of insects and crustaceans) is an attractive material for tissue-growth applications. To overcome chitosan's inherent lack of chemical functionality, O_2 plasma treatment has been used to activate the surface of chitosan membranes intended for bone-guided regeneration. Introduced polar groups were then used to control

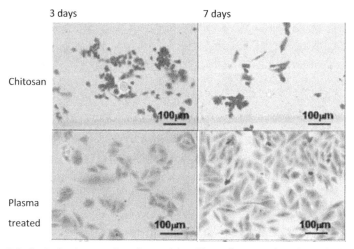

3 days   7 days

Chitosan

Plasma

treated

Figure 8.2 Optical micrographs of osteoblast-like cells stained with methylene blue cultured for three and seven days. *Reproduced with permission from [40].*

wettability and surface free energy [38], and for the immobilisation of biomolecules and control of protein absorption (which influences cell morphology and proliferation [39]). As shown in Fig. 8.2, this treatment was able to significantly improve the adhesion and proliferation of osteoblast-like cells (human osteosarcoma cell line SaOs-2) on chitosan membranes over a seven-day incubation period [40].

Fouling is a major challenge for membranes [29], leading to decreased flux or increased transmembrane pressure. In one successful attempt to produce antifouling membranes, poly(2-methoxyethyl acrylate) (PMEA) was plasma graft polymerised onto polyethylene membranes. The membranes were first treated with plasma and then exposed to air in order to introduce peroxide radicals, which subsequently facilitated the grafting of PMEA following immersion in monomer solution (MEA dissolved in MeOH aqueous solution). Polymeric membranes prepared in this fashion demonstrated a significant reduction in fouling when using bovine serum albumin as a model foulant substance. In contrast with pure plasma treatment processing, however, consideration must be given to the capacity for grafting and polymerisation facilitated by plasma treatment to introduce a surface layer that may reduce the physical dimensions of membrane pores (see Fig. 8.3), with possible implication for membrane flux and specificity.

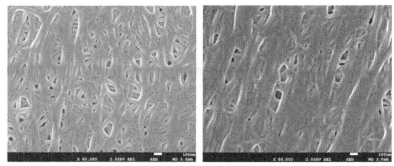

Figure 8.3 SEM images of the unmodified polyethylene membrane (left), and membrane grafted with 0.257 mg/cm^2 PMEA (right). *Reproduced with permission from [41].*

There exists a continuous drive to enhance the biocompatibility and antifouling capabilities of polymeric membranes destined for bio-applications and plasma treatment presents as an attractive technology for enabling such enhancement. Plasma treatment provides rapid, environmentally friendly, and selective modification of key surface characteristics that regulate biointerface interactions. Furthermore, this controlled engineering of surface chemistry and topography is compatible with masking techniques, can be applied to polymeric membranes with different geometries and sizes, and can be undertaken without altering the polymer's bulk mechanical properties.

8.3 MEMBRANE FOULING

Membrane fouling generally refers to the adherence/deposition of inorganic particles, colloids, and macromolecules, etc., at the membrane surface, pore walls, or inside the pore. Membrane flux decreases with increase in fouling, which causes higher operating pressure and requires more power to get the desired throughput [42]. The foulants not only deteriorate the membrane performance but also degrade the membrane material. It is indeed a serious problem in membranes used in high pressure processes such as reverse osmosis (RO), microfiltration (MF), ultrafiltration (UF) and nanofiltration (NF). Moreover, the attachment of foulants to membrane surfaces is a preliminary step in biofilm formation. Membrane fouling can be basically categorized into three main groups: reversible, irreversible, and irrecoverable, depending on the nature of foulant adhered onto membrane surface, as seen in Fig. 8.4. It is well-known

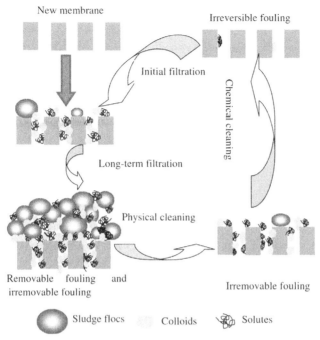

Figure 8.4 Reversible, irreversible, and irrecoverable fouling in membrane processes. *Reproduced with permission from [11].*

that membrane surface properties such as wettability, charge, and surface roughness highly affect their interaction with foulants.

Surface hydrophilicity. In the most general sense, fouling decreases with increases in the hydrophilicity of the polymeric material [43,44]. The hydrophobic membranes repel water molecules; as a result, hydrophobic organic molecules are driven towards the surface and get adsorbed on it. In contrast, hydrophilic surfaces are able to form hydrogen bonds with water molecules which results in the formation of a thin hydration layer between the membrane and bulk solution. This hydration layer acts as an energetic barrier and prevents the adhesion of fouling molecules.

The wettability of a surface is typically measured in terms of its contact angle. The contact angle of membranes depends on functional groups, porosity, roughness, as well as pore size distribution [45,46]. A highly porous membrane would exhibit a low contact angle even though it may not necessarily be hydrophilic. Increases in the density of hydrophilic groups such as $-OH$ and NH_2 has been identified to increase the hydrophilicity of membranes [47,48]. The contact angle value of a membrane with higher surface roughness is smaller if the material is hydrophilic and vice versa when the material is hydrophobic. However, the influence of chemical groups on hydrophilicity is higher than that of surface roughness.

Surface charge. Electrostatic repulsive forces between foulants and a membrane surface are known to decrease membrane fouling. The membrane charge is typically quantified in terms of zeta potential measurements. The surface of NF, UF, and RO membranes are usually negatively charged due to the presence of carboxylic/sulfonic acid groups [49,50]. Natural organic matter and most bacterial surfaces generally have a negative surface charge in the pH range of natural water and are less prone to be adsorbed onto a negatively charged surface. Likewise, a positively charged membrane has exhibited repulsion against positively charged proteins. The rise in pH of the feed solution increases the negative charge on the membrane, which is beneficial and reduces membrane fouling [51]. In this regard, Hadidi et al. fabricated charged, neutral and zwitterionic UF membranes with an identical pore size and compared their antifouling behavior. The fouling behavior of membranes was tested against the proteins of different charges at various pH. The highly hydrophilic zwitterionic membranes exhibited the least amount of fouling and showed a flux recovery ratio greater than 80 percent under all experimental conditions, even when the membranes and proteins were oppositely charged [52].

Surface roughness. A strong correlation exists between membrane fouling and its roughness. For example, RO hydrophilic cellulose acetate membrane showed less colloidal fouling compared to polyamide thin film composite on account of their smoother surface [53]. Higher membrane surface roughness increases the total surface area and colloidal particles are preferentially adsorbed on valleys of a rougher membrane causing valley clogging that leads to flux decline [54]. According to Bowen et al., higher membrane roughness reduces the electrostatic repulsion force between foulants and the surface, and the valley region experiences enhanced adhesion force [55]. Nonetheless, contradictory results showing a decrease in fouling with increasing membrane roughness have also been reported. For instance, Hasino et al. fabricated gear-shaped hollow cellulose acetate butyrate polymer membranes with varying surface roughness. The membrane with the highest roughness value exhibited the least biofilm formation due to an increase in surface area and foulants getting preferentially attaching in valleys, with the top of the membrane being kept relatively clean [56].

Despite the membrane hydrophilicity, roughness, and charge, the use of active molecules—such as grafted hydrophilic brushes and biocides—is also reported to importantly decrease fouling rate. Though the membrane properties such as hydrophilicity, roughness, and charge have shown to influence the degree of fouling, there are also studies showing fouling independent of these surface properties. Baek et al. studied the fouling in RO membranes in a laboratory scale cross-flow RO membrane system. It was found that the flux decline with time was independent of surface properties and was constant among all selected membranes [57]. Thus, it can be concluded that hydrophilization, smoothing, and the introduction of charged/biocidal groups are the main goals for membrane modification to reduce undesirable fouling.

8.3.1 Improve Membrane Antibiofouling Properties

In order to reduce the attachment of foulants, plasma treatment of a membrane significantly amends the chemical and physical structure of the surface. The pore size of the membrane can decrease/increase as a result of plasma treatment depending on competing processes of ablation or deposition. Various plasma modifications—such as treatment with air, oxygen, and carbon dioxide plasmas, grafting, and polymerization—have been effectively utilized to change polymer surfaces to a great extent.

Treatment in CO_2 plasma renders the membrane surface more hydrophilic due to the presence of oxygen incorporation. For example, poly(sulfone) and poly(ether sulfone) microporous membranes are most generally treated with CO_2 plasma, where the plasma is used to strongly etch the medium, which increases the pore diameter. Similarly, CO_2 treatment of a poly(propylene) membrane was reported to increase its hydrophilicity, pore size, and porosity and, thus, the membrane exhibited better fouling resistance and flux recovery in comparison to an unmodified surface [58]. By the same token, the application of air and oxygen plasmas leads to the formation of carboxyl groups. Surfaces treated with air plasma can exhibit amphoteric behavior depending on pH. For instance, when microporous poly(propylene) membranes were air plasma treated for 8 minutes, the hydrophilicity of the membranes considerably increased exhibiting less fouling compared to the unmodified surfaces. After continuous operation in a submerged membrane bioreactor for 110 hr, the flux recovery was 35 percent higher than that for the unmodified surface [59].

Furthermore, nitrogen moieties are incorporated into a membrane after treatment with nitrogen-based plasmas. Groups such as C-N, C = N and oxygen bearing functionalities are usually formed on the polymer surface. The rate of surface etching was found to be minimal in this type of plasma treatment, where the surface becomes more basic because of nitrogen groups [60,61]. Polyacrylonitrile membranes treated with ammonia plasma for 8 minutes showed improved antifouling property on account of their increased hydrophilicity due to the incorporation of nitrogen and oxygen groups. The flux recovery also increased about 49 percent and flux increased by 32 percent after plasma treatment. The membranes were used for oil−water emulsion separation [62].

In this regard, plasma-induced grafting and polymerization are also widely used to mitigate biofouling in membranes. For example, surface grafting to poly(vinylidene fluoride) (PVDF) membrane with zwitterionic copolymers significantly reduced fouling in static conditions [63]. Gancarz et al. modified polysulfone membranes with acrylic acid by plasma graft polymerization using three different approaches: (1) grafting in a solution, where the plasma modified membrane were exposed to air for 5 min and dipped in an aqueous solution of the monomer; (2) grafting in a vapor; where membranes were modified with the monomer vapor after treatment with Ar plasma; and (3) plasma polymerization of the monomer on PS surface. Membrane modification with grafting in the vapor phase revealed brush-like structures that increased the hydrophilicity of the membrane. Vapor-phase modified membranes showed the most efficient filtration property [64]. In another study, the polyacrylonitrile (PAN) membranes was prepared more hydrophobic by etching in Ar plasma, grafting hydrophobic groups by dipping in a fluorine-containing solution, followed by plasma irradiation. The contact angle increased from 101° to 132° after the plasma treatment. These membranes were subjected to a vacuum membrane distillation desalination test for 80 hours intermittently, where a steady flux recovery and almost 100 percent salt rejection were observed. This demonstrates the antifouling characteristic imparted to PAN membranes upon plasma modification [65]. In another report, the surface of RO membranes was nanostructured via atmospheric pressure plasma induced polymerization. The surface of the polyamide membrane was activated using atmospheric-pressure plasmas followed by graft polymerization of methacrylic acid and acrylamide monomers onto the membrane surface. Modification resulted in polyamide surfaces showing low mineral scaling propensity and higher permeability than the commercially available RO membranes [66].

In addition to the grafting method, plasma polymerisation can also be utilized to effectively reduce biofouling issues on membranes. For example, Zou et al. deposited hydrophilic triethylene glycol dimethyl ether onto a polyamide RO membrane via plasma polymerization to reduce organic fouling. Hydrophilicity of the membrane was significantly enhanced after the plasma deposition. After 210 min filtration of a protein solution, the modified membrane showed no flux decline, while 27 percent reduction in flux decline was observed for the untreated surface. Flux recovery after water cleaning of treated membranes was almost 100 percent whereas for the untreated ones it was 91 percent [67].

8.4 MEMBRANES SURFACE FUNCTIONALIZATION

Plasma surface grafting uses plasma for the formation of homogenously distributed covalently bonded functional groups and to induce surface-initiated polymerization on the membrane surface [68]. Indeed, plasma grafting can bring significant changes to the membrane characteristics, mainly to surface roughness, surface patterning, functionalities, hydrophilicity/hydrophobicity, and surface energy, thereby regulating crucial biocompatible parameters such as platelet adhesion [69], protein adsorption [70], cell adhesion and proliferation [71], and bacterial attachment [63], etc.

Plasma-assisted grafting can be broadly divided into plasma post-irradiation grafting and plasma syn-irradiation grafting [72,73]. Plasma post-irradiation grafting includes two steps, wherein an inert/reactive gas plasma is used initially to activate the membrane surface through the generation of free radicals and functional groups, and then subjected to liquid or gaseous phase monomers to initialize the grafting process [74]. On the other hand, plasma syn-irradiation grafting comprises three steps, where plasma is used in the first step for surface pre-activation, followed by the creation of a uniform pre-adsorbed layer of monomer by exposing with desired monomer, and, finally, treated with plasma again to induce surface-initiated grafting. At this point, free radicals stimulate the grafting process and are generated on the pre-adsorbed monomer layer through high-energy electronic/ionic collisions or UV radiation from the plasma [71,75]. In contrast to post-irradiation grafted surfaces, the grafted surfaces exhibit a highly crosslinked structure having less resemblance to the parent monomer due to highly reactive plasma exposure [72].

It has been demonstrated that post-irradiation grafting of polytetra-fluoroethylene (PTFE) membrane with poly(ethylene glycol)methacrylate (PEGMA) was very effective in improving the biofouling against plasma proteins and platelets [70]. Through PEGMA surface modification a highly hydrated hydrogel-like layer was formed on a hydrophobic PTFE membrane. It is also shown that biofouling was related to grafting coverage and degree of hydration which can be easily controlled by changing the PEGMA macromonomer concentration. As can be seen in Fig. 8.5, at lower concentration (Fig 8.5B) membrane pores were partially filled, but sequentially transformed into a nearly pore-free nature (Fig 8.5D) with respect to the increase in PEGMA percentage. It is worth noting that, compared to virgin PTFE, all the modified membranes showed good resistance to γ-globulin, fibrinogen, and HSA proteins and remarkably suppress platelet adhesion.

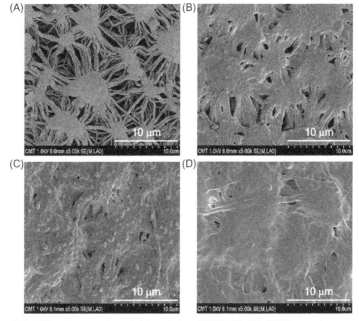

Figure 8.5 SEM micrographs of (A) pristine PTFE, (B)-(D) PTFE membrane grafted with different concertation of PEGMA macromonomer (5 to 20 wt%). The grafting coverage and hydrophilicity was found to increase with the increase in macromonomer concertation of PEGMA which is substantiated by the drastic reduction in water contact angle from 130° to 55°. Moreover, upon grafting the porous PTFE membrane became nearly pore-free and, thus, showed better biofouling properties. *Reproduced with permission from [70].*

Chang et al. used plasma syn-irradiation to graft zwitterionic poly(sulfobetaine methacrylate) (PSBMA) on poly(vinylidene fluoride) (PVDF) to improve the blood compatibility [69]. PVDF membranes were first treated with low pressure argon plasma, uniformly coated with SBMA by immersing in methanol solution containing 30 percent SBMA and finally subjected to atmospheric plasma to activate the grafting. During this stage, initiator radicals were formed on both the SBMA monomer and PVDF membrane which immobilized PSBMA on PVDF through copolymerization. At this phase, the plasma treatment time is very crucial which influences the grafting coverage, uniformity and charge neutrality. At optimal conditions, the PSBMA grafting has completely suppressed human blood platelet adhesion and improved the plasma clotting time (~14 min). Moreover, it reduced the blood cell disruption (3 percent) compared to the un-grafted PVDF. Another work done by Jhong and coworkers demonstrated the syn-irradiation grafting of zwitterionic PSBMA and PEGMA on PTFE at atmospheric pressure and its potential in skin tissue regeneration membrane applications [71]. The study revealed that after grafting both hydrophilicity and hydration capacity (ability to entrap water) of PTFE-g-PSBMA have significantly increased. Therefore, the grafted surface presented a very stable performance when in contact with protein solution, human blood, and bacterial medium. Furthermore, compared to the commercial hydrocolloid dressing, PTFE-g-PSBMA exhibited very fast wound-healing (within seven days). The in vivo wound-healing performance is depicted in Fig. 8.6. It is worth noting that surface grafting of highly polar zwitterionic monomers onto hydrophobic polymer surfaces is challenging and various techniques

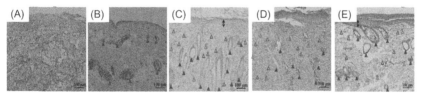

Figure 8.6 In vivo wound-healing performance of (A) gauze, (B) PTFE, (C), PTFE-g-PBSMA (D), PTFE-g-PEGMA, and (E) commercial hydrocolloid dressing after 14 days. It is clear that PTFE-g-PBSMA displayed excellent healing performance by closely resembling the normal skin tissue in terms of comparable granulation layer thickness (↕), well-proliferated fibroblast (△), collagen production, numerous hair follicles (▲), and blood vessels (Δ). Also, the healing performance was found to be very fast (within seven days) and even better than the commercial hydrocolloid dressing. *Reproduced with permission [71].*

(e.g., thermally induced molecular grafting and surface-initiated atom transfer radical polymerization) have been employed along with plasma-induced surface copolymerization. However, over these techniques, plasma-assisted grafting has the advantage of providing efficient grafting time, good control over the grafting coverage, better hydration capability, and the most promising biocompatible zwitterionic coatings [76,77].

The plasma-assisted grafting technique provides an efficient and easy way to tailor membrane properties, since a relatively large number of process parameters (e.g., pressure, plasma treatment time, reactive gas and concertation, etc.) can be altered to achieve the desired properties. It is highly appreciable that a wide variety of functional groups can be bonded to the membrane surface by suitably selecting the reactive gas discharges. For instance, hydroxyl, carboxylic, and amine groups can be attached to surfaces by exposure to oxygen, carbon dioxide, and ammonia plasmas respectively [67,78,79]. Besides, plasma-induced surface grafting is very fast and efficient in terms of both the process time and grafting yield. It was reported that plasma grafting of PEGMA on PVDF exhibited high grafting yields of 0.69 to 0.45 mg/cm^2 and a short grafting time of 120 s compared to the thermally-induced grafting (grafting period 48 h and yield of 0.12 mg/cm^2) [69]. In addition, the grafting was reconfigured from a network-like to brush-like structure by simply shifting the processes from low-pressure to atmospheric pressure. This structural variation was attributed to the higher degree of bond scission and chemical degradation of the pre-adsorbed PEGMA layer due to the higher penetration of plasma species at lower pressure. However, both structures were found to resist plasma protein and platelet adhesion; nonetheless the brush like-structure exceeds the network structure in performance. Fig. 8.7 illustrates the effect of plasma treatment time and pressure on grafting yield, hydrophilicity, and protein adsorption for PVDF-g-PEGMA membranes.

Despite grafting coverage and structure, plasma treatment time also has significant influence in controlling the surface polarity. It is a very crucial factor in determining hemocompatibility because even a slight charge imbalance causes electrostatic interactions with blood proteins. Chang et al. correlated plasma treatment time (0 to 120 s with 30 s interval) and surface polarity of syn-plasma grafted PBSMA membrane on PVDF with the hemocompatibility [77]. It was found that for ideal treatment time (90 s), the PBSMA layer maintained electrical neutrality, then became positively rich at lower exposure time and negatively rich with higher treatment time. Furthermore, the electrically neutral PBSMA layer

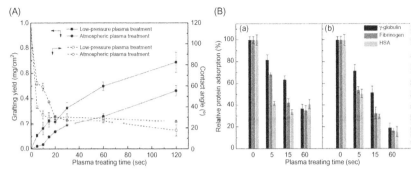

Figure 8.7 (A) Grafting yield and surface hydrophilicity versus plasma treatment time for PVDF-g-PEGMA membranes. The grafting process markedly increased grafting coverage and the surface hydrophilicity with drastic reduction of water contact angle from 120 to 20° within a short 120 s plasma treatment. (B) Human blood protein adsorption for (left) low pressure (network like structure) and (right) atmospheric pressure (brush like structure) deposited PVDF-g-PEGMA at different plasma treatment time (5 to 60 s). Both structures show good protein resistance characteristics but higher (80 percent reduction) for brush like PVDF-g-PEGMA. A decrease in protein adsorption with respect to plasma treatment time was also noticed due to the increase in grafting coverage. *Reproduced with permission from [69].*

showed virtually no platelet adhesion, relatively higher plasma clotting time (14 s), and low hemolytic activity.

The biocompatibility and cell−material interactions of membranes can be further improved by grafting followed by immobilizing bioactive molecules. In this approach, plasma is effectively used to create different functional groups on the membrane acting as potential binding sites for biomolecules. Covalent immobilization of collagen (CG) on plasma-treated poly(lactic acid) (PLLA) nanofiber membranes for tissue engineering has been reported by Chen and Su [80]. Oxygen plasma was used initially to create −COOH groups on the PLLA surface that promoted the attachment of collagen during the reaction with collagen solution in PBS. In vitro studies performed on pristine PLLA and CG-PLLA revealed that the latter displayed an improved viability, proliferation, and differentiation for rabbit articular chondrocytes. Likewise, a mixture of Ar, NH_3, and H_2 plasma has been used to graft heparin on to poly(L-lactide) [81]. The plasma generated primary $−NH_2$ groups on the PLA surface, which promote anchoring of a sufficiently high amount of heparin molecules. Enhanced infiltration of bovine aorta endothelial cells (BAECS) and decreased platelet adhesion was also achieved for heparin-modified PLA. Zhang et al. demonstrated that the hemocompatibility of a nonwoven

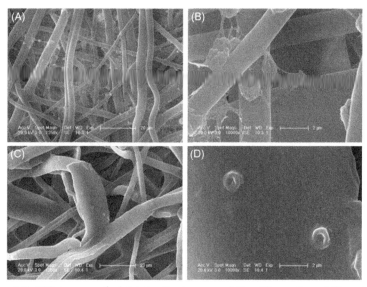

Figure 8.8 SEM images of platelet adhesion on pristine PP$_{NWF}$ (A and B) and BSA-modified PP$_{NWF}$ (C and D) membrane. SEM micrographs revealed that platelet adhesion is hardly found in BSA-PP$_{NWF}$ whereas it is predominant in pristine PP$_{NWF}$. *Reproduced with permission from [82].*

fabric poly propylene (PP$_{NWF}$) membrane was noticeably increased through surface immobilization with bovine serum albumin (BSA) [82]. A spacer layer of poly acrylic acid was made initially by combining plasma irradiation and UV-initiated graft polymerization to which BSA was successfully immobilized. It was also noticed that the antifouling properties were correlated to the amount of BSA adsorbed on the membrane surface. The modified membrane enhanced the anticoagulation properties with an improvement in blood clotting index from 5.7 percent to 23.9 percent. Moreover, the surface became hydrophilic and hence suppresses protein and platelet adhesion considerably. Fig. 8.8 represents SEM micrographs of platelet adhesion characteristics of bare and BSA-PP$_{NWF}$ membranes.

8.5 MEMBRANES EXPOSURE TO NONPOLYMERIZED PLASMA GASES

Plasma of nonpolymer-forming gases can be typically generated from noble gases (He, Ne, and Ar) or from reactive gases such as O$_2$, H$_2$, and N$_2$ in a vacuum system. It is widely used in laboratories for clearing polymeric surfaces owing to its capability to remove/destroy biological molecules such as proteins, cells, and peptides from material surfaces at low temperatures [83,84]. In many cases, exposure of a polymeric surface to a nonpolymerized plasma results in introducing new functional groups and/or creation of free radicals in the material and/or physical ablation.

Ar-plasma. Argon is a noble gas, thus, Ar-plasma is basically employed to change the chemical groups that already exist on the surface, rather than to introduce new groups onto the surface. Ar-plasma can potentially break existing molecular bonds, then form new ones; yet it is capable of introducing oxygen functionalities on a polymeric surface [85,86]. Aerts et al. studied the influence of plasmas generated in three different processing gases (Ar, Ar$-O_2$, and Ar$-H_2$) on laboratory-made polydimethylsiloxane membranes that possess excellent nanofiltration performance in organic solvents. In all cases, plasma led to an increase in oxygen-enriched species (C$-$O and C $=$ O) and consequently, in the hydrophilicity of the membranes, but the solvent permeability was affected. The decrease in permeability was attributed to crosslinking of the surface layer of the treated material and hydrophilicity modifications of the membranes [87].

At appropriate treatment conditions, argon plasma may significantly change surface parameters of various polymeric membranes. For instance, the water contact angles of chitosan treated-membranes was notably reduced (from 60.7° to 11.5°) and the surface energy increased (from 41.0 to 67.3 mJ/m^2) upon Ar-plasma treatments. In addition, the treated surface was more conducive to attachment and proliferation of human skin cells (fibroblasts) [88]. Similarly, low temperature Ar$^+$ ion irradiation has been applied (for 10 minutes) to improve cell adhesion of polymer-TiO$_2$ membranes. As a result of breaking of C$-$H bonds, the irradiated membranes were characterised by higher degree of crosslinking between C$-$C. The treatment also increased the surface energy and produced hydroxyl groups that prompted greater cell adhesion [89]. In the same way, the argon-plasma treatment led to a significant improvement in the surface and chemical properties of polymethyl methacrylate-Co membranes. The plasma action assisted in increasing the flux, porosity, and roughness of the surface as well as enhancing the crosslinking of the membrane, as shown by decreased the intensity of C$-$C and C$-$H bands [90].

Ar-plasma was successfully used to change physico-chemical properties of the first few molecular layers of polymeric membranes (polyethylene-terephthalate) prior to depositing metallic thin films (ZnO: Al) [91]. This enabled the researchers to achieve physically stable films with good adherence to the substrate. However, in the industrial filed, Ar plasma was reported to diminish undesirable plasticization effects of condensable fume (CO_2) in high pressure gas-separation system, notably reducing the gas permeability, and causing negligible impact on gas-separation properties of the membranes [92].

O_2-plasma. Oxygen-plasma is often used to clean polymeric membranes prior to bonding. Yet, it is efficient to introduce new functionalities to the surface. Oxygen plasma can generate a variety of oxygen species on the surface of polymers, including $C-O$, $C=O$, $O-C=O$, $C-O-O^-$, and CO_3 [93]. It is experimentally found that carbonyl, hydroxyl, and carboxyl groups are formed on a membrane surface (polysulfone) upon exposure to O_2-plasma, increasing O:C ratio in the polymer structure [94].

O_2-plasma treatment is widely used as an effective approach for improving wettability of a polymeric surface [95], and is recognised in many cases, to be more effective than Ar-plasma in producing hydrophilic surfaces [93]. Polypropylene, nonwoven fabric membranes were effectively modified with grafted chains by means of O_2-plasma pre-treatment combined with UV techniques. The biological performance (involving platelets adhesion and static protein adsorption) and antifouling activities of modified polypropylene were greatly enhanced when used in a microfiltration process [96]. Similarly, the irradiation of polyetheretherketone biomaterial by oxygen plasma is reported to increase the surface energy by incorporating extra functional groups into the polymeric surface which, in turn, modified protein adsorption that is a critical behavior in some medical applications (e.g., for eukaryotic cell adhesion) [97]. In the same way, polypropylene hollow fiber microporous membranes were subjected to several modifications, including O_2 and H_2 plasma, to enhance their hydrophilicity and blood compatibility. The O_2-plasma treatment was more effective at generating oxygenated functional groups on the surface of the treated material compared to H_2-plasmas. The plasma treatment clearly improved the wettability and enhanced blood compatibility of these microporous membranes; yet, some surface damage due to plasma-generated ion bombardment was also detected [98].

There are several reports that showed the positive influence of oxygen plasma treatments on membranes that can be utilized effectively in the industrial field. For instance, polyvinylidene fluoride (PVDF) membrane has not been employed for designing thin film composite membrane due to its hydrophobic nature. The surface properties of these membranes can be successfully tailored with O_2-plasmas and, thus, can be used to serve as a support substrate for the fabrication of films with good nanofiltration and reverse osmosis properties. The plasma treatment of PVDF can increase membrane water uptake and hydrophilic performance of these surfaces [99]. Furthermore, Modarresi et al. observed that important changes in the performance (CO_2/CH_4 separation) and surface properties of commercial membranes (polysulfone) occurred after O_2 plasma exposure. The separation properties of the tailored membranes were improved significantly with plasma treatment, and the selectivity was increased $\sim$ 50 percent for the treatment conditions of 6 W and 150 s. The water contact angle reduced with plasma treatment, by way of membrane surfaces showing the introduction of $C = O$-containing chemical groups onto the surface of the treated material [100].

However, it is important to mention that the degradation effects of an O_2-plasma treatment is much more prominent compared to that resulting from other gas plasmas, such as ammonia and nitrogen.

N_2-plasma. Nitrogen plasma has the capacity to introduce several functionalities (e.g., amine, nitrile, and imine) into a polymeric membrane that changes properties of the surface—such as polarity, reactivity, and wettability—whose density and distribution can be tuned by altering the plasma-processing conditions. According to Vesel et al., a few seconds of N_2-plasma exposure are sufficient to produce important changes in the chemical composition of a polymeric substance [101]. N_2-plasma is usually employed in the medical field to increase the biocompatibility and hydrophilicity of biomaterials. For example, the treatment of chitosan-based membranes with N_2-plasma considerably improves the adhesion/proliferation of L929 fibroblast-like cells; notably, no cytotoxic effects were generated as a result of the plasma treatment [102]. Similarly, the exposure of medically-used nanocomposite polycarbonate membranes to N_2-plasma produced significant enhancement in the cell growth and biocompatibility due to the formation of active sites on the surface (without changes to the bulk polymer structure) [103]. Furthermore, it has been shown that N_2-plasma can lead to a lasting increase in the number of functional groups on the surface of biomedical cellulose membranes which, in turn, can enhance the adhesion of endothelial and neuroblast cells. Furthermore, the treatment improved the porosity of the membranes without increasing their wettability. This is important in the development of materials for tissue engineering applications, since relatively low porosity is a key reason for poor cell penetration [104].

In the industrial field, low-temperature N_2-plasma treatment of commercial polypropylene hollow fiber-microporous membranes was reported to significantly increase the hydrophilicity and flux recovery, in addition to producing better filtration behavior, in a submerged membrane bioreactor compared to the unmodified material [105]. Likewise, upon exposure to N_2-plasma, surfaces of nonporous polyvinyl alcohol membranes were found to display a higher degree of cross-linking with relatively low etching rate, resulting in effective enhancement in separation performance of membrane. The gelation from self-crosslinking between hydroxyl functional groups in N_2-plasma treated membrane is the main reason for this increasing in separation function [106]. However, in some cases, N_2 was mixed with another gas, such as H_2 with the aim to control the dissociation of nitrogen molecules and, thereby, adjust the plasma treatment parameters and resultant concentrations of functional groups and, thus, material properties [16].

CO_2-plasma. CO_2-plasma treatment yields various polar groups (e.g., carbonyl, hydroxyl, and carboxylic) upon contact with polymeric surfaces. It is commonly used to modify membranes that possess hydrophobic properties to create hydrophilic surfaces. Morphological changes and enhanced hydrophilicities were demonstrated in CO_2-plasma-treated polyethersulfone ultrafiltration membranes. CO_2-plasma sharply reduced the membrane contact angles from $87°$ to $22.3°$ and decreased the surface roughness from of 42.8 nm to 15.7 nm (12 min exposure) [107]. Moreover, Yu et al. found that once subjected to CO_2 plasma, the presence of oxygen-containing polar groups on the polypropylene microporous membrane surfaces increased, which enhanced surface hydrophilicity and filtration performance [108]. Likewise, polysulfone ultrafiltration membranes were rendered more hydrophilic by a CO_2-plasma treatment due to the creation of surface polar functional groups (e.g., carbonyl and hydroxyl moieties). These modified membranes are significantly more hydrophilic, with the improvement being long lasting (a few months) and effective at significantly minimizing protein fouling and improving water flux [109].

Pal et al. showed the positive influence of CO_2-plasma on the overall properties of polyethersulfone membrane as the average roughness of membranes notably reduced from 111.25 nm to 19.8 nm and the permeability increased up to 30 percent (and subsequently remained unchanged for three months). These changes, interestingly, occurred without any alteration in the mechanical properties (tensile strength and breakdown forces) of the membranes [110].

NH₃-plasma. NH_3-plasma is broadly employed to produce selective high-density amino functionalization that is favourable in medical applications, in particular for biocompatibility improvement, cell adhesion strengthening, and metallization [111]. NH_3-plasma is capable of decreasing surface oxygen-containing groups owing to their interaction with chemically active species of NH_3-plasma (e.g., NH_2 and NH) [112]. It has been reported that NH_3-plasma greatly improves adhesion and survival of human cells on synthetic polymeric membranes (polyetheretherketone/polyurethane). The cells on the treated polymers were found to display higher metabolic activity (with regards to albumin synthesis and urea production). This probably relates to a significant increase in membrane hydrophilicity which improves adsorption of proteins and enhances surface interactions with attached cells [113]. Moreover, hydrophobic polypropylene microfiltration membranes were characterised to reveal desired surface changes by low temperature NH_3-plasma treatment method. The flux recoveries (after water and caustic cleaning) for the NH_3-plasma treated membranes for 1 min were higher (51.1 percent and 60.7 percent) than those of the virgin polymers. The adsorption of bovine serum albumin on the treated membranes was lower than that on the untreated membrane. The hydrophilicity of the membranes increased while the mechanical properties reduced as a result of plasma action [79].

8.6 CONCLUSIONS

Nonequilibrium low-temperature plasma treatment is a versatile technique for effectively modifying a variety of polymeric membranes. Plasma provides uniform surface modifications regardless of the geometry of the irradiated membranes, and is suitable for temperature-sensitive thin polymeric membranes. At appropriate processing conditions plasma has the potential to retain preferred bulk properties of membranes while significantly improve surface characteristics such as wettability, surface roughness, and biocompatibility. A wide range of new surface functionalities (e.g., carboxyl, hydroxyl, or amino) can also be induced depending on the selected plasma gases. Thus, successful utilization of plasma modifications will significantly increase the range of membranes applications.

REFERENCES

[1] S. Das, N.M. Ray, J. Wan, A. Khan, T. Chakraborty, M.B. Ray, Micropollutants in Wastewater: Fate and Removal Processes, Physico-Chemical Wastewater Treatment and Resource Recovery, ed, InTech, 2017.

[2] S. Yu, X. Liu, J. Liu, D. Wu, M. Liu, G. Gao, Surface modification of thin-film composite polyamide reverse osmosis membranes with thermo-responsive polymer (TRP) for improved fouling resistance and cleaning efficiency, Separation and Purification Technology 76 (2011) 283−291. 01/14/ 2011.

[3] R. Balamurugan, S. Sundarrajan, S. Ramakrishna, Recent trends in nanofibrous membranes and their suitability for air and water filtrations, Membranes 1 (2011) 232−248.

[4] P. Formoso, E. Pantuso, G. De Filpo, F.P. Nicoletta, Electro-Conductive Membranes for Permeation Enhancement and Fouling Mitigation: A Short Review, Membranes 7 (2017) 39.

[5] K. Hunger, N. Schmeling, H.B. Jeazet, C. Janiak, C. Staudt, K. Kleinermanns, Investigation of cross-linked and additive containing polymer materials for membranes with improved performance in pervaporation and gas separation, Membranes 2 (2012) 727−763.

[6] T. Nguyen, F.A. Roddick, L. Fan, Biofouling of water treatment membranes: a review of the underlying causes, monitoring techniques and control measures, Membranes 2 (2012) 804−840.

[7] N. Vandencasteele, F. Reniers, Plasma-modified polymer surfaces: Characterization using XPS, Journal of Electron Spectroscopy and Related Phenomena 178−179 (2010) 394−408. 5//.

[8] V.M. Kochkodan, V.K. Sharma, Graft polymerization and plasma treatment of polymer membranes for fouling reduction: A review, Journal of Environmental Science and Health, Part A 47 (2012) 1713−1727.

[9] A. Al-Jumaili, K. Bazaka, M.V. Jacob, Retention of Antibacterial Activity in Geranium Plasma Polymer Thin Films, Nanomaterials 7 (2017) 270.

[10] R. Morent, N. De Geyter, T. Desmet, P. Dubruel, C. Leys, Plasma surface modification of biodegradable polymers: a review, Plasma Processes and Polymers 8 (2011) 171−190.

[11] F. Meng, S.-R. Chae, A. Drews, M. Kraume, H.-S. Shin, F. Yang, Recent advances in membrane bioreactors (MBRs): Membrane fouling and membrane material, Water Res. 43 (2009) 1489−1512. /04/01/ 2009.

[12] T. Desmet, R. Morent, N.D. Geyter, C. Leys, E. Schacht, P. Dubruel, Nonthermal Plasma Technology as a Versatile Strategy for Polymeric Biomaterials Surface Modification: A Review, Biomacromolecules 10 (2009) 2351−2378. /09/14 2009.

[13] I. Gancarz, G. Poźniak, M. Bryjak, Modification of polysulfone membranes 1. CO2 plasma treatment, Eur. Polym J. 35 (1999) 1419−1428. /08/01/ 1999.

[14] K. Bazaka, M.V. Jacob, R.J. Crawford, E.P. Ivanova, Plasma-assisted surface modification of organic biopolymers to prevent bacterial attachment, Acta Biomater. 7 (2011) 2015−2028.

[15] M.A. Shannon, P.W. Bohn, M. Elimelech, J.G. Georgiadis, B.J. Marinas, A.M. Mayes, Science and technology for water purification in the coming decades, Nature 452 (2008) 301.

[16] C. Yanling, W. Yingkuan, P. Chen, S. Deng, R. Ruan, Non-thermal plasma assisted polymer surface modification and synthesis: A review, International Journal of Agricultural and Biological Engineering 7 (2014) 1.

[17] M.L. Lopez-Donaire, J.P. Santerre, Surface Modifying Oligomers Used to Functionalize Polymeric Surfaces: Consideration of Blood Contact Applications, J. Appl. Polym. Sci. 131 (2014) 15.

[18] L.D. Bartolo, S. Morelli, A. Bader, E. Drioli, Evaluation of cell behaviour related to physico-chemical properties of polymeric membranes to be used in bioartificial organs, Biomaterials. 23 (2002) 13.

[19] R. Zhang, X. Guo, X. Shi, A. Sun, L. Wang, T. Xiao, et al., Highly Permselective Membrane Surface Modification by Cold Plasma-Induced Grafting Polymerization of Molecularly Imprinted Polymer for Recognition of Pyrethroid Insecticides in Fish, Anal. Chem. 86 (2014) 9.

[20] J.D. Kittle, C. Wang, C. Qian, Y. Zhang, M. Zhang, M. Roman, et al., Ultrathin chitin films for nanocomposites and biosensors, Biomacromolecules 13 (2012) **5**.

[21] M.F. Rubner, R.E. Cohen, L. Zhai, M.C. Berg, Controlled drug release from porous polyelectrolyte multilayers, Biomacromolecules 7 (2006) 8.

[22] M.G. Bellino, I. Tropper, H. Duran, A.E. Regazzoni, G.J.A.A. Soler-Illia, Polymerase-Functionalized Hierarchical Mesoporous Titania Thin Films: Towards a Nanoreactor Platform for DNA Amplification, Small. 6 (2010) 5.

[23] Z. Tai, H. Ma, B. Liu, X. Yan, Q. Xue, Facile synthesis of Ag/GNS-g-PAA nanohybrids for antimicrobial applications, Colloids and Surfaces B: Biointerfaces 89 (2012) 5.

[24] S. Venkatraman, F. Boey, L.L. Lao, Implanted cardiovascular polymers: Natural, synthetic and bio-inspired, Progress in Polymer Science 33 (2008) 22.

[25] R.J. Crawford, H.K. Webb, V.K. Truong, J. Hasan, E.P. Ivanova, Surface topographical factors influencing bacterial attachment, Adv. Colloid. Interface. Sci. 179 (2012) 8.

[26] K. Bazaka, M.V. Jacob, R.J. Crawford, E.P. Ivanova, Efficient surface modification of biomaterial to prevent biofilm formation and the attachment of microorganisms, Appl. Microbiol. Biotechnol. 95 (2012) 13.

[27] B.D. Ratner, S.J. Bryant, Biomaterials: Where We Have Been and Where We Are Going, Annu. Rev. Biomed. Eng. 6 (2004) 27.

[28] Y.-Q. Song, J. Sheng, M. Wei, X.-B. Yuan, Surface modification of polysulfone membranes by low-temperature plasma-graft poly(ethylene glycol) onto polysulfone membranes, J. Appl. Polym. Sci. 78 (2000) 7.

[29] D. Rana, T. Matsuura, Surface modifications for antifouling membranes, Chem. Rev. 110 (2010) 24.

[30] K.S. Siow, L. Britcher, S. Kumar, H.J. Griesser, Plasma Methods for the Generation of Chemically Reactive Surfaces for Biomolecule Immobilization and Cell Colonization - A Review, Plasma Processes and Polymers 3 (2006) 27.

[31] M. Muller, C. Oehr, Plasma amino functionalisation of PVDF microfiltration membranes: comparison of the in plasma modifications with a grafting method using ESCA and an amino-selective fluorescent probe, Surface and Coatings Technology 116-119 (1999) 6.

[32] P. Favia, M.V. Stendardo, R. D'Agostino, Selective grafting of amine groups on polyethylene by means of NH3-H2 RF glow discharges, Plasmas and Polymers 1 (1996) 12.

[33] M. Tatoulian, F. Arefi-Khonsari, J. Amouroux, S.B. Rejeb, A. Martel, N.F. Durand, et al., Immobilization of Biomolecules on NH3, H2/NH3 Plasma-Treated Nitrocellulose Films, Plasmas and Polymers 3 (1998) 9.

[34] B.D. Tompkins, J.M. Dennison, E.R. Fisher, H$_2$O plasma modification of track-etched polymer membranes for increased wettability and improved performance, J. Memb. Sci. 428 (2013) 13.

[35] M.-S. Li, Z.-P. Zhao, N. Li, Y. Zhang, Controllable modification of polymer membranes by long-distance and dynamic low-temperature plasma flow: Treatment of PE hollow fiber membranes in a module scale, J. Memb. Sci. 427 (2013) 12.

[36] C.Y. Tang, Y.-N. Kwon, J.O. Leckie, Effect of membrane chemistry and coating layer on physiochemical properties of thin film composite polyamide RO and NF membranes. II. Membrane physiochemical properties and their dependence on polyamide and coating layers, Desalination 242 (2009) 15.

[37] K.R. Kull, M.L. Steen, E.R. Fisher, Surface modification with nitrogen-containing plasmas to produce hydrophilic, low-fouling membranes, J. Memb. Sci. 246 (2005) 13.

[38] H.-X. Sun, L. Zhang, H. Chai, H.-L. Chen, Surface modification of poly(tetra-fluoroethylene) films via plasma treatment and graft copolymerization of acrylic acid, Desalination 192 (2006) 9,

[39] L. Liu, S. Chen, C.M. Giachelli, B.D. Ratner, S. Jiang, Controlling osteopontin orientation on surfaces to modulate endothelial cell adhesion, Journal of Biomedical Materials Research - Part A 74 (2005) 9.

[40] P.M. Lopez-Perez, A.P. Marques, R.M.P.D. Silva, I. Pashkuleva, R.L. Reis, Effect of chitosan membrane surface modification via plasma induced polymerization on the adhesion of osteoblast-like cells, J. Mater. Chem. 17 (2007) 8.

[41] K. Akamatsu, T. Furue, F. Han, S.-I. Nakao, Plasma graft polymerization to develop low-fouling membranes grafted with poly (2-methoxyethylacrylate), Separation and Purification Technology 102 (2013) 157—162.

[42] I.C. Escobar, E.M. Hoek, C.J. Gabelich, F.A. DiGiano, Committee report: recent advances and research needs in membrane fouling, American Water Works Association. Journal 97 (2005) 79.

[43] A. Asatekin, S. Kang, M. Elimelech, A.M. Mayes, Anti-fouling ultrafiltration membranes containing polyacrylonitrile-graft-poly(ethylene oxide) comb copolymer additives, J. Memb. Sci. 298 (2007) 136—146. /07/20/ 2007.

[44] N.A. Ochoa, M. Masuelli, J. Marchese, Effect of hydrophilicity on fouling of an emulsified oil wastewater with PVDF/PMMA membranes, J. Memb. Sci. 226 (2003) 203—211. /12/01/ 2003.

[45] T. Humplik, J. Lee, S. O'hern, B. Fellman, M. Baig, S. Hassan, et al., Nanostructured materials for water desalination, Nanotechnology. 22 (2011) 292001.

[46] J. Gilron, S. Belfer, P. Väisänen, M. Nyström, Effects of surface modification on anti-fouling and performance properties of reverse osmosis membranes, Desalination 140 (2001) 167—179.

[47] C.Y. Tang, Y.-N. Kwon, J.O. Leckie, Effect of membrane chemistry and coating layer on physiochemical properties of thin film composite polyamide RO and NF membranes: I. FTIR and XPS characterization of polyamide and coating layer chemistry, Desalination 242 (2009) 149—167.

[48] H. Zou, Y. Jin, J. Yang, H. Dai, X. Yu, J. Xu, Synthesis and characterization of thin film composite reverse osmosis membranes via novel interfacial polymerization approach, Separation and Purification Technology 72 (2010) 256—262.

[49] S.S. Deshmukh, A.E. Childress, Zeta potential of commercial RO membranes: influence of source water type and chemistry, Desalination 140 (2001) 87—95.

[50] C. Bellona, J.E. Drewes, P. Xu, G. Amy, Factors affecting the rejection of organic solutes during NF/RO treatment—a literature review, Water Res. 38 (2004) 2795—2809.

[51] Y. Yoon, G. Amy, J. Cho, N. Her, J. Pellegrino, Transport of perchlorate (ClO 4 −) through NF and UF membranes, Desalination 147 (2002) 11—17.

[52] M. Hadidi, A.L. Zydney, Fouling behavior of zwitterionic membranes: Impact of electrostatic and hydrophobic interactions, J. Memb. Sci. 452 (2014) 97—103.

[53] M. Elimelech, X. Zhu, A.E. Childress, S. Hong, Role of membrane surface morphology in colloidal fouling of cellulose acetate and composite aromatic polyamide reverse osmosis membranes, J. Memb. Sci. 127 (1997) 101—109.

[54] E.M. Vrijenhoek, S. Hong, M. Elimelech, Influence of membrane surface properties on initial rate of colloidal fouling of reverse osmosis and nanofiltration membranes, J. Memb. Sci. 188 (2001) 115—128.

[55] W.R. Bowen, T.A. Doneva, Atomic force microscopy studies of membranes: effect of surface roughness on double-layer interactions and particle adhesion, J. Colloid. Interface. Sci. 229 (2000) 544—549.

[56] M. Hashino, T. Katagiri, N. Kubota, Y. Ohmukai, T. Maruyama, H. Matsuyama, Effect of surface roughness of hollow fiber membranes with gear-shaped structure on membrane fouling by sodium alginate, J. Memb. Sci. 366 (2011) 389−397.

[57] Y. Baek, J. Yu, S.-H. Kim, S. Lee, J. Yoon, Effect of surface properties of reverse osmosis membranes on biofouling occurrence under filtration conditions, J. Memb. Sci. 382 (2011) 91−99.

[58] H.-Y. Yu, Y.-J. Xie, M.-X. Hu, J.-L. Wang, S.-Y. Wang, Z.-K. Xu, Surface modification of polypropylene microporous membrane to improve its antifouling property in MBR: CO 2 plasma treatment, J. Memb. Sci. 254 (2005) 219−227.

[59] H.-Y. Yu, L.-Q. Liu, Z.-Q. Tang, M.-G. Yan, J.-S. Gu, X.-W. Wei, Surface modification of polypropylene microporous membrane to improve its antifouling characteristics in an SMBR: Air plasma treatment, J. Memb. Sci. 311 (2008) 216−224. 3/20/.

[60] M. Bryjak, I. Gancarz, G. Pozniak, Plasma-modified porous membranes, Chem. Papers 54 (2000) 496.

[61] H. Yu, M. Hu, K. Hu, K. Zh, J. Wang, Y. Wang Sh, Separation & Purif, Technol 45 (2005) 8.

[62] D. Pal, S. Neogi, S. De, Improved antifouling characteristics of acrylonitrile co-polymer membrane by low temperature pulsed ammonia plasma in the treatment of oil−water emulsion, Vacuum 131 (2016) 293−304.

[63] A. Venault, T.-C. Wei, H.-L. Shih, C.-C. Yeh, A. Chinnathambi, S.A. Alharbi, et al., Antifouling pseudo-zwitterionic poly (vinylidene fluoride) membranes with efficient mixed-charge surface grafting via glow dielectric barrier discharge plasma-induced copolymerization, J. Memb. Sci. 516 (2016) 13−25.

[64] I. Gancarz, G. Pozniak, M. Bryjak, A. Frankiewicz, Modification of polysulfone membranes. 2. Plasma grafting and plasma polymerization of acrylic acid, Acta polymerica 50 (1999) 317−326.

[65] L. Liu, F. Shen, X. Chen, J. Luo, Y. Su, H. Wu, et al., A novel plasma-induced surface hydrophobization strategy for membrane distillation: Etching, dipping and grafting, J. Memb. Sci. 499 (2016) 544−554.

[66] N.H. Lin, M.-M. Kim, G.T. Lewis, Y. Cohen, Polymer surface nano-structuring of reverse osmosis membranes for fouling resistance and improved flux performance, J. Mater. Chem. 20 (2010) 4642−4652.

[67] L. Zou, I. Vidalis, D. Steele, A. Michelmore, S. Low, J. Verberk, Surface hydrophilic modification of RO membranes by plasma polymerization for low organic fouling, J. Memb. Sci. 369 (2011) 420−428.

[68] F. Khelifa, S. Ershov, Y. Habibi, R. Snyders, P. Dubois, Free-radical-induced grafting from plasma polymer surfaces, Chem. Rev. 116 (2016) 3975−4005.

[69] Y. Chang, Y.-J. Shih, C.-Y. Ko, J.-F. Jhong, Y.-L. Liu, T.-C. Wei, Hemocompatibility of poly (vinylidene fluoride) membrane grafted with network-like and brush-like antifouling layer controlled via plasma-induced surface PEGylation, Langmuir 27 (2011) 5445−5455.

[70] Y. Chang, T.-Y. Cheng, Y.-J. Shih, K.-R. Lee, J.-Y. Lai, Biofouling-resistance expanded poly (tetrafluoroethylene) membrane with a hydrogel-like layer of surface-immobilized poly (ethylene glycol) methacrylate for human plasma protein repulsions, J. Memb. Sci. 323 (2008) 77−84.

[71] J.-F. Jhong, A. Venault, C.-C. Hou, S.-H. Chen, T.-C. Wei, J. Zheng, et al., Surface zwitterionization of expanded poly (tetrafluoroethylene) membranes via atmospheric plasma-induced polymerization for enhanced skin wound healing, ACS applied materials & interfaces 5 (2013) 6732−6742.

[72] N. De Geyter and R. Morent, *Non-thermal plasma surface modification of biodegradable polymers*: INTECH Open Access Publisher, 2012.

[73] T. Jacobs, R. Morent, N. De Geyter, P. Dubruel, C. Leys, Plasma surface modification of biomedical polymers: influence on cell-material interaction, Plasma Chemistry and Plasma Processing 32 (2012) 1039−1073.

[74] S.-Y. Lin, V.R. Parasuraman, S.L. Mekuria, S. Peng, H.-C. Tsai, G.-H. Hsiue, Plasma initiated graft polymerization of 2-methacryloyloxyethyl phosphorylcholine on silicone elastomer surfaces to enhance bio (hemo) compatibility, Surface and Coatings Technology 315 (2017) 342−349.

[75] S. Zanini, C. Riccardi, E. Grimoldi, C. Colombo, A.M. Villa, A. Natalello, et al., Plasma-induced graft-polymerization of polyethylene glycol acrylate on polypropylene films: chemical characterization and evaluation of the protein adsorption, J. Colloid. Interface. Sci. 341 (2010) 53−58.

[76] S.-H. Chen, Y. Chang, K.-R. Lee, T.-C. Wei, A. Higuchi, F.-M. Ho, et al., Hemocompatible control of sulfobetaine-grafted polypropylene fibrous membranes in human whole blood via plasma-induced surface zwitterionization, Langmuir 28 (2012) 17733−17742.

[77] Y. Chang, W.-J. Chang, Y.-J. Shih, T.-C. Wei, G.-H. Hsiue, Zwitterionic sulfobetaine-grafted poly (vinylidene fluoride) membrane with highly effective blood compatibility via atmospheric plasma-induced surface copolymerization, ACS applied materials & interfaces 3 (2011) 1228−1237.

[78] S. Pal, S.K. Ghatak, S. De, S. DasGupta, Characterization of CO_2 plasma treated polymeric membranes and quantification of flux enhancement, J. Memb. Sci. 323 (2008) 1−10.

[79] M.-G. Yan, L.-Q. Liu, Z.-Q. Tang, L. Huang, W. Li, J. Zhou, et al., Plasma surface modification of polypropylene microfiltration membranes and fouling by BSA dispersion, Chemical Engineering Journal 145 (2008) 218−224. 12/15/.

[80] J.-P. Chen, C.-H. Su, Surface modification of electrospun PLLA nanofibers by plasma treatment and cationized gelatin immobilization for cartilage tissue engineering, Acta Biomater. 7 (2011) 234−243.

[81] Q. Cheng, K. Komvopoulos, S. Li, Plasma-assisted heparin conjugation on electrospun poly (l-lactide) fibrous scaffolds, J. Biomed. Mater. Res. A. 102 (2014) 1408−1414.

[82] C. Zhang, J. Jin, J. Zhao, W. Jiang, J. Yin, Functionalized polypropylene non-woven fabric membrane with bovine serum albumin and its hemocompatibility enhancement, Colloids and Surfaces B: Biointerfaces 102 (2013) 45−52.

[83] K. Bazaka, M. Jacob, W. Chrzanowski, K. Ostrikov, Anti-bacterial surfaces: natural agents, mechanisms of action, and plasma surface modification, RSC Advances 5 (2015) 48739−48759.

[84] A. Al-Jumaili, S. Alancherry, K. Bazaka, M. Jacob, Review on the Antimicrobial Properties of Carbon Nanostructures, Materials 10 (2017) 1066.

[85] A.W. Mohammad, Y.H. Teow, W.L. Ang, Y.T. Chung, D.L. Oatley-Radcliffe, N. Hilal, Nanofiltration membranes review: Recent advances and future prospects, Desalination 356 (2015) 226−254. 1/15/.

[86] V. Vosmanska, K. Kolarova, S. Rimpelova, V. Svorcik, Surface modification of oxidized cellulose haemostat by argon plasma treatment, Cellulose 21 (2014) 2445−2456.

[87] S. Aerts, A. Vanhulsel, A. Buekenhoudt, H. Weyten, S. Kuypers, H. Chen, et al., Plasma-treated PDMS-membranes in solvent resistant nanofiltration: Characterization and study of transport mechanism, J. Memb. Sci. 275 (2006) 212−219. 4/20/.

[88] X. Zhu, K.S. Chian, M.B.E. Chan-Park, S.T. Lee, Effect of argon-plasma treatment on proliferation of human-skin−derived fibroblast on chitosan membrane in vitro, J. Biomed. Mater. Res. A. 73 (2005) 264−274.

[89] N.K. Agrawal, R. Agarwal, K. Awasthi, Y. Vijay, K. Swami, Surface Modification of Nano Composite Polymer Membranes by Ion Plasma Irradiation for Improving Biocompatibility of Polymers, Advanced Materials Letters 5 (2014) 639−644.

[90] N.K. Agrawal, R. Agarwal, A.K. Gautam, Y.K. Vijay, K.C. Swami, Surface Modification of Polymer Nanocomposites by Glow-Discharge Plasma Treatment, Materials Science 51 (2015) 68−75.

[91] S. Fernández, J.D. Santos, C. Munuera, M. García-Hernández, F.B. Naranjo, Effect of argon plasma treated polyethylene terephthalate on ZnO:Al properties for flexible thin film silicon solar cells applications, Solar Energy Materials and Solar Cells 133 (2015) 170−179. 2//.

[92] C.C. Hu, C.Y. Tu, Y.C. Wang, C.L. Li, K.R. Lee, J.Y. Lai, Effects of plasma treatment on CO_2 plasticization of poly (methyl methacrylate) gas-separation membranes, J. Appl. Polym. Sci. 93 (2004) 395−401.

[93] K.T. Lee, J.M. Goddard, J.H. Hotchkiss, Plasma modification of polyolefin surfaces, Packaging Technology and Science 22 (2009) 139−150.

[94] V. Kochkodan, D.J. Johnson, N. Hilal, Polymeric membranes: Surface modification for minimizing (bio)colloidal fouling, Adv. Colloid. Interface. Sci. 206 (2014) 116−140. 4//.

[95] K. Tsougeni, N. Vourdas, A. Tserepi, E. Gogolides, C. Cardinaud, Mechanisms of Oxygen Plasma Nanotexturing of Organic Polymer Surfaces: From Stable Super Hydrophilic to Super Hydrophobic Surfaces, Langmuir 25 (2009) 11748−11759. 2009/10/06.

[96] J. Zhao, Q. Shi, S. Luan, L. Song, H. Yang, H. Shi, et al., Improved biocompatibility and antifouling property of polypropylene non-woven fabric membrane by surface grafting zwitterionic polymer, J. Memb. Sci. 369 (2011) 5−12. 3/1/.

[97] E.T.J. Rochford, A.H.C. Poulsson, J. Salavarrieta Varela, P. Lezuo, R.G. Richards, T.F. Moriarty, Bacterial adhesion to orthopaedic implant materials and a novel oxygen plasma modified PEEK surface, Colloids and Surfaces B: Biointerfaces 113 (2014) 213−222. 1/1/.

[98] A.S. Abednejad, G. Amoabediny, A. Ghaee, Surface modification of polypropylene membrane by polyethylene glycol graft polymerization, Materials Science and Engineering: C 42 (2014) 443−450. 9/1/.

[99] E.-S. Kim, Y.J. Kim, Q. Yu, B. Deng, Preparation and characterization of polyamide thin-film composite (TFC) membranes on plasma-modified polyvinylidene fluoride (PVDF), J. Memb. Sci. 344 (2009) 71−81. 11/15/.

[100] S. Modarresi, M. Soltanieh, S.A. Mousavi, I. Shabani, Effect of low-frequency oxygen plasma on polysulfone membranes for CO2/CH4 Separation, J. Appl. Polym. Sci. 124 (2012).

[101] A. Vesel, I. Junkar, U. Cvelbar, J. Kovac, M. Mozetic, Surface modification of polyester by oxygen-and nitrogen-plasma treatment, Surface and interface analysis 40 (2008) 1444−1453.

[102] S.M. Luna, S.S. Silva, M.E. Gomes, J.F. Mano, R.L. Reis, Cell adhesion and proliferation onto chitosan-based membranes treated by plasma surface modification, J. Biomater. Appl. 26 (2011) 101−116.

[103] B. Bagra, P. Pimpliskar, N.K. Agrawal, C. Murli, D. Bhattacharyya, and S. Gadkari, "Bio-compatibility, surface & chemical characterization of glow discharge plasma modified ZnO nanocomposite polycarbonate," in *AIP Conference Proceedings*, 2014, pp. 189-191.

[104] R.A. Pertile, F.K. Andrade, C. Alves, M. Gama, Surface modification of bacterial cellulose by nitrogen-containing plasma for improved interaction with cells, Carbohydrate Polymers 82 (2010) 692−698.

[105] H.-Y. Yu, X.-C. He, L.-Q. Liu, J.-S. Gu, X.-W. Wei, Surface modification of polypropylene microporous membrane to improve its antifouling characteristics in an SMBR: N 2 plasma treatment, Water Res. 41 (2007) 4703—4709.

[106] D.J. Upadhyay, N.V. Bhat, Pervaporation studies of gaseous plasma treated PVA membrane, J. Memb. Sci. 239 (2004) 255—263. 8/15/.

[107] S. Pal, S.K. Ghatak, S. De, S. DasGupta, Evaluation of surface roughness of a plasma treated polymeric membrane by wavelet analysis and quantification of its enhanced performance, Appl. Surf. Sci. 255 (2008) 2504—2511. 12/30/.

[108] H.-Y. Yu, Y.-J. Xie, M.-X. Hu, J.-L. Wang, S.-Y. Wang, Z.-K. Xu, Surface modification of polypropylene microporous membrane to improve its antifouling property in MBR: CO2 plasma treatment, J. Memb. Sci. 254 (2005) 219—227. 6/1/.

[109] D.S. Wavhal, E.R. Fisher, Modification of polysulfone ultrafiltration membranes by CO2 plasma treatment, Desalination 172 (2005) 189—205.

[110] S. Pal, S.K. Ghatak, S. De, S. DasGupta, Characterization of CO2 plasma treated polymeric membranes and quantification of flux enhancement, J. Memb. Sci. 323 (2008) 1—10. 10/1/.

[111] B. Finke, K. Schröder, A. Ohl, Surface Radical Detection on NH3-Plasma Treated Polymer Surfaces Using the Radical Scavenger NO, Plasma Processes and Polymers 5 (2008) 386—396.

[112] D.A. Shutov, S.-Y. Kang, K.-H. Baek, K.S. Suh, K.-H. Kwon, Influence of Ar and NH3 plasma treatment on surface of poly (monochloro-para-xylylene) dielectric films processed in oxygen plasma, Jpn. J. Appl. Phys. 47 (2008) 6970.

[113] S. Salerno, A. Piscioneri, S. Laera, S. Morelli, P. Favia, A. Bader, et al., Improved functions of human hepatocytes on NH3 plasma-grafted PEEK-WC—PU membranes, Biomaterials. 30 (2009) 4348—4356. 9//.

CHAPTER 9

Selective Plasma Etching of Polymers and Polymer Matrix Composites

Harinarayanan Puliyalil[1,2], Gregor Filipič[3] and Uroš Cvelbar[3]
[1]National Institute of Chemistry (D-13), Ljubljana, Slovenia
[2]IMEC, Leuven, Belgium
[3]Jozef Stefan Institute, Ljubljana, Slovenia

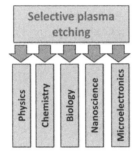

9.1 INTRODUCTION

After the revolutionary developments in the field of polymer science and technology, polymer materials are replacing many other classes of materials, such as wood or inorganics, due to their low-production cost, light weight, and simple fabrication. As evidently seen in many examples, surface modification of polymeric materials becomes essential for improving their performance level in many applications. Efficient and stable surface metallization, immobilization of biomolecules, and improved polymer-filler compatibility in composite fabrication are merely a few of them [1−3]. To date, the most commonly accepted surface modification techniques include wet-chemical treatments and plasma processing.

Plasma-assisted surface modification techniques (plasma functionalization or plasma etching) provide many benefits over the conventionally

Non-Thermal Plasma Technology for Polymeric Materials
DOI: https://doi.org/10.1016/B978-0-12-813152-7.00009-3

241

used wet-chemical methods, which include better material properties as well as being environmentally friendly [4]. Primarily, plasma etching allows us to control the etching precisely at very small dimensions (~ 10 nm) while wet-chemical treatment cannot provide precise etching on such a low dimension. Additionally, plasma etching overcomes many disadvantages associated with the wet-chemical methods, for example, the use of any hazardous solvents, toxic leftovers, surface contamination on the final products, and solvent absorption by the processed material during the modification process. A comparison of pros and cons of wet chemical etching versus plasma etching is listed in Table 9.1.

Plasma etching refers to the removal of a solid material by chemical transformation of it into small volatile molecules which proceed through the incorporation of various surface functional groups (Fig. 9.1) [5,6].

Table 9.1 A general comparison of wet chemical and plasma etching processes

	Wet chemical etching	Plasma etching
Etchants involved	Chemicals	Gas phase radicals/ions
Etch rate and selectivity:	High etch rate and lower control.	Good, controllable.
Advantageous:	1. Equipment cost is low 2. Easy to implement	1. Adequate to define small feature size around 10 nm 2. High surface sensitivity with no contamination issues 3. Ecologically benign technology 4. Low water/chemical consumption
Disadvantageous:	1. Inadequate to define features sizes below 1 μm 2. Hazardous chemicals involved 3. Contamination issues 4. Influences the bulk properties 5. Produces chemical waste	1. High initial investment needed. 2. Scaling up and continuous processing can face difficulties
Etching Directionality:	Provides only isotropic etching.	Provides both isotropic and anisotropic etching

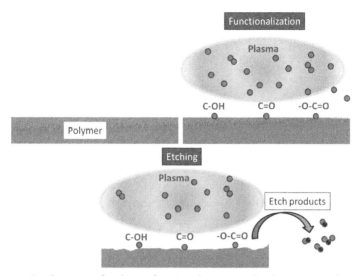

Figure 9.1 A schematic of polymer functionalization and subsequent etching in O_2 plasma.

Plasma is known to be used for etching of organic and inorganic materials; however, in this chapter, we will focus mainly on the etching of organic materials. Numerous studies revealed that organic materials undergo etching faster than most of the commonly known inorganic materials [7]. In the context of etch rate, the difference is found even between different polymers, based on their physical and chemical properties [8]. In this chapter, insight into various aspects of selective plasma etching related to the chemistries of materials and various gas plasmas will be given.

9.2 SELECTIVITY OF PLASMA ETCHING: ORIGIN AND INFLUENTIAL FACTORS

The selectivity of plasma etching originates due to multiple factors which can be divided mainly into (1) the properties of the material and (2) properties of the plasma/interacting plasma particles. Both physical and chemical properties of the polymer materials can significantly affect the plasma-etching rate. This includes the crystallinity, strength of various bonds on the polymer backbone, and the type of monomer involved in the polymer formation. As plasma properties are concerned, the predominantly important factor is the type and energy of the interacting plasma particles.

9.2.1 Crystallinity Properties of the Polymer Material

Among the physical properties, the crystallinity of polymers and their role in plasma-etching rate has been studied extensively. The comparison of polymer etching rate as a function of crystallinity reveals that the crystalline forms of a given polymer are etched at a lower rate compared to its amorphous forms. The lower etching rates of crystalline forms can be explained on the basis of the crystallization energy stabilization. Both theoretical and experimental arguments are available to validate these statements. Various studies reveal that the crystallinity of polyethylene terephthalate (PET), high-density polyethylene, and low-density polyethylene tend to increase after suitable plasma treatment, which is attributed to the higher removal rate of the amorphous form [9,10]. Thus, plasma etching selectivity can be utilized to increase the crystallinity concentration of a given polymer sample. In addition, such disparate etching rates of crystalline and amorphous forms can be utilized for rendering different surface morphologies after plasma etching. A typical example is well represented in Fig. 9.2.

Polymer in its crystalline form does not only provide higher stability, but the crystallinity also creates a difference in the plasma−polymer interactions. In an example, O_2-plasma treatment of various forms of PET foil revealed that the degree of polymer crystallinity can affect the recombination probability of neutral atoms on the surface [10]. For the amorphous form of the polymer, the recombination coefficient of neutral oxygen atoms on the surface is in the range from 2.1×10^{-4} to 8.3×10^{-4}, while for semi-crystalline form it is in the range from 2.5×10^{-5} to 4×10^{-4}. Such atomic recombination on the surface can dissipate energy in the form of heat and increase the etching rate. However, a clear explanation for such observations has not been provided yet.

9.2.2 Structural Features of the Polymer

9.2.2.1 Effect of the Aliphatic and Aromatic Moieties on the Polymer Backbone

The structural features of a polymer material has a significant role in defining its physical and chemical properties. Similarly, the structural features of a given polymer material can largely influence its stability towards various reactive plasmas. One of the primary structural features that influence the polymer-etching rate is the aliphatic and aromatic moiety on the polymer backbone. The etching rates for aliphatic polymers are

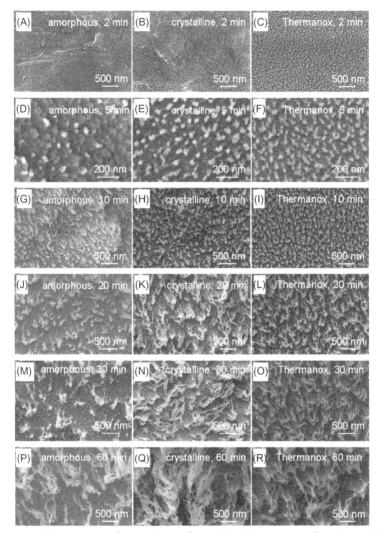

Figure 9.2 SEM images of amorphous (first column), semi-crystalline (second column), and third column, semi-crystalline biaxially oriented PET films (Thermanox) after different times of plasma treatments (time increasing down the columns). *Reproduced with permission from Macromolecules 2010, 43, 9908–9917 © American Chemical Society.*

moderately higher compared to aromatic ones due to the extra energy provided by the aromatic stabilization [7,11]. Another plausible explanation could be that the aliphatic moiety degrades easily to yield smaller volatile molecules. On the other hand, aromatics ring degradation can

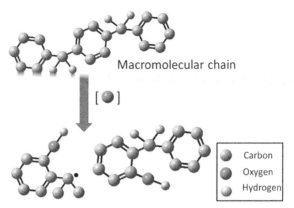

Macromolecular chain

[●]

●	Carbon
●	Oxygen
●	Hydrogen

Nonvolatile fragments, functionalized
at the aromatic rings

Figure 9.3 Interaction of plasma reactive species with aromatic polymer chain yielding nonvolatile fragments, functionalized at the aromatic rings.

result in the formation of heavier fragments which are nonvolatile. Further studies revealed that the extra stability of the aromatic polymers is due to their capability to quench the reactive plasma particles [11,12]. In this process, the ring functionalization is more probable compared to the ring rupture (Fig. 9.3); that is, the aromatic ring can quench the incoming reactive plasma species through the formation of functional groups on the rings and avoids ring fragmentation.

It is undoubtedly true that the etching of a polymer material commences as functionalization at the weakest bond junction [13]. In other words, regardless to the etching mechanism, the etching rate of a given polymer is associated with the energy required for breaking the bonds on the polymer chain. Experimental evidence has been provided by Taylor et al. in their studies related to the comparison of etching rate of different polymers (k_{rel}) with respect to the number of chain scissions per 100 eV of energy absorbed (G_s) (Table 9.2) [12]. The etching rate exhibited a linear trend with reference to the bond-breaking energy of the polymer chain.

9.2.2.2 Effect of Various Functional Groups on the Polymer Backbone

There are many scientific investigation reports for the comparison of the etching rates of polymeric materials bearing various functional groups on the polymer chain [14]. A comparison of etching rates for various polymers in Ar plasma is presented as an example in Fig. 9.4. It can be seen that the presence of heteroatoms on the polymer backbone increases the

Table 9.2 Relative rates of O_2 plasma removal (k_{rel}) and number of chain scissions per 100 eV of energy absorbed (G_s-values) for selected polymers

No.	Polymer	k_{rel}	G_s
1	Poly(α-methylstyrene)	1.11	0.3
2	Polyphenyl methacrylate	1.33	—
3	Polyvinyl methyl ketone	1.48	—
4	Polymethyl methacrylate (PMMA)	2.37	1.2
5	Polymethyl methacrylate-co-methacrylonitrile (94:6 mol %)	2.70	2.03
6	Polyisobutylene	3.56	4
7	Polybutene-1 sulfone	7.11	8

Source: Reproduced with permission from Polym. Eng. Sci. 1980, 20, 1087−1092 © John Wiley & Sons, 2004.

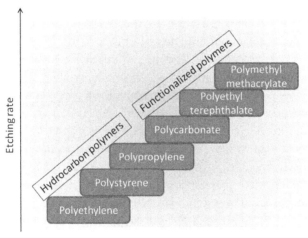

Figure 9.4 Comparison of trends in etching rates for various polymer substrates in Ar plasma at 300 W and 20 Pa. *Based on the results presented in Nuclear Instruments and Methods in Physics Research Section B: Beam Interactions with Materials and Atoms. 2003, 208, 281−6.*

etching rates of a given polymer when compared to its corresponding hydrocarbon analogue. This means, PET or polymethyl methacrylate (PMMA) can be etched in plasma at higher speeds compared to polystyrene (PS). This is due to the backward reaction between the highly reactive O atomic species or OH radical released into the plasma phase as a result of the polymer bond scission, which can further react with the polymer backbone and provide a higher etch kinetics. Nonetheless, the comparison of the presented etching rates brings a contradiction that the etching rates for PS is higher than that of polypropylene (PP). This can be attributed to the surface cross linkage on the exposed surface in the presence of ultraviolet (UV) irradiation from plasma and, thus, quenching the fast degradation of polymer by the reactive plasma particles [7,14,15].

9.2.3 Plasma Properties

For the treatment of delicate and heat-sensitive polymer materials, neutral atom rich plasmas with relatively low ion density are preferred in order to avoid any significant damage due to ion bombardment. In other words, for the successful surface functionalization or etching of the polymer materials, neutral atom-assisted processes are needed. Due to this reason, this section will discuss polymer etching only in neutral atom rich plasmas. Such plasmas are mostly generated by a high frequency discharge, which induces higher acceleration of the electrons due to their lower mass while the heavier ions are affected only to a smaller extent, providing higher dissociation of the feeding gas. Inductively coupled RF discharges or microwave discharges are examples which generate highly dissociated plasmas [16,17]. When the neutral radicals interact with a surface, the potential energy interactions with the surface are more significant than the kinetic energy interactions. For a plasma particle to react with an exposed surface, the incoming particle should have high enough energy to overcome the activation energy barrier for chemical reaction. Thus, the etching rate in plasma can be regulated by the energy of the particles generated in the system. This can be controlled by changing various discharge parameters such as feed gas flow, gas composition, plasma power, and operating pressure [18].

A systematic comparison of activation energies required for the gaseous phase reaction between hydrocarbon molecules with various neutral species reveals that O (1D) and F atoms do not need any activation energy for the chemical reaction to begin [19,20]. This means that such species react spontaneously with most of the organic polymer materials with little or no selectivity toward various bond types. On the other hand, the energy of activation for reaction of O (3P) species with a hydrocarbon molecule is found to be around 1 eV [11]. In other words, O (3P) can chemically differentiate between various bond types in contrast to O (1D) or F atom species. Similarly, highly reactive atomic Cl species reacts without any bond selectivity, whereas atomic Br species are less reactive and provide better etch selectivity.

The densities of various species can be changed by controlling the discharge parameters, such as inlet gas composition and discharge power. For example, adding trace amounts of noble gases to O_2 or CF_4 plasma tends to increase the dissociation rate of these gases and, thus, increase neutral atom density [21,22]. The addition of noble gases into the system increases plasma density due to their higher electron impact dissociation cross-section and consequently increase the dissociation [23].

9.3 APPLICATIONS OF SELECTIVE PLASMA ETCHING

The methodology of selective plasma etching has found to be an efficient tool for fabricating high-performing micro/nano devices derived from polymers and polymer matrix composites. A few of them are illustrated in this section with relevant examples.

9.3.1 Surface Nanostructuring of Polymer Materials

Surface nanostructuring of various polymeric materials has been achieved with the help of selective plasma etching for many applications, such as sensing, energy devices, and nanoelectronic devices. On a polymer surface, desired nanostructures can be fabricated either by template free or template assisted plasma etching. For example, by utilizing the ability of plasma to distinguish between different phases (crystalline/amorphous for example) of polymer materials, significantly important surface nanostructures are synthesized without using any templates [9]. Another adapted method is to use a thin coating of any metallic or inorganic plasma etch mask that can provide nonuniform etching and result in nano-dimensional growth on the surface [24]. One advantage of this methodology is that, we can control the surface characteristics by controlling the layer thickness and size of the particles involved in the mask formation.

Plasma-assisted selective etching of block copolymer materials is a promising methodology for achieving nanoscale featured polymer templates. Extensive application of conventional wet-chemical etching is limited due to the capillary forces needed for such applications, especially when the polymers possess lamellar arrangement. Indeed, selective plasma etching can function as an alternative tool in such cases. Polymers having low resistance against plasma reactive species can be etched away at higher rates while the latter one remains unaffected, leaving a nanostructured template pattern. Selective plasma etching of block copolymer of PMMA and PS (PS-b-PMMA) can be considered as a relevant example. Compared to PS, PMMA has a relatively higher etching rate in plasma. This characteristic feature is utilized for the preferential removal of PMMA from the block copolymer layer to leave behind nano-patterns of PS polymer [25]. Such structures are important and widely exploited in the semiconductor industry [26]. However, this application is not limited to (PS-b-PMMA) and is used for patterning many other block copolymers, where different polymer blocks involved in the block copolymer layer formation have distinct etching rates [27,28].

9.3.2 Reducing the Dimensionality of Carbon Allotropes

Carbon allotropes such as graphene, fullerene, or carbon nanotubes have already brought revolutionary developments in the field of nanoscience and nanotechnology. However, scientists have engaged in research towards further reducing their dimensionality to achieve increased specific surface area and strengthen the desired properties. A few examples are reducing the dimensionality of graphene nanosheet or carbon nanotubes to graphene nanoribbon and conversion of graphene into various forms such as graphene nanomesh or crumpled [29−32]. The importance of these materials lies in vast fields from nanoelectronics to catalysis or energy devices. Selective plasma etching has been employed as one of the crucial steps for achieving such targets and it is convenient to treat the composite materials derived from them for better process control. Further examples are illustrated below.

By utilizing the higher stability of the carbon atoms in the basal plane of the graphene sheet compared to the ones occupying the edge position, graphene sheets can be converted into graphene nanoribbon with the exposure of controlled plasma reactive species [33]. In other words, the formation of chemical bonds between plasma species and the nanostructures are energetically favorable at the edge positions. The plasma functionalization then leads to etching as the neutral gas molecules desorb (CO, CO_2) after further exposure [34]. However, this process might need longer plasma exposure time, especially for shortening the width of very large graphene nanosheets. Exposure of the graphene nanosheets to reactive plasmas for longer times might induce surface defects that may adversely affect the material properties. An alternative methodology for

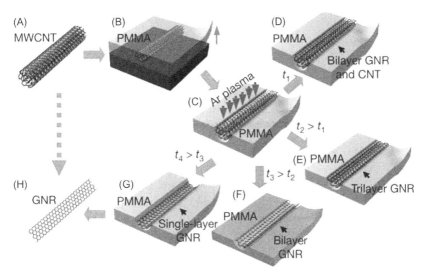

Figure 9.5 Making GNRs from CNTs. (A) A pristine MWCNT was used as the starting raw material. (B) The MWCNT was deposited on a Si substrate and then coated with a PMMA film. (C) The PMMA−MWCNT film was peeled from the Si substrate, turned over and then exposed to an Ar plasma. (D−G) Several possible products were generated after etching for different times: GNRs with CNT cores were obtained after etching for a short time t 1 (D) tri-, bi- and single-layer GNRs were produced after etching for times t 2, t 3 and t 4, respectively (t 4 > t 3 > t 2 > t 1; [E−G; H]) The PMMA was removed to release graphene nanoribbon. *Reproduced with permission from Nature 2009, 458, 877−880.*

the synthesis of graphene nanosheets would be to expose the carbon nanotubes embedded in a relatively less stable polymer matrix and expose it to reactive plasma [35,36]. One of the relevant examples is presented in Fig. 9.5. Nevertheless, a similar strategy has been utilized for the facile synthesis of various inorganic nanoribbons from their corresponding nanotube structures [37].

Graphene nanomesh, on the other hand, can be prepared by the controlled oxidation of carbon atoms in the graphene basal plane by reactive ion etching in the presence of suitable etch masks. By covering the graphene sheet with masking materials such as inorganic nanoparticles, structured polymer sheets, or both, followed by a reactive ion etching. The mask on the surface creates etching anisotropy, resulting in drilling of the graphene surface. During the plasma surface interactions, the presence of the mask is able to generate anisotropic etching to cause plasma drilling on the surface and yield of mesh-like graphene structure [38−40].

9.3.3 Applications in Biology and Medicine

Plasma surface modification of polymers has been emerging as one of the most important tools that can precisely control surface morphology and surface functionality, including the control of materials efficiency in biomedical applications, especially for designing artificial implants. This is done through polymer selective etching. For example, to attain a high compatibility between the polymer and exposed biological fluids, such as blood, both hydrophilic and hydrophobic surfaces are favored depending on the chemistry of the interacting molecule and surface morphology. By changing the surface hydrophilicity and morphology, it is possible either to increase or decrease the protein adsorption on the surface by changing the extend of chemical interactions between them [41].

Furthermore, plasma has been utilized widely for the successful bacterial inactivation, selectively etching layer-by-layer. Low-pressure as well as atmospheric-pressure plasma sources have been used for this process and more detailed information on the disinfectant ability of plasma can be found elsewhere [42−44]. However, the interesting aspect in this context is that various biological components, such as blood proteins or even some bacteria, are degraded in plasma at higher rates compared to thermoplastics or thermosetting plastics [45,46]. This difference in plasma etch selectivity can be utilized for the sterilization and recycling of various medical or surgical devices [47].

9.3.4 Tuning the Surface Wettability

Surface wettability of a polymer material is of great importance when it comes to different applications in material science. For example, achieving superhydrophilicility (surface water contact angle comes close to 0 degree) on nonpolar polymeric surfaces become essential to improve the bonding with the metals [48]. The best way to achieve superhydrophilic surfaces is to incorporate polar functional groups and, thereby, enhance the water−material surface attractive forces of interaction. Compared to conventional techniques, plasma can incorporate various functional groups and increase the surface roughness in a very short time, avoiding any changes in the bulk properties [5,49]. The combined effects of chemical and topographic modifications initiated in the course of plasma treatment can promote adhesion of the metal layer by improving the chemical compatibility between the polymer surface and the metal, and can even slightly increase the diffusion of the metal into the bulk polymer. It is been proven that the plasma pretreated samples exhibit significantly better metallic adhesion on the polymeric surfaces compared to the conventionally used industrial chemical bath process.

In contrast, achieving superhydrophobicity, mimicking naturally occurring water repellant surfaces (surface water contact angle above 150 degrees) is slightly more challenging. More details on the general methods adapted for creating superhydrophobic surfaces on polymeric surfaces can be found elsewhere [50]. By creating hydrophobicity on a surface, it is possible to slow down the aging of the surface, improve antifouling properties, and self-cleaning ability, etc. Nonetheless, surface morphology and surface chemistry are the deciding factors, irrespective of the fabrication method. Typically, fluorine-containing plasmas are the best candidates for providing suitable functional groups for creating water-repelling surfaces. For a detailed explanation of the effects of $C-F_x$ bond types on the surface wettability, readers are referred to [51].

Plasma-assisted isotropic and anisotropic etching has been efficiently used for fabricating superhydrophobic surfaces. In this manner, superhydrophobicity has been achieved on numerous polymeric substrates including PET, polyethylene (PE), polybutadiene, Teflon, and PMMA, etc. [52,53]. In general, fluorine-containing plasmas are used when the exposed polymer does not have any fluorine functionality on the polymer backbone. On the other hand, when the polymer backbone already contains fluorine (e.g., Teflon) any feed gas can be used to create the desired surface morphology and thereby achieve superhydrophobicity [54]. As an example, exposing a PS bead (mask) decorated Teflon film to O_2 plasma, nanocone arrays were generated on the surface due to the etching anisotropy (Fig. 9.6). Due to the effect of the newly created nanocone surface structure, the water contact angle increased to about 140 degrees [55]. Depending on the surface morphology, a water droplet can attain highly pinned state or they can easily roll off from the surface. This strategy of plasma anisotropic etching of polymer surface decorated with suitable masks has been utilized for fabricating natural mimicking superhydrophobic polymer surfaces. However, plasma etching assisted superhydrophilic or superhydrophobic surfaces face the challenge of aging of functional groups. This is one of the limiting factors for the applications where stable and durable surface chemistry is necessary.

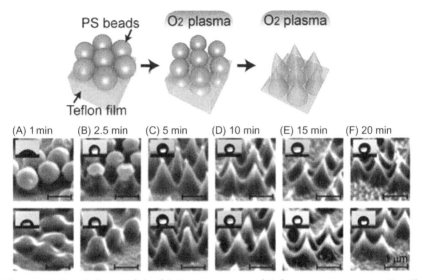

Figure 9.6 Top: Scheme of the fabrication process of Teflon nanocone arrays; (A − F) SEM images of Teflon nanocones with PS beads (top row) and after dissolution of PS beads (bottom row). The structures are created with different etching times; (A) 1, (B) 2.5, (C) 5, (D) 10, (E) 15, and (F) 20 min. All scale bars indicate 1 μm. The insets show the photos of water droplets (10 μL) on the corresponding samples. *Reproduced with permission from ACS Appl. Mater. Interfaces 2014, 6, 11110−11117. American Chemical Society.*

9.3.5 Fabrication of Materials for Extreme Environments

Selective plasma etching has been utilized for material fabrication in applications where the material has to stand extreme environments such as higher temperatures or electrical potential. Plasma pretreatments are already reported for reducing the flame retardant filler content in composite fabrication. In an example presented by Zeynep et al., plasma pretreatment of polyester fabric halved the flame-retardant concentration compared to the nontreated ones to achieve similar limiting oxygen index (LOI) [56]. This is because plasma treatments can improve the chemical compatibility between the polymer and the filler, and increase the surface area of the polymer by removing the least stable or amorphous surface residues.

In addition to this, plasma selective etching is used for improving the electrical insulation properties of composite materials for commercial applications according to International Electrotechnical Commission (IEC) standards. More details on the measurement methods, standard

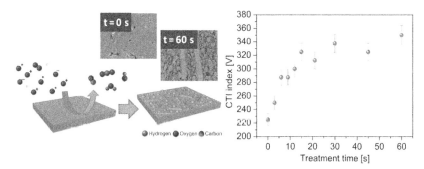

Figure 9.7 Plasma surface interaction of the glass-filled composite with corresponding SEM images for nontreated and plasma-treated samples for 60 s. The graph represents the variation of comparative tracking index (CTI) performance with plasma exposure time. *Reproduced with permission from RSC Adv. 2015, 5, 37853–37858.*

requirements for electrical insulators and associated issues can be found in [57,58] and the references cited therein. By using the disparate etching rate of various polymers and embedded organic or inorganic fillers, it is possible to expose the noncharring fillers onto the composite surface (Fig. 9.7) [5,21]. Presence of these fillers on the surface can act as a barrier towards surface polymer carbonization, and subsequent short circuits or fire. The method of selective plasma etching brings numerous advantages, such as fast processing, reducing the need of expensive filler for improving the electrical insulation properties and avoiding any changes to the bulk properties of the composite material.

9.4 CONCLUSIONS

Plasma can aggressively etch many materials. Furthermore, it has the ability to distinguish between various molecules (or chemical bonds) in the etchant based on its chemical stability as well as the material's physical properties. This has been validated by considering the energy of various bond types and the energy associated with various plasma reactive species. The characteristics of polymer material such as the type of monomer units, polymer crystallinity, and functionalities can largely influence the etching rate. Selectivity of plasma etching can thus be utilized for improving various surface characteristics of polymers or polymer matrix composites including nanostructuring, improved adhesion, surface wettability and ability to stand harsh environment. Thus, it represents one of the best and the most versatile tools for any material etching or surface modification applications. Nevertheless, significant improvement in the understanding of plasma chemistry and plasma system design is still necessary to achieve controlled plasma etching in the atom scale.

REFERENCES

[1] E.M. Liston, L. Martinu, M.R. Wertheimer, Plasma surface modification of polymers for improved adhesion: a critical review, J. Adhes. Sci. Technol. 7 (1993) 1091−1127. Available from: https://doi.org/10.1163/156856193X00600.

[2] R.A.N. Pertile, F.K. Andrade, C. Alves, M. Gama, Surface modification of bacterial cellulose by nitrogen-containing plasma for improved interaction with cells, Carbohydr. Polym. 82 (2010) 692−698. Available from: https://doi.org/10.1016/j.carbpol.2010.05.037.

[3] H. Kim, E. Oh, H.T. Hahn, K.H. Lee, Enhancement of fracture toughness of hierarchical carbon fiber composites via improved adhesion between carbon nanotubes and carbon fibers, Compos. Part A Appl. Sci. Manuf. 71 (2015) 72−83. Available from: https://doi.org/10.1016/j.compositesa.2014.12.014.

[4] V.M. Donnelly, A. Kornblit, Plasma etching: yesterday, today, and tomorrow, J. Vac. Sci. Technol. A Vacuum, Surfaces, Film. 31 (2013) 50825. Available from: https://doi.org/10.1116/1.4819316.

[5] H. Puliyalil, U. Cvelbar, G. Filipič, A.D. Petrič, R. Zaplotnik, N. Recek, et al., Plasma as a tool for enhancing insulation properties of polymer composites, RSC Adv. 5 (2015) 37853−37858. Available from: https://doi.org/10.1039/C5RA00304K.

[6] S.J. Pearton, D.P. Norton, Dry etching of electronic oxides, polymers, and semiconductors, Plasma. Process. Polym. 2 (2005) 16−37. Available from: https://doi.org/10.1002/ppap.200400035.

[7] H. Puliyalil, U. Cvelbar, Selective plasma etching of polymeric substrates for advanced applications, Nanomaterials 6 (2016) 108. Available from: https://doi.org/10.3390/nano6060108.

[8] X. Gu, Z. Liu, I. Gunkel, S.T. Chourou, S.W. Hong, D.L. Olynick, et al., High aspect ratio sub-15 nm Silicon trenches from block copolymer templates, Adv. Mater. 24 (2012) 5688−5694. Available from: https://doi.org/10.1002/adma.201202361.

[9] M.B. Olde Riekerink, J.G.A. Terlingen, G.H.M. Engbers, J. Feijen, Selective etching of semicrystalline polymers: CF4 gas plasma treatment of poly(ethylene), Langmuir 15 (1999) 4847−4856. Available from: https://doi.org/10.1021/la990020i.

[10] I. Junkar, U. Cvelbar, A. Vesel, N. Hauptman, M. Mozetič, The role of crystallinity on polymer interaction with oxygen plasma, Plasma. Process. Polym. 6 (2009) 667−675. Available from: https://doi.org/10.1002/ppap.200900034.

[11] S.J. Moss, A.M. Jolly, B.J. Tighe, Plasma oxidation of polymers, Plasma Chem. Plasma Process. 6 (1986) 401−416. Available from: https://doi.org/10.1007/BF00565552.

[12] G.N. Taylor, T.M. Wolf, Oxygen plasma removal of thin polymer films, Polym. Eng. Sci. 20 (1980) 1087−1092. Available from: https://doi.org/10.1002/pen.760201610.

[13] H. Puliyalil, G. Filipic, J. Kovac, M. Mozetic, S. Thomas, U. Cvelbar, Tackling chemical etching and its mechanisms of polyphenolic composites in various reactive

low temperature plasmas, RSC Adv. 6 (2016) 95120−95128. Available from: https://doi.org/10.1039/C6RA15923K.

[14] D. Hegemann, H. Brunner, C. Oehr, Plasma treatment of polymers for surface and adhesion improvement, in, Nucl. Instruments Methods Phys. Res. Sect. B Beam Interact. with Mater. Atoms (2003) 281−286. Available from: https://doi.org/10.1016/S0168-583X(03)00644-X.

[15] M. Hudis, Surface crosslinking of polyethylene using a hydrogen glow discharge, J. Appl. Polym. Sci. 16 (1972) 2397−2415. Available from: https://doi.org/10.1002/app.1972.070160918.

[16] F.D. Egitto, L.J. Matienzo, Plasma modification of polymer surfaces for adhesion improvement, IBM J. Res. Dev. 38 (1994) 423−439. Available from: https://doi.org/10.1147/rd.384.0423.

[17] J. Hopwood, Review of inductively coupled plasmas for plasma processing, Plasma Sources Sci. Technol. 1 (1992) 109−116. Available from: https://doi.org/10.1088/0963-0252/1/2/006.

[18] C. Lee, Global model of Ar, O2, Cl2, and Ar/O2 high-density plasma discharges, J. Vac. Sci. Technol. A Vacuum, Surfaces, Film. 13 (1995) 368. Available from: https://doi.org/10.1116/1.579366.

[19] S. Kondo, K. Tokuhashi, A. Takahashi, M. Kaise, Ab initio study of reactions between halogen atoms and various fuel molecules by Gaussian-2 theory, J. Hazard. Mater. 79 (2000) 77−86. Available from: https://doi.org/10.1016/S0304-3894(00)00266-1.

[20] C.L. Lin, W.B. DeMore, Reactions of atomic oxygen (1D) with methane and ethane, J. Phys. Chem. 77 (1973) 863−869. Available from: https://doi.org/10.1021/j100626a001.

[21] H. Puliyalil, G. Filipič, U. Cvelbar, Selective plasma etching of polyphenolic composite in O 2 /Ar plasma for improvement of material tracking properties, Plasma. Process. Polym. 13 (2016) 737−743. Available from: https://doi.org/10.1002/ppap.201600005.

[22] L.D.B. Kiss, J.P. Nicolai, W.T. Conner, H.H. Sawin, CF and CF2 actinometry in a CF4/Ar plasma, J. Appl. Phys. 71 (1992) 3186−3192. Available from: https://doi.org/10.1063/1.350961.

[23] J.T. Gudmundsson, E.G. Thorsteinsson, Oxygen discharges diluted with argon: dissociation processes, Plasma Sour. Sci. Technol. 16 (2007) 399−412. Available from: https://doi.org/10.1088/0963-0252/16/2/025.

[24] H. Fang, W. Wu, J. Song, Z.L. Wang, Controlled growth of aligned polymer nanowires, J. Phys. Chem. C. 113 (2009) 16571−16574. Available from: https://doi.org/10.1021/jp907072z.

[25] Y.-H. Ting, S.-M. Park, C.-C. Liu, X. Liu, F.J. Himpsel, P.F. Nealey, et al., Plasma etch removal of poly(methyl methacrylate) in block copolymer lithography, J. Vac. Sci. Technol. B Microelectron. Nanom. Struct. 26 (2008) 1684. Available from: https://doi.org/10.1116/1.2966433.

[26] K.W. Guarini, C.T. Black, K.R. Milkove, R.L. Sandstrom, Nanoscale patterning using self-assembled polymers for semiconductor applications, J. Vac. Sci. Technol. B Microelectron. Nanom. Struct. 19 (2001) 2784. Available from: https://doi.org/10.1116/1.1421551.

[27] Y.S. Jung, C.A. Ross, Orientation-controlled self-assembled nanolithography using a polystyrene - polydimethylsiloxane block copolymer, Nano. Lett. 7 (2007) 2046−2050. Available from: https://doi.org/10.1021/nl070924l.

[28] X. Zhang, C.J. Metting, R.M. Briber, F. Weilnboeck, S.H. Shin, B.G. Jones, et al., Poly(2-vinylnaphthalene)-block-poly(acrylic acid) block copolymer: self-assembled pattern formation, alignment, and transfer into silicon via plasma etching,

Macromol. Chem. Phys. 212 (2011) 1735–1741. Available from: https://doi.org/10.1002/macp.201100232.

[29] L. Jiang, Z. Fan, Design of advanced porous graphene materials: from graphene nanomesh to 3D architectures, Nanoscale. 6 (2014) 1922–1945. Available from: https://doi.org/10.1039/C3NR04555B.

[30] X. Zhang, H. Zhang, C. Li, K. Wang, X. Sun, Y. Ma, Recent advances in porous graphene materials for supercapacitor applications, RSC Adv. 4 (2014) 45862–45884. Available from: https://doi.org/10.1039/C4RA07869A.

[31] M. Terrones, A.R. Botello-Méndez, J. Campos-Delgado, F. López-Urías, Y.I. Vega-Cantú, F.J. Rodríguez-Macías, et al., Graphene and graphite nanoribbons: morphology, properties, synthesis, defects and applications, Nano Today. 5 (2010) 351–372. Available from: https://doi.org/10.1016/j.nantod.2010.06.010.

[32] L. Liu, X. Guo, R. Tallon, X. Huang, J. Chen, Highly porous N-doped graphene nanosheets for rapid removal of heavy metals from water by capacitive deionization, Chem. Commun. 53 (2017) 881–884. Available from: https://doi.org/10.1039/C6CC08515F.

[33] L. Xie, L. Jiao, H. Dai, Selective etching of graphene edges by hydrogen plasma, J. Am. Chem. Soc. 132 (2010) 14751–14753. Available from: https://doi.org/10.1021/ja107071g.

[34] H. Xiang, E. Kan, S.H. Wei, M.H. Whangbo, J. Yang, "Narrow" graphene nanoribbons made easier by partial hydrogenation, Nano. Lett. 9 (2009) 4025–4030. Available from: https://doi.org/10.1021/nl902198u.

[35] L. Jiao, L. Zhang, X. Wang, G. Diankov, H. Dai, Narrow graphene nanoribbons from carbon nanotubes, Nature 458 (2009) 877–880. Available from: https://doi.org/10.1038/nature07919.

[36] R. Thomas, K.P.S.S. Hembram, B.V.M. Kumar, G.M. Rao, High density oxidative plasma unzipping of multiwall carbon nanotubes, RSC Adv. 7 (2017) 48268–48274. Available from: https://doi.org/10.1039/C7RA04318J.

[37] H. Zeng, C. Zhi, Z. Zhang, X. Wei, X. Wang, W. Guo, et al., "White graphenes": boron nitride nanoribbons via boron nitride nanotube unwrapping, Nano. Lett. 10 (2010) 5049–5055. Available from: https://doi.org/10.1021/nl103251m.

[38] J. Bai, X. Zhong, S. Jiang, Y. Huang, X. Duan, Graphene nanomesh, Nat. Nanotechnol. 5 (2010) 190–194. Available from: https://doi.org/10.1038/nnano.2010.8.

[39] Y. Yang, X. Yang, X. Zou, S. Wu, D. Wan, A. Cao, et al., Ultrafine graphene nanomesh with large on/off ratio for high-performance flexible biosensors, Adv. Funct. Mater. 27 (2017). Available from: https://doi.org/10.1002/adfm.201604096.

[40] Y. Lin, Y. Liao, Z. Chen, J.W. Connell, Holey graphene: a unique structural derivative of graphene, Mater. Res. Lett. 5 (2017) 209–234. Available from: https://doi.org/10.1080/21663831.2016.1271047.

[41] C. Oehr, Plasma surface modification of polymers for biomedical use, Nucl. Instruments Methods Phys. Res. Sect. B Beam Interact. with Mater. Atoms (2003) 40–47. Available from: https://doi.org/10.1016/S0168-583X(03)00650-5.

[42] E. Stoffels, Y. Sakiyama, D.B. Graves, Cold atmospheric plasma: charged species and their interactions with cells and tissues, IEEE Trans. Plasma Sci. 36 (2008) 1441–1457. Available from: https://doi.org/10.1109/TPS.2008.2001084.

[43] M. Laroussi, Nonthermal decontamination of biological media by atmospheric-pressure plasmas: review, analysis, and prospects, IEEE Trans. Plasma Sci. 30 (2002) 1409–1415. Available from: https://doi.org/10.1109/TPS.2002.804220.

[44] Z. Ben Belgacem, G. Carre, E. Charpentier, F. Le-Bras, T. Maho, E. Robert, et al., Innovative non-thermal plasma disinfection process inside sealed bags: assessment of bactericidal and sporicidal effectiveness in regard to current sterilization norms,

PLoS One 12 (2017). Available from: https://doi.org/10.1371/journal. pone.0180183.

[45] A. Vesel, M. Kolar, N. Recek, K. Kutasi, K. Stana-Kleinschek, M. Mozetic, Etching of blood proteins in the early and late flowing afterglow of oxygen plasma, Plasma. Process. Polym. 11 (2014) 12–23. Available from: https://doi.org/10.1002/ ppap.201300067.

[46] G. Fridman, A.D. Brooks, M. Balasubramanian, A. Fridman, A. Gutsol, V.N. Vasilets, et al., Comparison of direct and indirect effects of non-thermal atmospheric-pressure plasma on bacteria, Plasma. Process. Polym. 4 (2007) 370–375. Available from: https://doi.org/10.1002/ppap.200600217.

[47] S. Lerouge, M.R. Wertheimer, L. Yahia, Plasma sterilization: a review of parameters, mechanisms, and limitations, Plasmas Polym. 6 (2001) 175–188. Available from: https://doi.org/10.1023/A:1013196629791.

[48] J. Drelich, E. Chibowski, D.D. Meng, K. Terpilowski, Hydrophilic and superhydrophilic surfaces and materials, Soft Matter. 7 (2011) 9804. Available from: https://doi. org/10.1039/c1sm05849e.

[49] J.P. Fernández-Blázquez, D. Fell, E. Bonaccurso, A. del Campo, Superhydrophilic and superhydrophobic nanostructured surfaces via plasma treatment, J. Colloid. Interface. Sci. 357 (2011) 234–238. Available from: https://doi.org/10.1016/j. jcis.2011.01.082.

[50] X. Zhang, F. Shi, J. Niu, Y. Jiang, Z. Wang, Superhydrophobic surfaces: from structural control to functional application, J. Mater. Chem. 18 (2008) 621. Available from: https://doi.org/10.1039/b711226b.

[51] V.H. Dalvi, P.J. Rossky, Molecular origins of fluorocarbon hydrophobicity, Proc. Natl. Acad. Sci. 107 (2010) 13603–13607. Available from: https://doi.org/ 10.1073/pnas.0915169107.

[52] T.S. Cheng, H.T. Lin, M.J. Chuang, Surface fluorination of polyethylene terephthalate films with RF plasma, Mater. Lett. 58 (2004) 650–653. Available from: https:// doi.org/10.1016/S0167-577X(03)00586-X.

[53] X.-M. Li, D. Reinhoudt, M. Crego-Calama, What do we need for a superhydrophobic surface? A review on the recent progress in the preparation of superhydrophobic surfaces, Chem. Soc. Rev. 36 (2007) 1350. Available from: https://doi.org/ 10.1039/b602486f.

[54] J. Zha, S.S. Ali, J. Peyroux, N. Batisse, D. Claves, M. Dubois, et al., Superhydrophobicity of polymer films via fluorine atoms covalent attachment and surface nano-texturing, J. Fluor. Chem. 200 (2017) 123–132. Available from: https://doi.org/10.1016/j.jfluchem.2017.06.011.

[55] M. Toma, G. Loget, R.M. Corn, Flexible Teflon nanocone array surfaces with tunable superhydrophobicity for self-cleaning and aqueous droplet patterning, ACS Appl. Mater. Interfaces. 6 (2014) 11110–11117. Available from: https://doi.org/ 10.1021/am500735v.

[56] Z. Ömeroğulları, D. Kut, Application of low-frequency oxygen plasma treatment to polyester fabric to reduce the amount of flame retardant agent, Text. Res. J. 82 (2012) 613–621. Available from: https://doi.org/10.1177/0040517511420758.

[57] N. Yoshimura, S. Kumagai, B. Du, Research in Japan on the tracking phenomenon of electrical insulating materials, IEEE Electr. Insul. Mag. 13 (1997) 8–19. Available from: https://doi.org/10.1109/57.620513.

[58] K.J.L. Paciorek, F.J. Perzak, F.F.C. Lee, J.H. Nakahara, D.H. Harris, Moist tracking investigation of organic insulating materials, IEEE Trans. Electr. Insul. EI-17 (1982) 423–428. Available from: https://doi.org/10.1109/TEI.1982.298485.

CHAPTER 10

Wettability Analysis and Water Absorption Studies of Plasma Activated Polymeric Materials

Jorge López-García
Faculty of Technology, Tomas Bata University, Zlín, Czech Republic

10.1 POLYMER MATERIALS: A REAL NEED IN THE MODERN WORLD

Despite the undeniable environmental problems that our planet is facing right now and the near-future solution that must be addressed, polymeric materials are vital assets in the modern world, and whether we like them or not, we may not be able to live without using polymers. According to Statista, a specialized statistics website, polymeric materials are utilized in diverse realms. For example, the European plastic material demand by sectors in 2016 was the following: A total demand of 49 million tons where 39.9% went to packaging, building and construction 19.7%, automotive 8.9%, electrical and electronic 5.8%, agriculture 3.3%, and others (i.e., sports, healthy, safety, aeronautics, household, furniture, and goods, etc.) 22.4%. These figures represent a crucial part in European industry; in fact, the plastics industry gives direct employment to over 1.5 million people, close to 60,000 companies are related to this business with a trade balance of over 16.5 billion euros in 2015 with a turnover for the European plastics industry for more than 340 billion euros. Worldwide in 2015, plastics production reached approximately 322 million tons, where China was the biggest producer with 27.8% of the planet's production followed by the NAFTA countries and Europe with 18.5% each, 16.7% from southern Asia, 7.3% from the Middle East and Africa, 4.4% South America, 4.3% Japan, and 2.6% of the former Soviet Union countries. Table 10.1 summarizes the top 10 most produced plastic in Europe in 2015.

Likewise, Table 10.2 highlights the main usages of these top 10 materials.

Non-Thermal Plasma Technology for Polymeric Materials
DOI: https://doi.org/10.1016/B978-0-12-813152-7.00010-X

261

Table 10.1 The most produced plastics in Europe.

Material	European production (million tons)
Polypropylene (PP)	≈ 8.7
Low density polyethylene (LDPE)	≈ 8.0
High density polyethylene (HDPE)	≈ 5.2
Poly vinyl chloride (PVC)	≈ 4.8
Polystyrene (PS)	≈ 4.2
Polyurethane (PU)	≈ 3.5
Polyethylene terephthalate (PET)	≈ 3.0
Acrylonitrile butadiene styrene (ABS) and Styrene-acrylonitrile copolymer (SAN)	≈ 0.8
Polyamide polymers (PA)	≈ 0.5
Other plastics	≈ 5.5

Source: PlasticsEurope

Table 10.2 European plastics demand of the most-used polymers in Europe.

Material	Main applications
Polypropylene (PP)	Food packaging, sweet and snack wrappers, hinged caps, microwave-proof containers, pipes, automotive parts, bank notes, etc.
Low density polyethylene (LDPE)	Reusable bags, trays and containers, agricultural film, food packaging film, etc.
High density polyethylene (HDPE)	Toys, milk bottles, shampoo bottles, pipes, houseware, etc.
Poly vinyl chloride (PVC)	Window frames, profiles, floor and wall covering, pipes, cable insulation, garden hoses, inflatable pools, etc.
Polystyrene (PS)	Eyeglasses frames, plastic cups, egg trays, building insulation, etc.
Polyurethane (PU)	Building insulation, pillows and mattresses, insulating foams for fridges, etc.
Polyethylene terephthalate (PET)	Bottles for water, soft drinks, juices, cleaners, etc.
Acrylonitrile butadiene styrene (ABS) and Styrene-acrylonitrile copolymer (SAN)	Containers, kitchenware, computer products, high quality packaging, battery cases, swimming pool components, shower screens, greenhouse glazing, autoclavable devices, optical fibers, etc.

(Continued)

Table 10.2 (Continued)

Material	Main applications
Polyamide polymers (PA)	Automotive, textiles, packaging, building and construction, distinct kind of fibers, etc.
Other plastics	Other uses, such as telecommunications, aeronautics, medical implants, surgical devices, membranes, valves and seals, protective coatings, etc.

Source: PlasticsEurope

The central concern of plastic production is its devastating role in pollution. As far as recycling is concerned, in 2014, 25.8 million tons of postconsumer plastic waste ended up in the official waste streams, where 69.2% was recovered through recycling (29.7%) and energy recovery processes (39.5%); whereas 30.8% still went to landfill. Within the different plastic applications, plastic packaging reached the highest recycling rate with 39.5%, and represented more than 80% of the total recycled quantities. Furthermore, it has been noticed that countries with landfill bans (i.e., Switzerland, Austria, The Netherlands, Germany, Sweden, Luxembourg, Denmark, Belgium, and Norway) achieve higher recycling rates [1,2].

10.2 CONTACT ANGLE MEASUREMENT AND ITS IMPORTANCE ON POLYMER STUDIES: SURFACE FREE ENERGY AND MEASURING METHODS

Before starting, it is imperative to remind ourselves of some basic definitions. According to the Oxford dictionary, a surface is the uppermost part of something or a continuous set of points that has length and breadth, but no thickness [3]. In this regard, surface science is defined as the study of physical and chemical phenomena that occur at the interface of two phases, such as solid—liquid, solid—gas, solid—vacuum, and liquid—gas. The determination of surface free energy of a solid is essential in a wide range of challenges in applied science. Due to the difficulties of measuring the surface free energy of a solid phase directly, indirect techniques have been used and contact angles measurements are considered the simplest indirect method for the estimation of surface free energy. The determination of the surface free energy of solids from contact angles relies on Young's equation which is the equation related to when a liquid comes

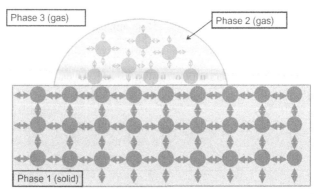

Figure 10.1 Graphic representation of the solid, liquid, and gas interfaces. *Krüss GmbH.*

into a contact with a solid in bulk. According to Young's equation, there is a relationship between the contact angle θ, the surface tension of the liquid σ_{lg}, the interfacial tension σ_{sl} between liquid and solid, and the surface free energy σ_{sg} of the solid:

$$\sigma_{sg} = \sigma_{sl} + \sigma_{lg} \cdot \cos\theta \tag{10.1}$$

The equation is valid for three-phase systems in thermodynamic equilibrium for ideal (smooth and chemically homogenous) solids and pure liquids. It is the basis of all models for determining the surface free energy of solids by means of contact angle measurements, Fig. 10.1.

This equation and the obtained value of the contact angle are still used for calculating the surface free energy of polymeric materials. In fact, there are various methods for surface free energy calculation based on the Young's equation, such as Equation of the state, Owens-Wendt-Raeble-Kaeble, Zisman theory, Fowkes theory, Van Oss-Chaudhury-Good, an Acid/Base theory, among others. All of these approaches have been developed as a result of the studies on the phenomena occurring at the interface, such as adsorption, catalysis, and wetting [4,5].

There are four techniques which are widely utilized for contact angle measurement. Perhaps, the most universal and widely used is the Drop Shape Analysis.

10.2.1 Drop Shape Analysis

The analysis of drop profile is commonly dedicated for the contact angle measurement between the liquid drop and sample surface. The contact

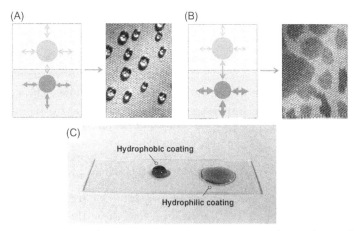

Figure 10.2 Contact angle measurement: (A) Water on A polypropylene Fabric, no wetting. (B) Cleaning solvent on Cotton, complete wetting. (C) Hydrophilic versus hydrophobic coatings.

angle is monitored by a goniometer; the measurement may also be made with photographs or video images. The image of the drop is recorded with the help of a camera and transferred to the drop shape analysis software. A contour recognition is initially carried out based on a gray–scale analysis of the image. In the second step, a geometrical model describing the drop shape is fitted to the contour. Drop size and volume, liquid density, liquid vapor pressure, surface quality, time of equilibrium, solubility, and laboratory temperature may considerably influence the assessment.

The contact angle measurement might be easily performed by establishing the tangent angle of a liquid drop with a solid surface. The contact angle, θ, is the angle formed by the solid–liquid and the liquid–vapor interfaces as may be seen in Fig. 10.2.

When the contact angle is $0°$, the liquid spreads over the solid surface. In contrast, if θ is $180°$, the liquid does not wet the solid surface. A solid surface is considered wettable if the contact angle is less than $90°$; whereas the surface is not wettable when θ is above $90°$. Solid surfaces may be classified into two basic groups, hydrophilic (wettable with water and high-surface energy) and hydrophobic (not wettable with water and low-surface energy). Another two key definitions to be recalled are wetabilitty and wetting: the former is the ability of a solid surface to reduce the surface tension of a liquid in contact with it, spreads over the surface and wets it. The latter is the ability of a liquid to maintain contact with a solid surface.

10.2.2 The Wilhelmy Plate Method

This technique is based on measuring the force, F, acting on a rod or a plate, which is partially or completely immersed in a probe liquid, and thus, from this force measurement the contact angle may be determined. The plate is attached to a balance with a thin metal wire. The force on the plate due to wetting is measured using a tensiometer or microbalance and used for calculating surface tension σ. The basic equation is as follows:

$$\sigma = F/L \cdot \cos\theta \qquad (10.2)$$

where σ is the surface tension, F the force, L the wetted length, which is equal to its perimeter ($2w + 2d$) being w and the d the plate width and the thickness (depth) respectively. θ is the contact angle between the liquid phase and the plate. Platinum is chosen as the plate material since it is chemically inert, easy to clean, and because it may be optimally wetted on account of its very high-surface free energy, hence, generally forming a contact angle θ of 0 degree (cos $\theta = 1$) with liquids. To appraise the contact angle, the solid sample (in plate form) is immersed in a liquid with known surface tension. The contact angle may be calculated from the measured force by transposing the Wilhelmy equation:

$$\theta = \arccos\left(F/(L \cdot \sigma)\right) \qquad (10.3)$$

A dynamic contact angle is normally measured by slowly immersing (advancing) and then withdrawing (receding) the solid. The advancing angle is determined during the wetting process, while the receding angle is during the de-wetting process (Fig. 10.3) [6,7].

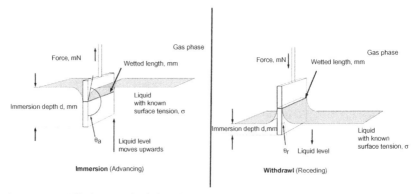

Figure 10.3 Wilhelmy method description.

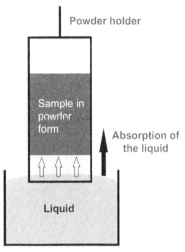

Figure 10.4 Washburn method description.

10.2.3 Washburn Sorption Method

The wettability of powders may be monitored by using the Washburn method where the contact angle is calculated from the weight increase over time when the powder is in contact with the liquid (Fig. 10.4).

In a Washburn measurement, a glass tube with a filter base is filled with powder that comes into contact with the testing liquid. The liquid is drawn up as a result of capillary action. The increase in mass of the tube, which is suspended from a force sensor, is determined with respect to time of the experiment. The basic equation of this method is as follows:

$$m^2/t = c \cdot \rho^2 \cdot \sigma \cdot \cos \theta/\eta \qquad (10.4)$$

where m is the mass; t the flow time; σ is the surface tension of the liquid; c corresponds to the capillary constant of the powder; ρ is the density of the liquid; θ the contact angle, and η is the viscosity of the liquid [8].

10.2.4 Top-View Distance Method

This technique is dedicated to making measurements either when solid samples have a curved shape or in depressions which are not easily accessible by means of the classical drop shape analysis. This is also recommended for small contact angles, since it has very good resolution in low measuring ranges. The systems consist of an optical device capable of

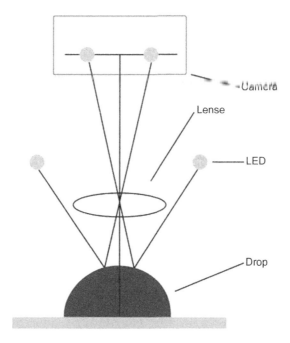

Figure 10.5 View distant method. *Krüss GmbH.*

measuring the contact angle. The curvature of the surface of a drop, which correlates with the contact angle, is determined from the distance of reflected light spots in a video image. The light spots originate from small light sources, which are arranged in a plane above the drop, as may be seen in Fig. 10.5.

These are some parameters that must be known before carrying out the analysis, such as the working distance, the distance of the LEDs from one another, the enlargement, and the dosed drop volume.

10.3 ON THE CORRELATION OF PLASMA TREATMENT IN SURFACE WETTABILITY MODIFICATION

10.3.1 Plasma Surface Modification

One straightforward strategy to change certain surface properties and, thus, overcome physico-chemical difficulties is by using plasma-based technologies, such as corona, dielectric barrier, radiofrequency, and microwave discharges. Plasma treatments are efficient and economical tools in the field of surface modification, pertaining to the fact that they are environmentally friendly, easy to operate, and require relatively inexpensive equipment. The primary effect of plasma treatments is to impart surface reactivity via electron, ion, radical, and UV radiation collisions, thereby confining the treatment effects to the outermost layers. The technique is versatile since various carrier gases (air, carbon dioxide, halogen-containing gases, hydrogen, nitrogen, nitrogen-containing gases, noble gases, oxygen, ozone, oxygen-containing gases, and steam) may be utilized for producing unique surface properties. For instance, oxygen, oxygen-containing and ozone plasmas are conventionally devoted to increasing the surface energy and to introducing polar (O-containing) moieties. Ozone is known as a strong oxidant and is a powerful tool to promote adhesion. Noble gases are inert and do not chemically react with the treated sample, but transfer reactivity giving rise to bond breakage and the subsequent origin of free radicals, which may endure for several days and undergo many reactions. In fact, helium, neon, and argon are employed for cleaning and sputtering, while argon, krypton, and xenon have found applications in implantation and deposition. Argon is the typical noble gas used in plasma treatment owing to its relatively low cost, availability, and high yield in sputtering processes. Inert gases are also dedicated for cleaning before treating the substrate with a reactive gas. Nitrogen is deemed as a low reactive gas on account of its electron configuration ($1s^2$ $2s^2$ $2p_x^1$ $2p_y^1$ $2p_z^1$). However, oxygen functionalities are usually incorporated in polymer surfaces after nitrogen and other non-oxygen plasma treatments. This phenomenon is a consequence of breaking bonds and free radicals formation that once the samples are withdrawn from the plasma reactor trigger the reaction between atmospheric oxygen and free radicals. Fluorine-containing plasma is set for surface etching. Besides, it has the opposite effect to oxidative plasmas. Consequently, this treatment prevents the inclusion of oxygen functionalities and decreases surface energy. Although hydrogen plasmas may be applied to raise hydrophobicity, this technique is not effective, since atomic hydrogen reacts with atmospheric oxygen forming oxygen-containing groups on the surface [9−11]. Surface wettability, printability, adherence, and biocompatibility may be ameliorated by using nonthermal plasma treatments.

Owing to the relevance and convenience of nonthermal plasma technologies, there are plenty of studies about this crucial topic. For example, Table 10.3 lists the water contact angle of some of the most-used polymers in the world.

Table 10.3 Water contact angle of some of the most sold polymer in the world

Polymeric material	Water contact angle
Ethylene propylene diene monomer (EPDM)	≥ 111
Poly(tetrafluoro-ethylene) (PTFE)	≥ 105
Polypropylene (PP)	≥ 95
Polyurethane (PU)	≥ 94
Poly vinyl chloride (PVC)	≥ 92
Polyethylene	≥ 90
Poly vinyl chloride (PVC)	≥ 92
Polystyrene (PS)	≥ 85
Polycarbonate	≥ 84
Polymethyl methacrylate (PMMA)	≥ 80
Stryrene-acrylonitrile copolymer (SAN)	≥ 78
Polyethylene terephthalate (PET)	≥ 70
Polyacrylonitrile	≥ 60

10.3.2 Hydrophilicity

Solid surfaces may be classified into two basic groups: hydrophilic (wettable with water and high-surface energy) and hydrophobic (not wettable with water and low-surface energy). The values in Table 10.3 may drastically plummet after a few seconds of air, nitrogen, noble gases, or oxygen-containing plasma treatment. Some polymers, such as Polypropylene, Polymethyl metacrylate, Polystyrene, and Stryrene-Acrylonitrile Copolymer have a distinctive, the so-called hydrophobic recovery, aging. Fig. 10.6 displays the aging behavior of polypropylene. This phenomenon may be ascribed to the desorption of low molecular weight oxidized materials (LMWOM) in the atmosphere, the diffusion of surface functionalities from the surface into the bulk, as polymer surfaces try to minimize their surface energy, molecular rearrangements, and finally the further reaction of free radicals trapped inside the polymer with unsaturated hydrocarbon, and reactive species from the atmosphere. Several authors have observed LMWOM species agglomerate as globular features at the surface. Using capillary electrophoresis some researchers found that the main constituents were in fact low-weight oxidized products such as organic acids, containing one or two carboxylic acid groups, and sometimes even hydroxyl or vinyl functions [13,14]. Fig. 10.6 depicts how the total surface energy of a nitrogen plasma-treated sample is slightly higher than its untreated counterpart 30 days after the plasma treatment.

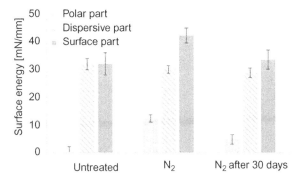

Figure 10.6 Aging behavior of the BOPP samples treated in nitrogen with the laboratory DBD system. *Seidelmann L.J.W., Bradley J.W., Ratova M., Hewitt J., Moffat J., Kelly P. Reel-to-reel atmospheric pressure dielectric barrier discharge (DBD) plasma treatment of polypropylene films. Appl. Sci. 7 (4) (2017) 1–18 [12].*

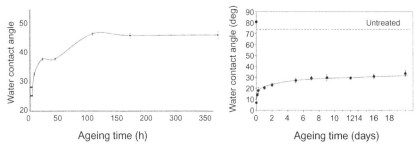

Figure 10.7 Ageing of SAN (on the left) and PS (on the right) after using oxygen nonthermal plasma treatment. *Vesel A. Modification of polystyrene with a highly reactive cold oxygen plasma. Surf Coat Technol 2010; 205(2): 490-497 [15] and Lopez-Garcia J., Primc G., Junkar I., Lehocky M., Mozetic M. On the hydrophilicity and water resistance effect of styrene-acrylonitrile copolymer treated by CF4 and O2 plasmas. Plasma. Process. Polym. 12 (10) (2015) 1075–1084 [16].*

In the case of polymers with aromatic rings, these molecular rearrangements might be also connected with a possible restore of aromaticity. Fig. 10.7 shows the ageing time of two polymers with aromatic rings after being treated with oxygen plasma. It is necessary to emphasize that after using plasmas that increase surface energy, the water contact angle of any untreated sample will always be higher than the treated one no matter the ageing time [15,16]. Ageing also depends on the storage conditions. There are studies on polymethyl methacrylate, polyethylene, and polyethylene terephthalate which state that ageing in water is slightly faster than in air. Besides, another important factor that impinges on the rate of

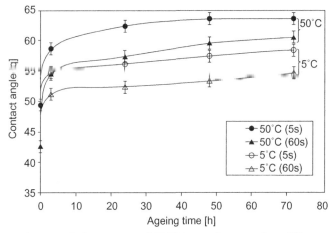

Figure 10.8 Ageing of plasma-treated PMMA samples stored at different temperatures in ambient air. *Vesel A., Mozetic M. Surface modification and ageing of PMMA polymer by oxygen plasma treatment. Vacuum 86 (6) (2012) 634−637.*

ageing is temperature. A higher storing temperature may lead to higher chain mobility and, therefore, faster ageing. In contrast, lower temperature may reduce ageing effects, as may be observed in Fig. 10.8 [17].

10.3.3 Hydrophobicity

Fluorine-containing plasma is set for rising hydrophobicity and diminishing surface energy, having the opposite effect to oxidative plasmas. The research on hydrophobicity receives much attention in diverse fields; for example, material self-cleaning properties, outdoors, nonwoven fabrics, or even cell adhesion [18−21]. The typical carrier gases employed for improving hydrophobicity are tetraflouromethane, CF_4, triflouromethane also known as flouroform, CHF_3, and sulfur hexafluoride, SF_6. Stryrene-Acrylonitrile Copolymer (SAN) is a polymer with growing demand, it is used for food containers, kitchenware, computer products, battery cases, swimming pool components, and shower screens, among others; nonetheless, this material has a tricky disadvantage, its water absorption is higher than polystyrene. This drawback may be solved after few seconds of CF_4 plasma treatment, where the water contact angle may perfectly rise from $80°$ to over $100°$. Another case is polymethyl methacrylate (PMMA) whose water contact angle is around $80°$ and may be increase to over $150°$. Fig. 10.9 is an indubitable example of the extent of hydrophobicity after using CF_4 plasma treatment. It should be noted, that

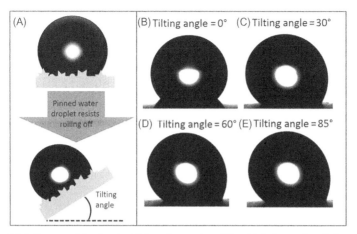

Figure 10.9 Extent of hydrophobicity after using CF₄. (A) Schematic of the pinned water droplet resistance to rolling off from the composite surface. (B–D) The images of the water droplet of volume 10 μl at various sample tilting angles. *Puliyalil H., Recek N., Filipic G., Cekada M., Jerman I., Mozetic M., et al. Mechanisms of hydrophobization of polymeric composites etched in CF4 plasma. Surf. Interface Ana. 49 (4) (2017) 334–339 [22].*

fluorine-containing gases are strong etchant gases which concomitantly alter surface topography. The fluorination implies the incorporation of C-F₃, C-F₂ and C-F functionalities onto the treated surface. It is a stable process that remains over time; however, it is well-established that the amount of C−F₃ and C−F₂ bonds diminishes, whereas the amount of C−F bond rises, where C−F₃ bond types contribute more towards the hydrophobicity of the surface.

The use of chemicals capable of repelling water, oil, and other liquids that cause stains is a relevant issue in the textiles industry. Plasma treatment has advantages over wet-treatment, since it is environmentally friendly and does not require expensive machines for its operation. Fluorinated resins are employed to make a fiber hydrophobic, but the use of substances by wet means drastically impact on the fiber bulk properties; furthermore, fluorine compounds may cause long-term ecological damage on account of its slow degradation which is making their use more restricted. In contrast, plasma treatment does not generate any waste liquid and does not modify the bulk properties. Organosilicons are dedicated for textile hydrophobization and oleophobization since many of them are volatile close to room temperature, nonflammable, inexpensive, and commercially available. Hexamethyldisiloxane, n-propyltrimethoxysilane, and hexadecyltrimethoxysilane are often preferred as a precursor in plasma polymerization processes because of its highly organic character as well as its high vapor pressure. Hexamethyldisiloxane plasma-polymerized thin silicon oxide films may be utilized in a large number of applications, such as protective antiscratch coatings, barrier films for food and pharmaceutical packaging, coatings for biomaterials, textiles (including woven polyesters, rayon filaments, cotton, and Dacron) and in wood manufacturing. For instance, the silica nanoparticle treatment itself does not change the hydrophilic surface of cotton; nevertheless, fluorinated silanes or alkylated silanes are reported to be sufficient to transform that hydrophilic character into a hydrophobic one, increasing the repellence dramatically [23,24].

The standard method ISO 9073-6 (Textiles-Test methods for nonwovens-Part 6: Absorption) is used for determining the liquid absorbency time and liquid absorptive capacity and, thus, it may be employed for evaluating wettability changes. The liquid absorbency test measures the time required for the complete wetting of a specimen strip loosely rolled into a cylindrical wire basket and immersed into a liquid for a specified period of time (test intervals include 10, 30, and 60 seconds). The masses of samples must be measured before and after absorption of water. The liquid absorption capacity (LAC) is calculated as follows:

$$\text{LAC} = m\text{-}m_0/m_0 \cdot 100[\%] \tag{10.5}$$

where LAC(%) is the liquid absorption capacity, m_0 corresponds to the mass in grams of the dry sample, and m is the mass in grams of the sample after water absorption [25].

10.3.4 Polymer Printability

One of the most important topics associated with increasing the surface energy of polymers is devoted to improving the printability of films. It has been comprehensively examined that ink adheres more firmly to polymer surfaces that have been plasma-treated than to untreated ones. The improvement in ink adhesion is mainly ascribed to two factors: surface polarity and roughness. The former because the incorporation of polar groups (hydrophilicity) onto the surface renders good adhesion, the latter since a rise in surface roughness leads to a higher effective area, and consequently more available sites for ink-polymer interaction. Surface chemistry and topography are of paramount importance for printability [26]. Fig. 10.10 provides convincing evidence about how plasma treatment is able to ameliorate ink adhesion.

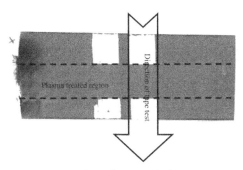

Figure 10.10 This image exhibits the influence of plasma treatment on ink adhesion after sticking and then lifting the tape off with a consistent force. *Seidelmann L.J.W., Bradley J.W., Ratova M., Hewitt J., Moffat J., Kelly P. Reel-to-reel atmospheric pressure dielectric barrier discharge (DBD) plasma treatment of polypropylene films. Appl. Sci. 7 (4) (2017) 1–18.*

For instance, progresses on Polyethylene and PVC printability have been evinced after carrying out the tape-peeling test. This experiment is employed for estimating the extent of ink adhesion on polymeric surfaces. These results indicated that only by using simple natural eye-observation, ink on the printed-untreated samples seems to be completely removed after five peels and this observation was confirmed by means of spectroscopic techniques. Contrariwise, the data clearly proves that plasma treatment improves ink adhesion since all the UV-VIS spectra of all the peeled-untreated material are significantly different than the treated specimens, where the absorbances of the plasma-treated samples were approximately two times higher than the untreated ones even after 5, 10, 20, and 40 peels. Investigations on ethylene propylene diene monomer (EPDM) rubber compositions reveal that adhesion is notoriously enhanced after using plasma treatment without damaging its elastic and mechanical properties [27]. In the textiles industry, plasma treatment has been used for improving the dye uptake in wool and other types of fabrics [28]. It is possible to conclude that plasma treatment is a cost-effective way for modification of polymeric materials while at the same time strengthening its printability [29,30].

10.4 WETTABILITY AND ITS ROLE ON PLASMA TREATMENT FOR MEDICAL APPLICATIONS

Any biomaterial must possess a suitable mechanical strength, malleability, and functionality. These parameters are regulated by bulk properties, while biological responses are controlled by surface composition. Nonthermal plasmas are extensively assigned to biomaterial science, since surface properties and biocompatibility may improve confining the treatment to the top layer without affecting the bulk properties. Although there is a broad range of values, many authors claim that plasma penetration depth is within the mesoscopic scale (1—1,000 nm) and hinges on the substrate and on the operating parameters. Plasma treatments have been utilized for numerous applications in the biomedical area, for example, contact lens materials, as these materials need high oxygen permeability and surface hydrophilicity. The commercially available lens material is a copolymer of an alkyl acrylate and a siloxane. The siloxane component in the copolymer increases oxygen permeability but reduces the lens surface, resulting in poor patency of the tear film/lens interface. Oxygen-plasma treatment on the lens material increases hydrophilicity, making the lenses more comfortable for wearing. Plasma deposition (including both grafting and polymerization) may create a new surface setting which promotes cell attachment and cell proliferation on tested surfaces, it is also possible to create resilient surfaces against bacterial strains or improving biocompatibility. Furthermore, plasma etching treatments are applied for cleaning, sterilization, and/or wettability changes. Table 10.4 lists some of the search areas of plasma treatment in biomedicine [31,32].

Table 10.4 Main applications of plasma treatment in biomaterial engineering

Barriers coatings
Drug-release, gas-exchange membranes, device protection, protection from corrosion

Biosensors
Biomolecules immobilized on surfaces

Blood-compatible surfaces
Vascular grafts, catheters, stents, heart valves, membranes, filters

Nonfouling surfaces
Contact lenses, wound healing, catheters, biosensors

Tissue engineering and cell culture
Cell growth, antibody production

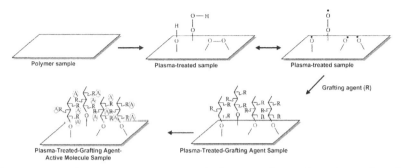

Figure 10.11 Schematic representation of plasma grafting where R signifies the grafting agent and A the active molecule respectively.

Owing to the reactivity that plasma confers to any material surface, there are two systems which are probably the most used techniques of nonthermal plasmas for biomedical purposes. The first one is plasma-based coatings, as these coatings have a fundamental advantage over other traditional processes; they are prepared by solventless and eco-friendly plasma deposition methods. Fig. 10.11 is a schematic illustration of plasma deposition. Some polymeric materials possess low-surface energy and lack proper bonding interfaces, making them suitable for surface alteration and suitable for tissue engineering or any other medical application.

Currently, there are studies focus on plasma deposition in order to give the polymer surfaces properties to counteract bacterial strains or for improving biocompatibility Table 10.5 highlights some investigations associated with this topic.

As may be noticed on the table above, there are some gases of certain mixture of gases such as ammonia and sulfur-containing gases, which may be devoted for plasma deposition instead of utilizing grafting agents [33−43]. Plasma deposition has been successfully used for preparing bio-sensors, where both, oxygen-rich and nitrogen-rich regions are located in distinct parts of the sample [44,45]. Apart from plasma deposition, another possibility is by introducing polar entities straight to the surfaces which conveys reactivity via plasma species. Cell attachment and proliferation may be ameliorated by using diverse carrier gases, such as air, oxygen, noble gases, ammonia, and hydrogen sulfide. Numerous cell lines have been seeded on plasma-treated surfaces of either synthetic polymers or biological ones. Cell attachment is significantly better on plasma-treated substrate in comparison with untreated ones, and it may be estimated by means of quantitative colorimetric techniques like the MMT

Table 10.5 Recent approaches on plasma-deposition for medical applications

Polymeric substrate	Plasma carrier gas	Grafting agent	Active molecule
Low density polyethylene (LDPE)	• Air • Oxygen	• Allylamine • N-allylmethylamine • N,N-dimethylallylamine • Acrylic acid • Allylalcohol • Allylamine • Hydroxyethyl methacrylate	• Chondroitin sulfate • Fucoidan • Heparin • Fluoroquinolones (such as norfloxacin, ofloxacin and ciprofloxacin) • Alginic acid • Benzalkonium Chloride • Bronopol • Triclosan
Poly vinyl chloride (PVC)	• Air	• Acrylic acid	• Irgasan • Chitosan • Pectin
Polyethylene terephthalate (PET)	• Argon • Nitrogen • Ammonia • Hydrogen	• Allylamine	• Heparin • Chitosan
Poly (lactid acid) PLA	• Air	• Acrylic acid	• D-Glucosamine

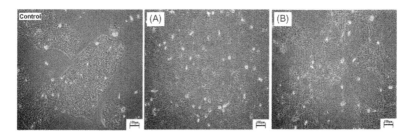

Figure 10.12 Micrographs of human skin HaCaT keratinocytes in culture on atecollagen films. (A) untreated, (B) argon plasma-treated.

assay and also qualitatively by light micrographs as the ones shown in Fig. 10.12.

Despite the plethora of studies related to the plasma treatment of biomaterials, hitherto, there is no a categorical statement about whether biocompatibility is better on hydrophilic or hydrophobic surfaces; the fact is that surface wettability (generally referred to as hydrophilicity/hydrophobicity) is a pivotal parameter affecting the biological response of any biomaterial. Although there is no a strict protocol to obtain an ideal biomaterial, biocompatibility relies upon surface crystallinity, the hydrophilic/hydrophobic character of the material, the topographical patterning, and chemical compositions [46−50].

10.5 OTHER APPROACHES TO ASSESS WETTABILITY AND SURFACE ENERGY

There are other ways to appraise wettability and surface energy. The Wilhelmy method indirectly measures the contact angle on a solid sample. The process may be summed up as follows: The sample approaches the liquid, and the force/length is zero. Thereupon, the sample is in contact with the liquid surface, forming a contact angle θ <90°; the liquid rises up, causing a positive wetting force. When the specimen is immersed further, the increase of buoyancy causes a decrease in the force detected on the balance and then the force is measured for the advancing angle. Finally, the sample is pulled out of the liquid after having reached the desired depth; the force is measured for the receding angle. The method has some advantages, for example, the task of measuring an angle, which is indeed a critical aspect of the optical systems, is reduced to the measurements of weight and length, giving higher accuracy without any kind of subjectivity. On the other hand, it also holds drawbacks, viz., the specimen must have a uniform cross section and is not suitable for irregular geometries; the sample requires the same composition and geometry at all sides. This accurate technique embraces in a big spectrum of materials including polymer films, wood, fibers and asphalts [51−54].

Contact angle evaluation of powders or granules is a tricky process and almost impossible when using conventional means. The powders or granules have to be compressed into a flat cake form where the liquid drops are applied and the contact angles are measured. Nevertheless, porosity and liquid penetration may lead to dubious results. It has been substantiated that the contact angles on a porous surface is higher than on a smooth surface with the same composition.

The Washburn method is intended to determine the wettability of powders. This involves wettability determination by using liquid penetration into a powder bed as a function of time. Wettability analysis using contact angle determination for materials that absorb the liquid in contact or become soluble in the test liquid is a major problem for the goniometric systems. Nonetheless, it does not represent a problem for the Washburn method. If the sample is soluble in a test liquid then the test liquid is saturated with the soluble material and it may be employed for the liquid penetration. With regard to polymer powders, the Washburn method has been used on both untreated and plasma-treated polymers, either biological or synthetic ones, such as starch, polyethylene, and polyethylene terephthalate [55—58].

Wettability and surface energy determination are overriding and indeed the main theme of this chapter; notwithstanding, there are different approaches for a complete surface analysis. Electrokinetic measurements are sensitive, accurate, and reliable techniques for surface characterization. The ζ-potential is an indicator for charge formation at the solid/liquid interface and the surface charge is generated by the interaction of the solid surface with the electrolyte solution. The surface charge formation hinges on the pH value of the electrolyte solution; thereby, the ζ-potential gives insight into some physico-chemical features of a solid surface. These assessments are performed when the liquid is in contact with a solid sample and moves through a capillary system. The liquid carries a net charge whose flow gives rise to a streaming current because of the hydraulic pressure, which in turn generates a potential difference. Fig. 10.13 depicts the ζ-potential versus pH for untreated, oxygen-treated and CF$_4$-treated Styrene-Acrylonitrile Copolymer (SAN)

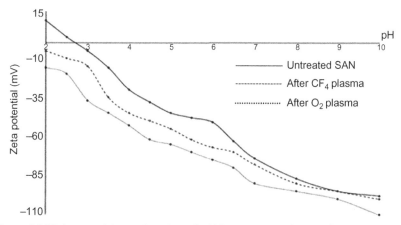

Figure 10.13 ζ-potential as a function of pH in aqueous solution of 0.001 M potassium chloride. Samples: SAN sheets untreated and after CF$_4$ and oxygen plasma treatment.

sheets. The values at neutral pH are negative, which denotes the basic nature of the examined specimens. The negatively surface charge increased after the treatments, which may be a consequence of the dissociation of some plasma-introduced moieties. Concomitantly, the isoelectric points (IEPs, which are defined as the pH at which a substance has a net charge of zero, or at which it is at its minimum ionization) of the treated samples shifted towards lower pH values. This information plainly confirms the incorporation of distinct organic functional groups which were deprotonated by the pH variation. The new surface functionalities act as electron donors, which may explain why the curves dropped to negative numbers; in fact, previous reports describe the same phenomenon after using either oxidative or fluorine-containing plasma treatments. The ζ-potential is a good complementary analysis, which is currently utilized in polymer science [59,60].

10.6 FUTURE PROSPECTS AND CONCLUSIONS

Polymer science and material innovation are undeniably a path towards creativity with eminent progress in the past few years and with great challenges to overcome. The overwhelming related projects, patents, and scientific publications are the best evidence we may get. Nonetheless, Polymer industry may not ignore the environmental challenges that we, as a global society must confront and solve, hence, it is imperative to use our plastics, either thermoplastics or thermosets, responsibly. Plastic are not waste, quite the contrary, they are a source and a meaningful asset in the modern world, providing solutions, creating plenty of jobs, and thriving businesses worldwide. Likewise, it is essential to strengthen the use of biopolymers and other kind of materials with fast degradation rates. With respect to nonthermal plasma technologies, it has been extensively demonstrated that this is an efficient tool for any kind of surface modification, which circumscribes numerous materials in several domains. From material sterilization to tissue engineering, from the most commercial and ordinary polymer in the world to the most sophisticated biopolymer, covering a vast range of industrial realms (e.g., automotive, electronics, photography, packaging, textiles, sports, rubber, wood, and bioengineering, etc.). Every polymer, either biological or synthetic, behaves differently under the presence of water or any other liquid, therefore, hydrophilicity and hydrophobicity are highly relevant parameters of treated surfaces and material customization. With respect to material customization, nonthermal plasma treatment is an outstanding alternative either to improve materials surface hydrophilicity or to repel water and oils; it all depends upon the carrier gas that is used for it and the goal to achieve. Regardless of how competent a plasma reactor is, it is indispensable to always heed to experimental parameters, such as gas pressure or gas flow rate, on the plasma reactors. Moreover, there are machines that are more robust than other ones, with different efficiency factors, that may render more uniform treatments. Finally, it is worth pointing out that wettability analysis may be performed by different methods, it all depends on the sample and the needs the researcher and the study require. There is no a better technique, they have pros and cons, and complement each other. In the same sense, wettability analysis is just a part of a bigger package, a group that ought to include at least, spectroscopic and optical evaluations in order to have succinct and reliable surface analysis. Only a few areas of technology demand more interdisciplinary teamwork than polymers or have the impact on quality and length of life, therefore, researchers are facing untold challenges in materials, biology, and engineering sciences, among others.

REFERENCES

[1] PlasticsEurope. Plastics statistics report for EU, <http://www.plasticseurope.org/documents/document/20161014113313-plastics_the_facts_2016_final_version.pdf>; 2016.
[2] K. Chidambarampadmavathy, O.P. ArthibaKarthikeyan, K. Heimann, Sustainable bio-plastic production through landfill methane recycling, Renew. Sust. Energ. Rev. 71 (5) (2017) 555−562.
[3] Oxford dictionary. Surface definitions <http://en.oxforddictionaries.com/definition/surface>; 2017.
[4] V. Bursikova, P. Stahel, Z. Navratil, J. Bursik, J. Janca, Surface Energy Evaluation of Plasma Treated Materials by Contact Angle Measurement, Masaryk University, Brno, 2004.
[5] M. Zenkiewicz, Methods for the calculation of surface free energy of solids, J. Ach. Mater. Manuf. Eng. 24 (1) (2007) 137−145.
[6] Y. Yuan, L. Randall, Contact angle and wetting properties, in: G. Bracco, B. Holst (Eds.), Surface Science Techniques., Springer-Verlag, Berlin, 2013, pp. 3−34.
[7] X. Wang, Q. Min, Z. Zhang, Y. Duan, Y. Zhang, J. Zhai, Influence of head resistance force and viscous friction on dynamic contact angle measurement in Wilhelmy plate method, Colloids Surf. A 527 (1) (2017) 115−122.
[8] A. Alghunaim, S. Kirdponpattara, N.B. Zhang, Techniques for determining contact angle and wettability of powders, Powder Technol. 287 (1) (2016) 201−215.
[9] J. Lopez-Garcia, Surface Treatment of Collagen-Based Biomaterials in Medical Applications, Association of Czech Booksellers and Publishers, Zlin, 2012.
[10] M. Mozetic, K. Ostrikov, D.N. Ruzic, D. Curreli, U. Cvelbar, A. Vesel, et al., Recent advances in vacuum sciences and applications, J. Phys. D: Appl. Phys. 47 (1) (2014) 153001.

[11] A. Vesel, M. Mozetic, New developments in surface functionalization of polymers using controlled plasma treatments, J. Phys. D: Appl. Phys. 50 (1) (2017) 293001.
[12] L.J.W. Seidelmann, J.W. Bradley, M. Ratova, J. Hewitt, J. Moffat, P. Kelly, Reel-to-reel atmospheric pressure dielectric barrier discharge (DBD) plasma treatment of polypropylene films, Appl. Sci. 7 (4) (2017) 1–18.
[13] S. Guimond, I. Radu, G. Czeremuszkin, D.J. Carlsson, M.R. Wertheimer, Biaxially Oriented polypropylene (BOPP) surface modification by nitrogen atmospheric pressure glow discharge (APGD) and by air corona, Plasma Polym. 7 (1) (2002) 71–88.
[14] S.M. Mirabedini, H. Arabi, A. Salem, S. Asiaban, Effect of low-pressure O$_2$ and Ar plasma treatments on the wettability and morphology of biaxial-oriented polypropylene (BOPP) film, Prog. Org. Coat. 60 (2) (2007) 105–111.
[15] A. Vesel, Modification of polystyrene with a highly reactive cold oxygen plasma, Surf. Coat. Technol. 205 (2) (2010) 490–497.
[16] J. Lopez-Garcia, G. Primc, I. Junkar, M. Lehocky, M. Mozetic, On the hydrophilicity and water resistance effect of styrene-acrylonitrile copolymer treated by CF$_4$ and O$_2$ plasmas, Plasma. Process. Polym. 12 (10) (2015) 1075–1084.
[17] A. Vesel, M. Mozetic, Surface modification and ageing of PMMA polymer by oxygen plasma treatment, Vacuum 86 (6) (2012) 634–637.
[18] C. Chaiwong, P. Rachtanapun, P. Wongchaiya, R. Auras, D. Boonyawan, Effect of plasma treatment on hydrophobicity and barrier property of polylactic acid, Surf. Coat. Technol. 204 (18-19) (2010) 2933–2939.
[19] D.C. Bastos, A.E.F. Santos, M.D. Da Fonseca, R.A. Simao, Inducing surface hydrophobization on cornstarch film by SF$_6$ and HMDSO plasma treatment, Carbohydr. Polym. 91 (2) (2013) 675–681.
[20] K. Matsubara, M. Danno, M. Inoue, H. Nishizawa, Y. Honda, T. Abe, Hydrophobization of polymer particles by tetrafluoromethane (CF$_4$) plasma irradiation using a barrel-plasma-treatment system, Appl. Surf. Sci. 284 (1) (2013) 340–347.
[21] S. Park, J. Kim, C.H. Park, Influence of micro and nano-scale roughness on hydrophobicity of a plasma-treated woven fabric, Text Res. J. 87 (2) (2017) 193–207.
[22] H. Puliyalil, N. Recek, G. Filipic, M. Cekada, I. Jerman, M. Mozetic, et al., Mechanisms of hydrophobization of polymeric composites etched in CF$_4$ plasma, Surf. Interface Anal. 49 (4) (2017) 334–339.
[23] A. Montarsolo, M. Periolatto, M. Zerbola, R. Mossotti, F. Ferrero, Hydrophobic sol-gel finishing for textiles:Improvement by plasma pre-treatment, Text Res. J. 83 (11) (2013) 1190–1200.
[24] K. Gotoh, E. Shohbuke, G. Ryu, Application of atmospheric pressure plasma polymerization for soil guard finishing of textiles, Text Res. J. (2017). Available from: https://doi.org/10.1177/0040517517698988.
[25] N. Radic, M.O. Obradovic, M. Kostic, B. Dojcinovic, M. Hudcova, M. Cernak, Deposition of gold nanoparticles on polypropylene nonwoven pretreated by dielectric barrier discharge and diffuse coplanar surface barrier discharge, Plasma Chem. Plasma Process. 33 (1) (2013) 201–218.
[26] M. Naebe, P.G. Cookson, J.A. Rippon, X.G. Wang, Effects of leveling agent on the uptake of reactive dyes by untreated and plasma-treated wool, Text Res. J. 80 (4) (2010) 312–324.
[27] R.P. Dos Santos, E. Mattos, M.F. Diniz, R.C.L. Dutra, Caracterização por FT-IR da Superfície de Borracha EPDM Tratada via Plasma por Micro-ondas [FT-IR characterization of EPDM rubber surface treated by microwave plasma], Polimeros 11 (2011) 1–7 [in Portuguese].
[28] R. Morent, N. De Geyter, J. Verschuren, K. De Clerck, P. Kiekens, C. Leys, Non-thermal plasma treatment of textiles, Surf. Coat. Tech. 202 (14) (2008) 3427–3449.

[29] M. Sowe, I. Novak, A. Vesel, I. Junkar, M. Lehocky, P. Saha, et al., Analysis and characterization of printed plasma-treated polyvinyl chloride, Int. J. Polym. Anal. Charact. 14 (7) (2009) 641−651.

[30] J. Lopez-Garcia, F. Bilek, M. Lehocky, I. Junkar, M. Mozetic, M. Sowe, Enhanced printability of polyethylene through air plasma treatment, Vacuum 95 (2013) 43−49.

[31] T. Desmet, R. Morent, N. De Geyter, C. Leys, E. Schacht, P. Dubruel, Nonthermal plasma technology as a versatile strategy for polymeric biomaterials surface modification: A review, Biomacromolecules 10 (9) (2009) 2351−2378.

[32] A. Asadinezhad, M. Lehocky, P. Saha, M. Mozetic, Recent progress in surface modification of polyvinyl chloride, Materials 5 (12) (2012) 2937−2959.

[33] A. Asadinezhad, I. Novak, M. Lehocky, F. Bilek, A. Vesel, I. Junkar, et al., Polysaccharides coatings on medical PVC: A probe into surface characteristics and the extent of material adhesion, Molecules 15 (2) (2010) 1007−1027.

[34] R. Morent, N. De Geyter, T. Desmet, P. Dubruel, C. Leys, Plasma surface modification of biodegradable polymers: a review, Plasma. Process. Polym. 8 (3) (2011) 171−190.

[35] E. Karbassi, A. Asadinezhad, M. Lehocky, P. Humpolicek, A. Vesel, I. Novak, et al., Antibacterial performance of alginic acid coating on polyethylene film, Int. J. Mol. Sci. 15 (8) (2014) 14684−14696.

[36] E. Karbassi, A. Asadinezhad, M. Lehocky, P. Humpolicek, P. Saha, Bacteriostatic activity of fluoroquinolone coatings on polyethylene films, Polym Bull. 72 (8) (2015) 2049−2058.

[37] E. Stoleru, M.C. Baican, A. Coroaba, G.E. Hitruc, M. Lungu, C. Vasile, Plasma-activated fibrinogen coatings onto poly(vinylidene fluoride) surface for improving biocompatibility with tissues, J. Bioact. Compat. Pol. 31 (1) (2016) 91−108.

[38] K. Ozaltin, M. Lehocky, P. Humpolicek, J. Pelkova, P. Saha, A new route of fucoidan immobilization on low density polyethylene and its blood compatibility and anticoagulation activity, Int. J. Mol. Sci. 17 (6) (2016) 1−12.

[39] A. Vesel, R. Zaplotnik, M. Modic, M. Mozetic, Hemocompatibility properties of a polymer surface treated in plasma containing sulfur, Surf. Interface Anal. 48 (7) (2016) 601−605.

[40] E. Stoleru, T. Zaharescu, E.G. Hitruc, A. Vesel, E.G. Ioanid, A. Coroaba, et al., Lactoferrin-immobilized surfaces onto functionalized PLA assisted by the gamma-rays and nitrogen plasma to create materials with multifunctional properties, ACS Appl. Mater. Interfaces 46 (8) (2016) 31902−31915.

[41] A.E. Swilem, M. Lehocky, P. Humpolicek, Z. Kucekova, I. Junkar, M. Mozetic, et al., Developing a biomaterial interface based on poly(lactic acid) via plasma-assisted covalent anchorage of d-glucosamine and its potential for tissue regeneration, Colloid Surface B 148 (12) (2016) 59−65.

[42] K. Ozaltin, M. Lehocky, Z. Kucekova, P. Humpolícek, P. Saha, A novel multistep method for chondroitin sulphate immobilization and its interaction with fibroblast cells, Mater Sci. Eng. C 70 (1) (2017) 94−100.

[43] T.V. Ivanova, R. Krumpolec, R. Homola, E. Musin, G. Baier, K. Landfester, et al., Ambient air plasma pre-treatment of non-woven fabrics for deposition of antibacterial poly (L-lactide) nanoparticles, Plasma. Process. Polym. (2017). Available from: https://doi.org/10.1002/ppap.201600231.

[44] G. Mishra, C.D. Easton, G.J.S. Fowler, S.L. McArthur, Spontaneously reactive plasma polymer micropatterns, Polymer. (Guildf). 59 (9) (2011) 1882−1890.

[45] M. Vandenbossche, L. Bernard, P. Rupper, K. Maniura-Weber, M. Heuberger, G. Faccio, et al., Micro-patterned plasma polymer films for bio-sensing, Mater. Des. 114 (1) (2017) 123−128.

[46] J.L. Garcia, A. Asadinezhad, J. Pachernik, M. Lehocky, I. Junkar, P. Humpolicek, et al., Cell proliferation of HaCaT keratinocytes on collagen films modified by argon plasma treatment, Molecules 15 (4) (2010) 2845−2856.

[47] J.L. Garcia, J. Pachernik, M. Lehocky, I. Junkar, P. Humpolicek, P. Saha, Enhanced keratinocyte cell attachment to atelocollagen thin films through air and nitrogen plasma treatment, Prog. Colloid. Polym. Sci. 138 (1) (2011) 89−94.

[48] M. Lee, Y.-G. Ko, J.B. Lee, W.H. Park, D. Cho, O.H. Kwon, Hydrophobization of silk fibroin nanofibrous membranes by fluorocarbon plasma treatment to modulate cell adhesion and proliferation behavior, Macromol. Res. 22 (7) (2014) 746−752.

[49] N. Recek, M. Resnik, H. Motaln, T. Lah-Turnsek, R. Augustine, N. Kalarikkal, et al., Cell adhesion on polycaprolactone modified by plasma treatment, Int. J. Polym. Sci. (2016). Available from: https://doi.org/10.1155/2016/7354396.

[50] N. Recek, M. Resnik, R. Zaplotnik, M. Mozetic, H. Motaln, T. Lah-Turnsek, et al., Cell proliferation on polyethylene terephthalate treated in plasma created in SO_2/O_2 mixtures, Polymers 9 (3) (2017) 2−16.

[51] M.S. Moghaddam, P.M. Claesson, M.E.P. Walinder, A. Swerin, Wettability and liquid sorption of wood investigated by Wilhelmy plate method, Wood Sci. Technol. 48 (1) (2014) 161−176.

[52] K. Abe, Y. Kusaka, M. Fujita, N. Yamamoto, H. Ushijima, Wilhelmy plate method of polymer film surfaces for printed electronics, Mol. Cryst. Liq. Cryst. 622 (1) (2015) 19−24.

[53] M.S. Moghaddam, M.E.P. Walinder, P.M. Claesson, A. Swerin, Wettability and swelling of acetylated and furfurylated wood analyzed by multicycle Wilhelmy plate method, Holzforschung 70 (1) (2016) 69−77.

[54] A. Habal, D. Singh, Comparison of Wilhelmy plate and Sessile drop methods to rank moisture damage susceptibility of asphalt − aggregates combinations, Constr. Build. Mater. 113 (1) (2016) 351−358.

[55] J. Pichal, J. Hladik, P. Spatenka, Atmospheric-air plasma surface modification of polyethylene powder, Plasma. Process. Polym. 6 (2) (2009) 148−153.

[56] M. Thakker, V. Karde, D.O. Shah, P. Shukla, C. Ghoroi, Wettability measurement apparatus for porous material using the modified Washburn method, Meas. Sci. Technol. 24 (12) (2013) 125902.

[57] A. Alghunaim, S. Kirdponpattara, B.Z. Newby, Techniques for determining contact angle and wettability of powders, Powder Technol. 287 (1) (2016) 201−215.

[58] G. Oberbossel, C. Probst, V.R. Giampietro, P.R. Von Rohr, Plasma afterglow treatment of polymer powders: Process parameters, wettability improvement, and aging effects, Plasma. Process. Polym. 14 (3) (2017) 1−10.

[59] J. Lopez-Garcia, M. Lehocky, P. Humpolicek, I. Novak, On the correlation of surface charge and energy in non-thermal plasma-treated polyethylene, Surf. Interface Anal. 46 (9) (2014) 625−629.

[60] P. Vanek, Z. Kolzka, T. Luxbacher, J.A.L. Garcia, M. Lehocky, M. Vandrovcova, et al., Electrical activity of ferroelectric biomaterials and its effects on the adhesion, growth and enzymatic activity of human osteoblast-like cells, J. Phys. D: Appl. Phys. 49 (17) (2016) 175403.

CHAPTER 11

Microscopic Analysis of Plasma-Activated Polymeric Materials

Jemy James[1,2], Blessy Joseph[1,2], Ashin Shaji[3], Parvathy Nancy[4], Nandakumar Kalarikkal[1,4], Sabu Thomas[1,5], Yves Grohens[2] and Guillaume Vignaud[2]

[1]International and Inter University Centre for Nanoscience and Nanotechnology (IIUCNN), Mahatma Gandhi University, Kottayam, Kerala, India
[2]FRE CNRS 3744, IRDL, University of Southern Brittany, Lorient, France
[3]Institute of Physics, Slovak Academy of Sciences, Slovak, Slovak Republic
[4]School of Pure and Applied Physics, Mahatma Gandhi University, Kottayam, Kerala, India
[5]School of Chemical Science, Mahatma Gandhi University, Kottayam, Kerala, India

11.1 INTRODUCTION

This chapter provides a microscopic analysis of plasma-activated polymeric materials. The microscopic analysis gives first-hand information of the nature and extend of plasma activation, thereby providing an understanding of the activation and efficiency of the technique.

Polymers play an integral role in our daily lives. The advancements in the field of polymers have grown steeply from normal household materials to intelligent and smart polymer systems that can be tailored for specific applications. Polymers are generally categorized as synthetic or naturally occurring materials. Many polymers exist in nature, like cotton, starch, and silk, etc., while many of them have been synthesized in laboratory. Polymers were not discovered overnight, but came forth from the hard work and persistence of inspiring and motivated scientists, chemists, and research scholars whose efforts have benefited the world and are still helping us in our daily lives [1] (Figs. 11.1–11.3).

It is amazing to note some of the developments in the polymeric materials. One of the latest innovations in polymeric materials is their use as a polymer foam, which, when injected into the abdominal cavity, expands and puts pressure on the internal organs thereby reducing internal bleeding [2]. An injectable scaffold made out of polymer materials can enhance the healing ability of our bodies [3]. The injected scaffold reaches the target site and regains its required shape due to the shape memory effect. Another development in this field is soft robots or flexible

Non-Thermal Plasma Technology for Polymeric Materials
DOI: https://doi.org/10.1016/B978-0-12-813152-7.00011-1

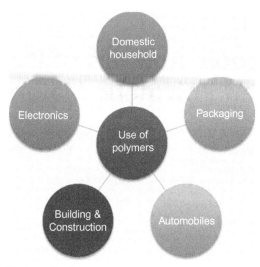

Figure 11.1 Application of polymers.

robots. They can act as camouflage and can assist doctors in surgery or help in search and rescue operations [4]. It has also been found that some kinds of injectable polymers have the ability to curtail tumor growth [5]. In most cases, blindness is caused by corneal diseases, and corneal transplant is an option, but in most cases, the matching between the cornea and availability of the donor cornea limits treating this issue and blindness cannot be solved. Our eyes are the windows to the world, and giving vision is a best ever gift. Researchers have developed artificial cornea so that patients suffering from corneal blindness can benefit [6]. Kelvar and Nomex are significant polymers these days. Kelvar is a bullet-proof material whereas Nomex is a fire-resistant polymer. Polymers are fascinating in the fact that they are light-weight, cost-effective, and easy to fabricate [7].

However, the surface properties of polymers seldom meet adequate scratch-resistance, wettability, biocompatibility, gas transmission, adhesion, or friction [8]. In order to improve their property and their versatility, surface modifications are often necessary. Surface modifications can be physical or chemical. Physical treatments like UV, gamma-ray, electron beam irradiations, ion beam, plasma, and laser treatments offers the possibility of precise modification without the use of harsh organic solvents [9]. Plasma treatment of polymers is one of the emerging technologies of modifying/activating polymers. This chapter presents the microscopic analysis of plasma-activated polymers. Most of the literature reflects the vast possibility of plasma modification of polymers for the use in the biomedical arena. Though the diverse application of plasma-modified polymers needs to be appreciated, the major focus of this chapter is on the microscopic analysis of the surfaces of the plasma-activated polymers.

11.2 PLASMA ACTIVATION OF POLYMERS

Plasma treatment of polymers is a process which improves the properties of the polymers which make them more suitable to the customized applications. Wettability is one of the important properties of polymer materials which is required for the polymer when it is subjected to coating with some other materials for the enhancement of the properties. One of the possibilities is to functionalize the polymer with polar functional groups so as to enhance the wettability. If the polymer surface is rich in polar components, higher wettability can be achieved thereby making the surface more compatible for coatings [7]. Though there are numerous methods to induce these polar groups, plasma is usually used due to its distinct advantages.

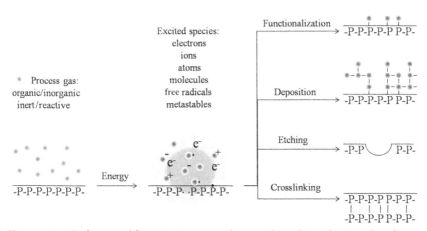

Figure 11.2 Surface modification processes that can be achieved using the plasma technique. *Reprinted from Bazaka, K., Jacob, M. V., Crawford, R. J., Ivanova, E. P., Plasma-assisted surface modification of organic biopolymers to prevent bacterial attachment, Actabiomaterialia, 7 (5), 2015–2028. Copyright (2011), with permission from Elsevier.*

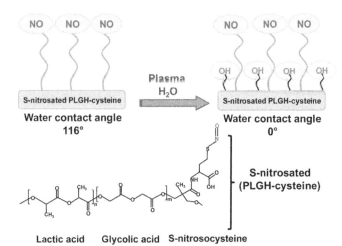

Figure 11.3 Modification of the surface of a polymer by H₂O plasma without disturbing the active NO groups. *Reproduced from Hetemi, D., Pinson, J., Surface functionalisation of polymers. Chem. Soc. Rev. 2017, 46, 5701–5713 with permission of The Royal Society of Chemistry.*

In the plasma modification process, the substrate is exposed to a reactive environment of partially ionized gas comprising large concentrations of excited atomic, molecular, ionic, and free radical species [10]. Further, the type and degree of the chemical and physical modifications that will take place depends on the nature of the interactions between the excited species and the solid surface.

Until recently, low-pressure plasma has been used to treat the surface of the polymer to functionalize them, but nowadays, atmospheric plasma is used to functionalize the polymers as atmospheric plasma treatment does not require highly sophisticated vacuum chambers and pumps.

Super-hydrophilic or super-hydrophobic polymeric surfaces can be achieved using plasma modifications by etching organic polymers, such as poly(methyl methacrylate) (PMMA) and polyether ether ketone (PEEK) [11,12]. The surface properties of the biomaterial determine its nature of interaction with a biological environment. This has major applications in the biomedical field where the polymer has to interact with proteins or other biologically active moieties. In this direction the effect of plasma modification to enhance protein adsorption and to support cell growth has been studied extensively [13,14]. A recent article on surface functionalization of polymers schematically shows the effect of plasma treatment on polymer surfaces [15]. In one case of water plasma exposure, a highly hydrophobic polymer was modified into a highly hydrophilic polymer through calculated exposure to H_2O plasma.

11.3 MICROSCOPIC ANALYSIS

The extend and efficiency of the plasma functionalization on polymers can be understood from an analysis of the surface. As a common observation, the surface of the polymer can be viewed through an optical microscope.

11.3.1 Optical Microscopy

The most common and the cheapest analysis tool, the optical microscope, has an important role to play. The formation of microphase domains when plasma modification is carried out can be imaged using an optical microscope [16]. In most of the cases, the effect of plasma modification on the properties of the polymer is being observed using microscopy.

López-Perez et al. studied the plasma modification of chitosan films on the osteoblast-like cells adhesion [17]. The chitosan membrane surface was plasma modified and then the vinyl monomer graft polymerization was carried out. The chitosan membranes were prepared by solvent casting and later on this was plasma treated using radio frequency plasma equipment, followed by an exposure of the membrane to oxygen plasma. The plasma treatment was carried out to activate the surface of the chitosan membrane through free radical formation and graft polymerization was initiated by the formed free radicals. Fig. 11.4 shows the optical micrograph of untreated and treated chitosan membrane, and it can be seen that osteoblast-like cells has better adhesion towards the plasma-activated membranes.

The studies on plasma-assisted graft co-polymerization of the acrylic acid on the surface of polyurethane was carried out by the research group headed by Dr. Byung-Cheol Shin [16]. Oxygen plasma was used to introduce carboxylic acid groups onto the surface of the polyurethane. The morphology of the material was analyzed using an optical microscope (OLYMPUS BX60, Japan) with a TOSHIBA CCD color camera. The transmission light technique was used to image the samples and the examination was carried on samples spread on the glass slides and at a magnification of 1000X. Fig. 11.5 shows the optical micrographs of the polyurethane and acrylic acid grafted polyurethane films. Fig. 11.5A demonstrates that microphase separation of the polyurethane films leads to the domain formation. The white spots represent the hard segments,

3 days   7 days

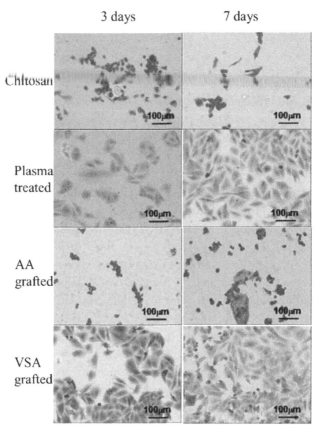

Figure 11.4 Optical micrographs of osteoblast-like cells stained with methylene blue cultured for 3 and 7 days. The plasma-activated membrane shows improved cell adhesion. *Reproduced with permission from López-Pérez, P.M., Marques, A.P., Silva, R.M. P.D., Pashkuleva, I., Reis, R.L., Effect of chitosan membrane surface modification via plasma induced polymerization on the adhesion of osteoblast-like cells. J. Mater. Chem. 2007, 17 (38), 4064 DOI: 10.1039/b707326g © Royal Society of Chemistry.*

whereas the black phase indicates the soft segments. Fig. 11.5B represents an uneven surface which demonstrates that the grafting of the acrylic acid on to the polyurethane surface was completed. The three dimensional images show the results much more clearly.

Anna Cifuentes and Salvador Borrós compared two different plasma modification techniques for the immobilization of protein monolayers [18]. The immobilization becomes necessary in the case for fabrication of smart bioactive membranes and surfaces. Strips of polystyrene were modified by the monomer, pentafluorophenyl methacrylate (PFM), using two

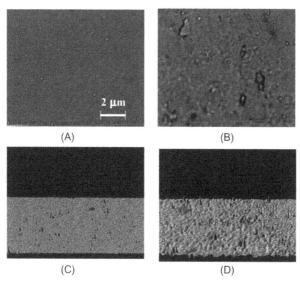

Figure 11.5 The optical micrographs of the polyurethane surface (A) and plasma-assisted co-polymerized polyurethane surface with acrylic acid (B), the 3D images of the respective surfaces are shown in (C) and (D). *Reprinted from Choi, H. S., Kim Young-Sun, Y., Zhang, Y., Tang, S., Myung, S. W., Shin, B. C., Plasma-induced graft co-polymerization of acrylic acid onto the polyurethane surface, Surf. Coatings Technol., 182 (1), 55−64. Copyright (2004), with permission from Elsevier.*

techniques: plasma enhanced chemical vapor deposition (PECVD) and plasma grafting using argon plasma. Fluorescence microscopy (AxioVs40, Zeiss Imaging Solutions) was used to image the surfaces of the polystyrene samples for comparison. A fluorescent dye, fluorescein-5-tiosemicarbazide (Fluka, FTSC), was used in the fluorescent imaging.

Fig. 11.6 shows the fluorescent microscopy images of the polystyrene sample surfaces which was modified using two techniques: PECVD (Fig. 11.6A) and argon plasma (Fig. 11.6B). The polystyrene samples onto which the PFM thin film was prepared using the PECVD technique had a highly reactive surface. This is attributed to the PFM thin film being reactive due to the number of exposed PFM groups and this leads to the hydrophobic nature of the thin films. When the PFM was grafted onto the polystyrene surface by plasma grafting, an etching effect is formed along with the formation of free radicals and thereby rendering the surface hydrophilic in nature.

In another study, the surface plasma modification of polyurethane co-polymer was carried out for enhanced cell attachment [19]. The plasma

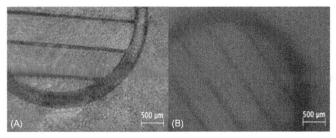

Figure 11.6 The fluorescent microscopy images of the polystyrene samples functionalized with FTSC, PFM polymerization using PECVD(A) and PFM grafting using argon plasma. *Reproduced with permission from Cifuentes, A., Borrós, S., Comparison of two different plasma surface-modification techniques for the covalent immobilization of protein monolayers. Langmuir 2013, 29 (22), 6645−6651, American Chemical Society.*

modification enhanced the biocompatibility of the polymer. Polyurethane and its blends are used in a wide variety of clinical applications, like cardio-vascular implants and vein replacement, etc. Materials like Elast-Eon are comprised of a group of low modulus, flexible polyurethanes due to the incorporation of silicon-based chain extenders to the polymer. Plasma treatment of polyurethane has found to increase the cellular interactions of the human coronary artery endothelial cells [20] and other kinds of human cells.

Phase contrast microscopy was carried out and Fig. 11.7 shows that the plasma-treated samples have an increased cellular adhesion. In the absence of protein coating, a comparatively low level of human dermal fibroblasts (HDF) spreading was observed on the untreated Elast-Eon, E2A, with a specific increase in the spreading on the plasma-treated samples. The confocal microscopy was carried out using a Zeiss LSM 510 meta confocal microscope. Fig. 11.8 shows the confocal microscopy images of the cell adhesion on the treated and untreated polymers. The images show that in the absence of protein coating, the human dermal fibroblasts were rounded with no cytoskeletal assembly, whereas the cells were spread on the plasma treated with the presence of the actin cytoskeletal assembly in some cells.

Poly(3-hydroxybutyrate-*co*-hydroxyvalerate) (PHBV) is a less crystalline, more flexible and, thus, more readily processable polymer which can be used for surgical sutures, wound dressings, and tissue engineering due to its biodegradability, good biocompatibility, and nontoxicity. The main hindrance to use PHBV is its hydrophobic nature and it needs to be made hydrophilic so as to increase cell adhesion. Plasma treatment can enhance

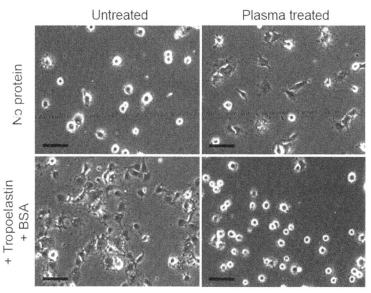

Figure 11.7 Phase contrast microscopy of the human dermal fibroblasts (HDFs) cell adhesion on untreated and plasma treated polymer (Elast-Eon E2A) in the absence of (top) and in the presence (bottom) of Tropoelastin coating and BSA blocking [19] Scale bars indicate 200 μm. *Reprinted from , Bax, D.V., Kondyurin, A., Waterhouse, A., McKenzie, D.R., Weiss, A.S., Bilek, M.M.M, Surface plasma modification and tropoelastin coating of a polyurethane co-polymer for enhanced cell attachment and reduced thrombogenicity, Biomaterials, 35 (25), 6797–6809. Copyright (2014), with permission from Elsevier.*

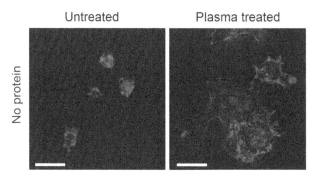

Figure 11.8 Actin cytoskeleton visualization (red) in HDF on untreated (left), plasma treated (right) Elast-Eon E2A in the absence of tropoelastin coating and BSA blocking [19] Scale bars indicate 50 μm. *Reprinted from Bax, D.V., Kondyurin, A., Waterhouse, A., McKenzie, D.R., Weiss, A.S., Bilek, M.M.M., Surface plasma modification and tropoelastin coating of a polyurethane co-polymer for enhanced cell attachment and reduced thrombogenicity, Biomaterials, 35 (25), 6797–6809. Copyright (2014), with permission from Elsevier.*

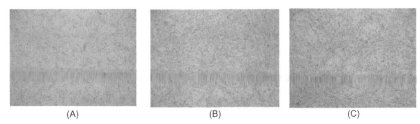

(A)   (B)   (C)

Figure 11.9 Optical micrographs of BMSCs cultured on PHBV films for 1 week (A) untreated PHBV; (B) O₂ plasma treated PHBV, and (C) N₂ plasma treated PHBV [21]. *Reproduced from Wang, Y., Lu, L., Zheng, Y., Chen, X., Improvement in hydrophilicity of PHBV films by plasma treatment, J. Biomed. Mater. Res. - Part A, 76 (3), 589–595, Copyright (2006), with permission from John Wiley and Sons.*

the hydrophilic nature of the polymer. Wang et al., used oxygen plasma and nitrogen plasma to improve the hydrophilicity of the PHBV films [21].

The dog bone marrow stromal cells (BMSCs) were seeded on to the plasma treated and untreated PHBV films to illustrate the cell affinity onto the PHBV films. The optical micrographs of the BMSCs (Fig. 11.9) cultured for a week on the PHBV neat film as well as oxygen and nitrogen plasma treated PHBV films were analyzed. Though the BMSCs spread on untreated and treated films, the cells proliferated better on the plasma-treated PHBV film.

11.3.2 Atomic Force Microscopy

Plasma treatments are environmentally safe as they do not release toxic organic solvents and are easy to control; current, pressure and voltage being the main parameters. Moreover it is reproducible. However, in spite of this, plasma-aided technology is not considered as a viable one mainly because of the complexity of the processes involved during the plasma treatment [22] and ageing of the polymer after treatment. Different strategies have been explored for improving the wettability of polymers, and plasma treatment has a major role among others. Atomic Force Microscopy has emerged as a valuable tool providing insight into the structure and properties of polymers. Plasma treatments lead to the formation of radicals that promotes surface crosslinking or functionalization thereby improving the surface properties [23]. A comparison of surface modifications of carbon fiber using various techniques is presented by Sharma et al. [24]. The atomic force microscopic images (Fig. 11.10) show the effect of the various surface modification techniques on the surface of the carbon fibers.

Slepicka et al. used argon plasma as a tool to modify the surface properties of different polymer materials (poly (ethyleneterephthalate) (PET),

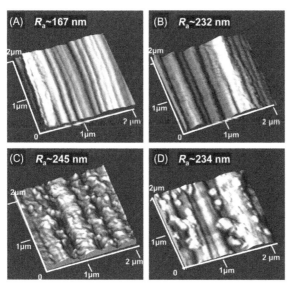

Figure 11.10 AFM images of CF: (A) untreated CF. (B) Plasma treated CF with 1% O2. (C) HNO3 treated CF for 90 min. (D) Nano YBF3 treated CF with 0.3 wt% dose (0.3% YbF3). *Reprinted from Sharma, M., Gao, S., Mäder, E., Sharma, H., Wei, L.Y., Bijwe, J. Carbon fiber surfaces and composite interphases Compos. Sci. Technol., 102, 35—50, Copyright (2014), with permission from Elsevier.*

high-density polyethylene (HDPE), poly(tetrafluoro–ethylene) (PTFE) and poly(L-lactic acid) (PLLA) with a view to enhance their biocompatibility [25]. It was concluded that plasma treatment could improve the surface polarities to a varied extent. Surface morphology and roughness were studied by atomic force microscopy (AFM). It could be observed that the surface morphology of PET and HDPE significantly changed after plasma modification, with an increase in the surface roughness. Whereas, lamellar structure was achieved for the PTFE after plasma modification, but no significant change in surface roughness was seen (Fig. 11.11). Plasma-modified PLLA had a dramatic increase in surface roughness when compared to pristine PLLA.

In order to improve the cell adhesion properties of biocompatible poly(L-lactide acid) (PLLA) and poly-4-methyl-1-pentene (PMP), argon plasma was used for modification [26]. The AFM was used for the surface morphology and roughness study of pristine PLLA, PMP, and plasma-modified surfaces. Surface roughness increased for PLLA at the discharge powers 3 and 8 W and for an exposure time of 0−240 s when compared to unmodified PLLA. Pits were also formed on the modified surface. The surface morphology and roughness of PMP did not vary significantly due

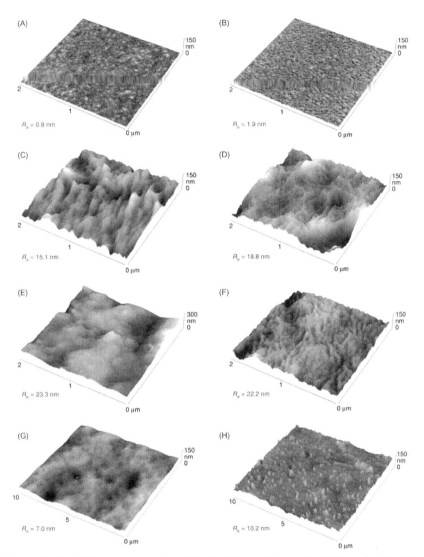

Figure 11.11 AFM images: pristine PET (A), PET modified with 10 W and 240 s (B), pristine HDPE (C), HDPE modified with 10 W and 240 s (D), pristine PTFE (E), PTFE modified with 10 W and 240 s (F), pristine PLLA (G), PLLA modified with 10 W and 240 s (H). Ra is the average surface roughness in nm. *Reproduced with permission from Slepička, P., Kasálková, N.S., Stránská, E., Bačáková, L., & Švorčík, V. Characterization of Plasma Treated Polymers for Applications as Biocompatible Carriers. Express Polym. Lett. 2013, 7 (6) Copyright (2013) Express Polymer Letters.*

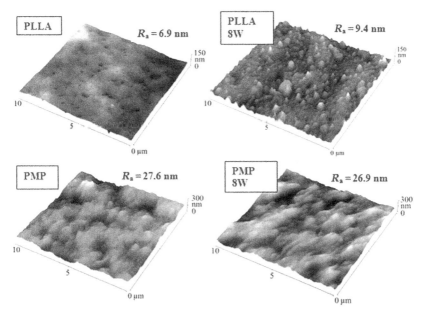

Figure 11.12 AFM images of pristine PLLA (PLLA) and PLLA irradiated with Ar plasma [power 8 W, time 240 s (PLLA/8 W)], pristine PMP (PMP) and PMP irradiated with Ar plasma [8 W (PMP/8 W)]. Ra is average surface roughness in nm. *Reproduced from Slepička, P., Trostová, S., Slepičková Kasálková, N., Kolská, Z., Sajdl, P., Švorčík, V, Surface Modification of Biopolymers by Argon Plasma and Thermal Treatment,Plasma Process. Polym., 9 (2), 197–206, Copyright (2012), with permission from John Wiley and Sons.*

to plasma exposure and the granular morphology was preserved after plasma treatment as evident from Fig. 11.12.

A similar study was done on perfluoroethylenepropylene (FEP) polymer for improving the surface properties [27]. The treatment of FEP by the powers of 3 and 8 W with the duration of 20 and 240 s induced an increase in average surface roughness when compared to pristine FEP. The surface morphology of treated FEP samples is shown in Fig. 11.13. The wrinkled pattern was enhanced and, moreover, crystalline lamellas were visible on the modified FEP surface which could be attributed to the different ablation rate of crystalline and amorphous phases.

Membrane fouling has become a major problem in ultra-filtration technology. Extensive investigations have shown that hydrophobicity is a major factor that influences membrane fouling. Subsequent studies have shown that membranes with lower surface roughness and higher hydrophilicity are less prone to fouling [28] (Figs. 11.14 and 11.15).

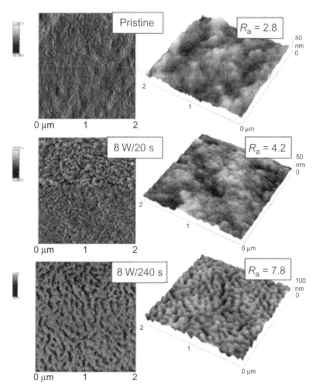

Figure 11.13 AFM images of pristine FEP (pristine), FEP treated with Ar plasma (the power of 8 W, time 20 s (8 W/20 s) and polymer treated with 8 W and 240 s (8 W/240 s). The left column represents 2D phase images, right column represents 3D morphology images. Ra is the average surface roughness in nm [27]. *Reprinted from Slepička, P., Peterková, L., Rimpelová, S., Pinkner, A., Kasálková, N.S., Kolská, Z., Ruml, T., Švorčík, V., Plasma activated perfluoroethylenepropylene for cytocompatibility enhancement, Polym. Degrad. Stab, 130, 277–287. Copyright (2016), with permission from Elsevier.*

Pal et al. evaluated changes in surface hydrophilicity and surface roughness of polyethersulfon (PES) membranes by CO_2 plasma treatment [29]. From the AFM analysis, it's very clear that a significant reduction in surface roughness was achieved after plasma treatment when compared to the pristine ones. Roughness decreased with increase in exposure duration and even with 1 min of plasma treatment, a substantial decrease of surface roughness was achieved.

The effects of radio frequency argon plasma on the surface properties of the Poly(p-phenylene benzobisoxazole) (PBO) fibers was investigated by Liu et al [30]. PBO fibers are well-known for their excellent tensile

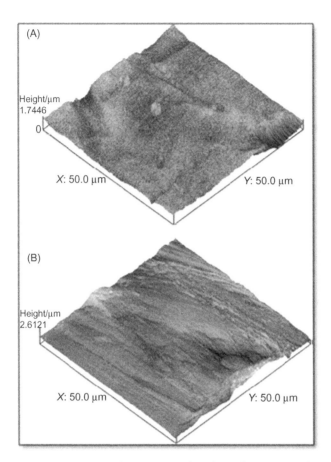

Figure 11.14 3D AFM images of (A) untreated and (B) plasma treated (12 min) PES membrane [29]. *Reprinted from Pal, S., Ghatak, S.K., De, S., DasGupta, S., Evaluation of surface roughness of a plasma treated polymeric membrane by wavelet analysis and quantification of its enhanced performance, Appl. Surf. Sci., 255 (5), 2504–2511, Copyright (2008), with permission from Elsevier.*

strength, modulus, and thermal stability. But the presence of high crystallizing surface layers present them with a smooth and inert surface restricting its further applications. In this work, the fibers were treated at a pressure of 80 Pa for 7 min with varying input powers of 100, 150, 200, 250, and 300 W, respectively. It is well-understood from the AFM images that plasma teratment yielded fibers with many notches, spots, and cone-like protuberances compared with untreated fibers having smooth morphology. The fiber surface layers were destroyed when subjected to

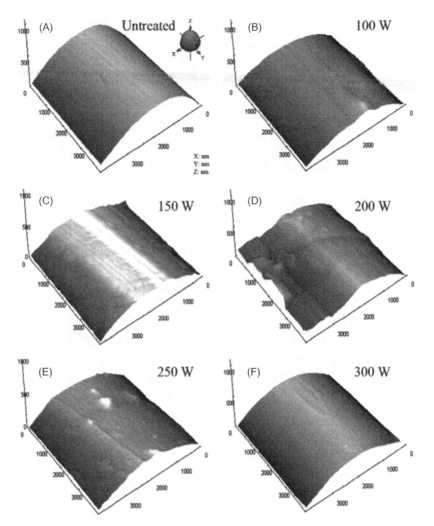

Figure 11.15 Surface morphologies of fibers by AFM [30]. *Reprinted from Liu, D., Chen, P., Chen, M., Liu, Z., Surface modification of high performance PBO fibers using radio frequency argon plasma, Surf. Coatings Technol, 206 (16), 3534−3541. Copyright (2012), with permission from Elsevier.*

plasma due to the heavy collisions between the fiber and the high-energy particles. Surface degradation occurred as a result.

11.3.3 Scanning Electron Microscopy

Scanning electron microscopy has become an indispensable tool for the characterization of surface analysis of polymer materials. The ease of specimen preparation makes SEM a valuable characterization technique to study larger and even smaller polymer structures [31].

H$_2$O vapor plasma modification of track etched membranes, PC-TE (etched polycarbonate) and PET-TE (polyethylene terephthalate) was carried out by Tompkins et al. in order to increase the hydrophilicity of the membranes [32]. A home-built inductively coupled RF plasma reactor using H$_2$O vapor plasma was employed for this purpose.

Fig. 11.16 shows representative SEM images of plasma-treated PC-TE and PET-TE membranes. When the reactor pressure was increased to 350 mtorr, no significant change was observed for both PC-TE and PET-TE membranes. Whereas increasing the exposure time with an elevated plasma power, altered PC-TE and PET-TE surface structures. Increasing the time to 30 min and an 87 percent increase in pore diameter was seen

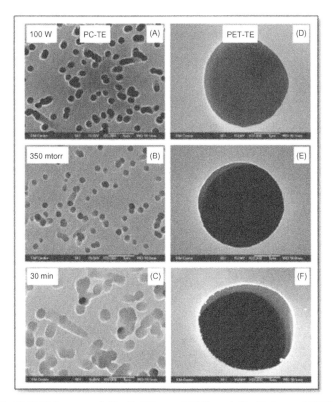

Figure 11.16 SEM images of (A—C) 0.2 mm PC-TE and (D—F) 3 mm PET-TE treated to determine the effect of plasma treatment parameters on surface morphology. Images in (A) and (D) are membranes treated using P¼100 W (50mtorr, 2 min); (B) and (E) are membranes treated using 350 m torr reactor pressure (25 W, 2 min); (C) and (F) are membranes treated for 30 min (25 W, 50 mtorr) [32]. *Reprinted from Tompkins, B.D., Dennison, J.M., Fisher, E.R., H$_2$O plasma modification of track-etched polymer membranes for increased wettability and improved performance, J. Memb. Sci., 428, 576—588, Copyright (2013), with permission from Elsevier.*

in PC-TE membranes (c), whereas for PET-TE membranes, only a slight increase in pore size was observed.

Polysulfone (PSf) is widely used as an ultra-filtration membrane due to its excellent thermal and chemical stability. Its major drawback is that it suffers from irreversible fouling. With a view to overcome membrane fouling, polysulfone ultra-filtration membranes were modified using water vapor plasmas [33]. Morphology of the modified membranes were analyzed using SEM (Fig. 11.17). Both sides of the membrane were analyzed so as to verify

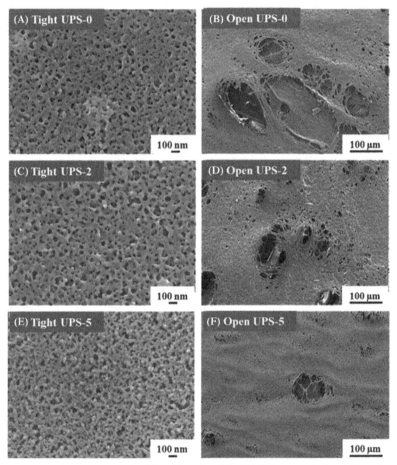

Figure 11.17 SEM images of UPS-0 (A) tight side (50,000 ×, scale bar 100 nm) and (B) open side (200 ×, scale bar 100 mm); UPS-2 (C) tight side and (D) open side; and UPS-5 (E) tight side, and (F) open side. *Reproduced from Pegalajar Jurado, A., Mann, M.N., Maynard, M.R., Fisher, E.R., Hydrophilic modification of polysulfone ultrafiltration membranes by low temperature water vapor plasma treatment to enhance performance, Plasma Process. Polym, 13 (6), 598—610. Copyright (2016), with permission from John Wiley and Sons.*

the penetration of the plasma throughout the cross-section of the UF membrane. The open side of the membrane faced the plasma glow and was oriented upstream, and that which was oriented downstream was the tight side.

Open and tight sides of the membranes which were exposed to the plasma treatment for 2 min (open UPS-2 and tight UPS-2) didn't exhibit any significant change in bulk architecture when compared to the untreated ones (Open UPS-0 and tight UPS-0). However, the tight side of UPS-5 (membranes exposed to plasma for 5 min) displayed a slight shrinkage of the membrane pores when compared to the untreated ones.

It was polysulfone earlier and now its polyether sulfone. FESEM imaging was used to study the effect of plasma treatment on the impregnation of polyether sulfone on carbon fibers [24]. It was observed that the plasma treatment enhanced the adherence of the polymer on the carbon fibers by around three times, as seen from the FESEM images (Fig. 11.18).

Electrospun polyvinyl alcohol/ Chitosan (PVAC) nanofibers were modified using dielectric barrier discharge (DBD) plasma using oxygen and argon gases, as reported in a recent article by Punamshree et al [34]. Field Emission Scanning Electron Microscopy was used to observe the surface of the nanofibers. It was observed from the FESEM images that fibers obtained after plasma treatment were without much surface defects or visible surface damage (Fig. 11.19).

Untreated and plasma treated PVA/C nanofibres were incubated with 10 ml of human blood to carry out hemocompatibility studies. The FESEM imaging of the fibers revealed that there is not much effect of

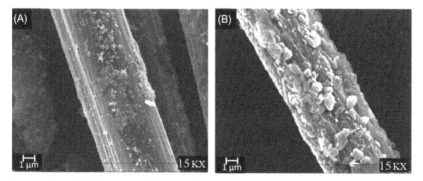

Figure 11.18 FESEM images (X 15 K) of CF impregnated with polyether sulfone: (A) before and (B) after plasma modification shows an incremental matrix pickup for plasma modified CF. *Reprinted from Sharma, M., Gao, S., Mäder, E., Sharma, H., Wei, L. Y., Bijwe, J. Carbon fiber surfaces and composite interphases, Compos. Sci. Technol., 102, 35−50. Copyright (2014), with permission from Elsevier.*

Figure 11.19 FESEM images of (A) untreated PVA/Cs nanofibers and PVA/Cs nanofibers treated with DBD plasma with (B) Ar and (C) O$_2$ gases, respectively (FESEM images, scale:1 μm). *Reprinted from Das, P., Ojah, N., Kandimalla, R., Mohan, K., Gogoi, D., Dolui, S.K., Choudhury, A.J, Surface modification of electrospun PVA/chitosan nanofibers by dielectric barrier discharge plasma at atmospheric pressure and studies of their mechanical properties and biocompatibility. Int. J. Biol. Macromol, DOI: 10.1016/j.ijbiomac.2018.03.115. Copyright (2018), with permission from Elsevier.*

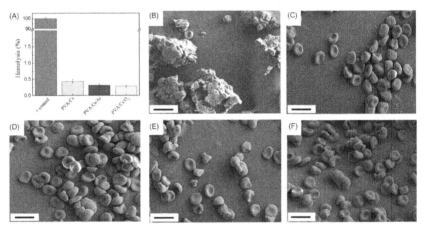

Figure 11.20 (A) Hemolysis (%) of human red blood cells after incubation with PVA/Cs, PVA/Cs/Ar and PVA/Cs/O2 nanofibers. Values represent mean $\pm$ SD of three batches. + ve control: RBC in distilled water. Surface morphologies of RBC subjected to hemolysis test with (B) DW, (C) saline, (D) PVA/Cs and (E) PVA/Cs/Ar, and (f) PVA/Cs/O2 nanofibers observed under FESEM (FESEM images, scale: 10 μm). *Reprinted from Das, P., Ojah, N., Kandimalla, R., Mohan, K., Gogoi, D., Dolui, S.K., Choudhury, A.J., Surface modification of electrospun PVA/chitosan nanofibers by dielectric barrier discharge plasma at atmospheric pressure and studies of their mechanical properties and biocompatibility, Int. J. Biol. Macromol, DOI: 10.1016/j.ijbiomac.2018.03.115. Copyright (2018), with permission from Elsevier.*

plasma-treated and untreated nanofibres on the morphology of the red blood cells (RBC). The measured hemolytic activities of the plasma-treated and untreated PVA/C nanofibres were found to be well below the permissible standards or limits of the hemolysis for biomaterials (Fig. 11.20).

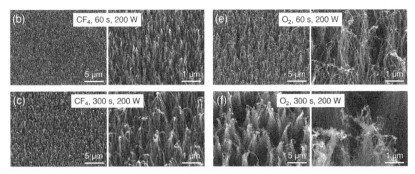

Figure 11.21 (B), (C) FESEM images taken at 35_ tilt of CF4 plasma treated VACNTs for 60 s and 300 s at 200 W, respectively. (E), (F) FESEM images taken at 35_ tilt of O2 plasma treated VACNTs for 60 s and 300 s at 200 W, respectively. *Reprinted from Yung, C.S., Tomlin, N.A., Heuerman, K., Keller, M.W., White, M.G., Stephens, M., Lehman, J.H., Plasma modification of vertically aligned carbon nanotubes: superhydrophobic surfaces with ultra-low reflectance, Carbon N. Y, 127, 195—201, Copyright (2017), with permission from Elsevier.*

In another recent article, the vertically aligned carbon nanotubes (VACNT) were modified using O_2 and CF_4 plasma to study the effect of plasma modification on these carbon nanotubes [35]. FESEM imaging was used to study the surface morphology of the carbon nanotubes. It has been observed that oxygen plasma treatment for 300 s has rendered a different morphology on the as-grown VACNT. The development of conical bundles with clumped features is evident from the FESEM images as shown in Fig. 11.21.

Several studies have shown that plasma treatments using Helium (He) discharge induce both functionalization and crosslinking of the uppermost layers of the polymer film [36]. Reis et al. treated commercial TFC membranes with pure He and H_2O gas plasma to evaluate the effect of parameters like morphology, surface energy, and microstructure on membrane performance [37].

SEM images (Figs. 11.22 and 11.23) showed that with an increase in exposure times, the surface of the H_2O-treated membranes was significantly roughened, with the effect intensified for higher excitation power. In contrast, SEM analysis for Helium plasma showed that at 10 W, the modified membranes were only slightly smoothed for short plasma treatments, while roughening occurred for a treatment time of 5 min, and the same effect was also found at 80 W.

Oxygen/Ar plasma was employed to enhance the mechanical properties of poly-p-phenylene benzobisoxazole/bismaleimide (PBO/BMI)

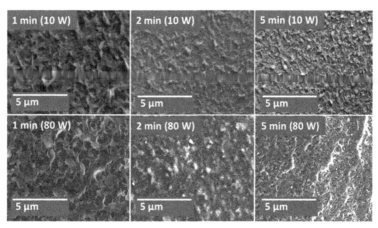

Figure 11.22 SEM images from H$_2$O plasma modified membranes [37]. *Reprinted from Reis, R., Dumée, L.F., Merenda, A., Orbell, J.D., Schütz, J.A., Duke, M.C. Plasma-induced physicochemical effects on a poly (amide) thin-film composite membrane, Desalination, 403, 3–11, Copyright (2017), with permission from Elsevier.*

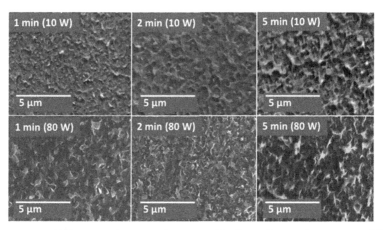

Figure 11.23 SEM images from He plasma modified membranes [37]. *Reprinted from Reis, R., Dumée, L.F., Merenda, A., Orbell, J.D., Schütz, J.A., Duke, M.C,. Plasma-induced physicochemical effects on a poly (amide) thin-film composite membrane, Desalination, 403, 3–11, Copyright (2017), with permission from Elsevier.*

composite [38]. Shear fracture morphology was observed by scanning electron microscopy (SEM). The SEM images of failed PBO/BMI composites are shown in Fig. 11.24. From Fig. 11.24A, it is evident that the fracture was initiated by the PBO fibers being pulled out from the matrix. The fibers had a smooth, clean surface with almost no resin adhering to

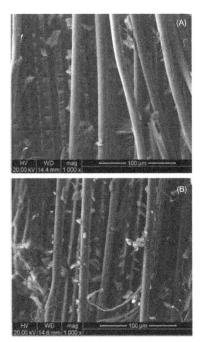

Figure 11.24 SEM picture of the failure surface: (A) untreated; and (B) treated for 7 min [38]. *Reprinted from Chen, M., Liu, D., Chen, P., Liu, Z., Ding, Z, The interfacial adhesion of poly-p-phenylene benzobisoxazole/bismaleimide composites improved by oxygen/argon plasma treatment and surface aging effects, Surf. Coatings Technol, 207, 221−226, Copyright (2012), with permission from Elsevier.*

them. Thus, it could be concluded that the interfacial adhesion between the PBO fibers and the BMI resins was weak in the case of unmodified fibers. Whereas in Fig. 11.24B it is clear that the resin fragments still adhered to the fiber surface and this was due to enhanced adhesion between fibers and resins in case of treated fibers. The authors concluded that after plasma treatment the failure mode changed from adhesive failure to partial cohesive failure.

The impact of nonthermal plasma (NTP) treatment on conductive polymer polypyrrole was evaluated by Galar et al [39]. In this work, polypyrrole samples were prepared as globular (PPy-G) and also nanostructured (PPy-NT) forms and were treated by nonthermal plasma for varied time intervals ranging from 1 to 60 min (Fig. 11.25). The nondestructive nature of NTP was evident from the SEM micrographs. There was no effect on the morphological properties of the polymer after treatment.

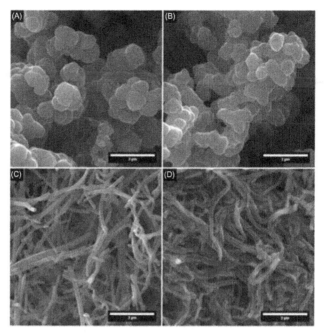

Figure 11.25 Morphology of PPy observed by scanning electron microscopy. (A) as-prepared PPy-G; (B) PPy-G after 50 min of plasma treatment; (C) as-prepared PPy-NT; and (D) after 50 min of plasma treatment [39]. *Galář, P., Khun, J., Kopecký, D., Scholtz, V., Trchová, M., Fučíková, A., . . . & Fišer, L. Influence of non-thermal plasma on structural and electrical properties of globular and nanostructured conductive polymer polypyrrole in water suspension. Sci. Rep. 2017, 7 (1), 15068. Copyright (2017), Springer Nature.*

From further studies it was found that NTP influenced the elemental composition and physical properties of the polymer.

Plasma treatment has been found to be an effective technique for the surface modification of biomaterials thereby leading to their enhanced performance in terms of biocompatibility and mechanical properties [40]. Parameters like roughness, morphology, charge, chemical composition, surface energy, and wettability affect the interactions of the biomaterial with the biological environment [41]. Nitrogen and argon plasma treatments were explored in order to improve the biocompatibility of chitosan membranes [42]. In vitro studies were performed to access the viability of L929 fibroblast-like cells on the membrane (Fig. 11.26). It could be seen that nitrogen and argon plasma treated membranes promoted higher cell viability than untreated chitosan membranes. Moreover, the fibroblast like cells were able to attach and stretch on nitrogen and argon plasma

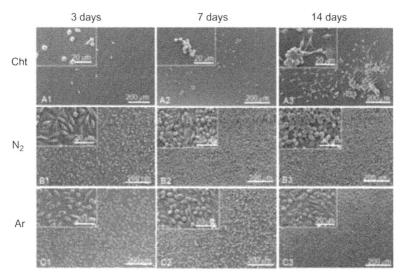

Figure 11.26 SEM micrographs of L929 fibroblast-like cells cultured on: (A) Cht (untreated membrane control); (B) ChtP2 (chitosan membranes modified by nitrogen plasma); (C) ChtP6 (chitosan membranes modified by argon plasma), after 3, 7 and 14 d of culture. *Reprinted from Silva, S.S., Luna, S.M., Gomes, M.E., Benesch, J., Pashkuleva, I., Mano, J.F., Reis, R.L, Plasma surface modification of chitosan membranes: characterization and preliminary cell response studies, Macromol. Biosci, 8(6), 568–576, Copyright (2008), with permission from John Wiley and Sons.*

modified membranes whereas poor cell attachment was seen for the untreated membranes.

11.3.4 Transmission Electron Microscopy of Plasma Activated and Plasma Treated Polymeric Materials

Transmission electron microscopy (TEM) is one of the versatile techniques which are commonly implemented in the study of polymeric materials. TEM utilizes electrons with high energy to provide compositional, morphological, and crystallographic information about the polymeric samples. TEMs are the most powerful microscopes which can make a maximum potential magnification of one nanometer and which can also produce 2D high-resolution images which can be used for a wide range of scientific, educational, and industrial applications (Figs. 11.27 and 11.28).

In plasma-activated polymers, one can produce only a very small number of solid products and along which are insoluble in organic solvents due to their high degree of crosslinking. These characteristics of plasma polymers cause complications in several analytical methods which are generally

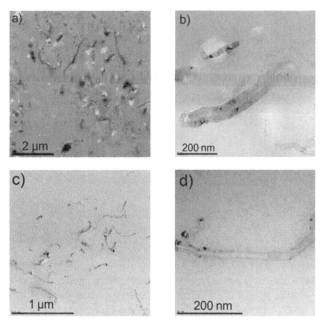

Figure 11.27 TEM images of microtomed epoxy composite reinforced by (A, B) untreated 60−100 nm diameter MWCNTs, and (C, D) plasma nanocoated 10−30 nm diameter MWCNTs at various magnifications [44]. Acknowledgments. *Copyright (2011), Polymers.*

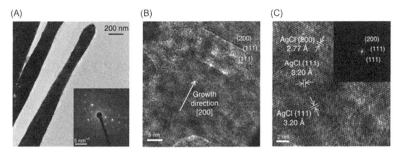

Figure 11.28 (A) transmission electron microscopy (TEM) images of single AgCl nanorod and selected area electron diffraction (SAED) patterns (inset). SAED pattern recorded along [011] zone axis. (B, C) High-resolution TEM (HR-TEM) image of AgCl nanorod. Nanorod grows along [200] direction. (111) and (200) planes of AgCl are noted [45]. Acknowledgments. *Copyright (2017), Springer Nature.*

used for conventional polymers. As matter of fact, we need to implement more sophisticated tools like TEM [43]. TEM has a crucial role in the direct evaluation of composites morphology of plasma polymers. There is no other technique except TEM which can provide direct information with high accuracy about the structure, dispersion of state, and adhesion state between particles the polymers in the nanometer regime.

The main advantage of implementing TEM for the analysis of plasma-treated polymers is its powerful magnification. Apart from this, information regarding the elemental and compound structure of the polymeric structure can be extracted from the TEM technique.

Compared to other analysis methods, TEM provides high quality and detailed images of the materials which are able to provide information about the surface properties, shape, size, and structure. Thus, TEM analysis is an important approach in most of the research reported in journals recently. Andrew C. Ritts et al. reported the implementation of TEM analysis to show the uniform dispersion of plasma-treated multiwalled carbon nanotubes (MWCNTs) in an epoxy matrix [44]. The average length of the MWCNTs are also predicted in this experiment with the help of TEM analysis.

Jae Yong Park et al. [45] reported on the characterization of the crystal structure of single-crystal silver chloride nanorods which is fabricated using a C-l2 plasma on Ag-coated polyimide. HR-TEM images of the AgCl NRs show well-aligned rock salt single crystal formation and the selected area electron diffraction (SAED) patterns show the defects formed inside during the structure formation.

Apart from these advantages, there are also some disadvantages for TEM analysis which eventually affect the implementation of the TEM analysis in plasma polymers. The main reasons for this are the size and the operating cost of the equipment, laborious sample preparation which sometimes results in the sample structure alteration during preparation process, potential artefacts from sample preparation, requirement of special training for the analysis and operations, and the analysis is limited to only those kinds of samples that are electron transparent, are able to tolerate the vacuum chamber and also small enough to fit inside the chamber, and, finally, the images are black and white. Choon-Sang Park et al. [46] reported one of these drawbacks in their work. The work demonstrates the study on the fabrication of the plasma-polymerized aniline (pPANI) nanofibers and nanoparticles by an intense plasma cloud-type atmospheric pressure plasma jets (iPC-APPJ) device with a single bundle of three glass tubes. The particle sizes monitored using TEM are observed to be smaller than those measured using scanning electron microscope (SEM), which is clearly due to the sample preparation method used for the TEM analysis.

Instead of all these kind of fallbacks, TEM analysis is one of the most powerful microscopic tools for plasma polymers available till date which is capable of producing high-resolution detailed images in an order of 1 nm.

11.4 CONCLUSION

The plasma treatment, or modification, of polymers have a plethora of applications. Microscopic analysis takes track of the extend of the modification of polymers. The technology is ever growing and recent additions to microscopic techniques are coming out of the industry which gives new insight into plasma modification. Moreover, the improvements in the present technology overcome the hurdles faced by the microscopes and high-end modern gadgets are becoming more handy and user-friendly. This, in turn, will boost the practical application of polymers into our daily lives in the benefit of humanity for a better tomorrow.

REFERENCES

[1] V.R. Gowariker, N.V. Viswanathan, J. Sreedhar, Polymer Science, ninth ed., Newage international Limited, Delhi, 1986.
[2] Injectable Foam Could Prevent Fatal Blood Loss in Wounded Soldiers. http://releases.jhu.edu/2014/07/10/injectable-foam-could-prevent-fatal-blood-loss-in-wounded-soldiers/.
[3] S.A. Bencherif, R.W. Sands, D. Bhatta, P. Arany, C.S. Verbeke, D.A. Edwards, et al., Injectable preformed scaffolds with shape-memory properties, Proc. Natl. Acad. Sci. (2012).
[4] P. Reuell, Soft Robots Go for Color, The Harvard Gazette, Camouflage, 2012.
[5] W. Liu, J. McDaniel, X. Li, D. Asai, F.G. Quiroz, J. Schaal, et al., Brachytherapy using injectable seeds that are self-assembled from genetically encoded polypeptides in situ, Cancer Res. 72 (22) (2012). 5956 LP-5965.
[6] J. Storsberg, Artificial cornea gives the gift of vision https://www.fraunhofer.de/en/press/research-news/2012/october/artificial-cornea-gives-the-gift-of-vision.html.
[7] A. Vesel, M. Mozetic, New developments in surface functionalization of polymers using controlled plasma treatments, J. Phys. D. Appl. Phys. 50 (29) (2017). Available from: https://doi.org/10.1088/1361-6463/aa748a.
[8] D. Hegemann, H. Brunner, C. Oehr, Plasma treatment of polymers for surface and adhesion improvement, Nucl. Instrum. Methods Phys. Res. Sect. B Beam Interact. Mater. Atoms 208 (2003) 281−286.
[9] M. Ozdemir, C.U. Yurteri, H. Sadikoglu, Physical polymer surface modification methods and applications in food packaging polymers, Crit. Rev. Food. Sci. Nutr. 39 (5) (1999) 457−477.
[10] K. Bazaka, M.V. Jacob, R.J. Crawford, E.P. Ivanova, Plasma-assisted surface modification of organic biopolymers to prevent bacterial attachment, Actabiomaterialia 7 (5) (2011) 2015−2028.

[11] K. Ellinas, A. Tserepi, E. Gogolides, From superamphiphobic to amphiphilic polymeric surfaces with ordered hierarchical roughness fabricated with colloidal lithography and plasma nanotexturing, Langmuir 27 (7) (2011) 3960−3969.

[12] K. Tsougeni, N. Vourdas, A. Tserepi, E. Gogolides, C. Cardinaud, Mechanisms of oxygen plasma nanotexturing of organic polymer surfaces: from stable super hydrophilic to super hydrophobic surfaces, Langmuir 25 (19) (2009) 11748−11759.

[13] C. Oehr, Plasma surface modification of polymers for biomedical use, Nucl. Instrum. Methods Phys. Res. Sect. B Beam Interact. Mater. Atoms 208 (2003) 40−47.

[14] S. Yoshida, K. Hagiwara, T. Hasebe, A. Hotta, Surface modification of polymers by plasma treatments for the enhancement of biocompatibility and controlled drug release, Surf. Coat. Technol. 233 (2013) 99−107.

[15] D. Hetemi, J. Pinson, Surface functionalisation of polymers, Chem. Soc. Rev. 46 (2017) 5701−5713.

[16] H.S. Choi, Y. Kim Young-Sun, Y. Zhang, S. Tang, S.W. Myung, B.C. Shin, Plasma-induced graft co-polymerization of acrylic acid onto the polyurethane surface, Surf. Coat. Technol. 182 (1) (2004) 55−64.

[17] P.M. López-Pérez, A.P. Marques, R.M.P.D. Silva, I. Pashkuleva, R.L. Reis, Effect of chitosan membrane surface modification via plasma induced polymerization on the adhesion of osteoblast-like cells, J. Mater. Chem. 17 (38) (2007) 4064. Available from: https://doi.org/10.1039/b707326g.

[18] A. Cifuentes, S. Borrós, Comparison of two different plasma surface-modification techniques for the covalent immobilization of protein monolayers, Langmuir 29 (22) (2013) 6645−6651. Available from: https://doi.org/10.1021/la400597e.

[19] D.V. Bax, A. Kondyurin, A. Waterhouse, D.R. McKenzie, A.S. Weiss, M.M.M. Bilek, Surface plasma modification and tropoelastin coating of a polyurethane co-polymer for enhanced cell attachment and reduced thrombogenicity, Biomaterials 35 (25) (2014) 6797−6809. Available from: https://doi.org/10.1016/j.biomaterials.2014.04.082.

[20] S. De, R. Sharma, S. Trigwell, B. Laska, N. Ali, M.K. Mazumder, Plasma treatment of polyurethane coating for improving endothelial cell growth and adhesion, J Biomater. Sci. Polym. 16 (8) (2005).

[21] Y. Wang, L. Lu, Y. Zheng, X. Chen, Improvement in hydrophilicity of PHBV films by plasma treatment, J. Biomed. Mater. Res. - Part A 76 (3) (2006) 589−595. Available from: https://doi.org/10.1002/jbm.a.30575.

[22] N. Vandencasteele, F. Reniers, Surface characterization of plasma-treated PTFE surfaces: an OES, XPS and contact angle study, Surf. Interface Anal. 36 (8) (2004) 1027−1031.

[23] E.F. Castro Vidaurre, C.A. Achete, F. Gallo, D. Garcia, R. Simão, A.C. Habert, Surface modification of polymeric materials by plasma treatment, Mater. Res. 5 (1) (2002) 37−41. Available from: https://doi.org/10.1590/S1516-14392002000100006.

[24] M. Sharma, S. Gao, E. Mäder, H. Sharma, L.Y. Wei, J. Bijwe, Carbon fiber surfaces and composite interphases, Compos. Sci. Technol. 102 (2014) 35−50.

[25] P. Slepička, N.S. Kasálková, E. Stránská, L. Bačáková, V. Švorčík, Characterization of plasma treated polymers for applications as biocompatible carriers, Express Polym. Lett. 7 (6) (2013).

[26] P. Slepička, S. Trostová, N. Slepičková Kasálková, Z. Kolská, P. Sajdl, V. Švorčík, Surface modification of biopolymers by argon plasma and thermal treatment, Plasma. Process. Polym. 9 (2) (2012) 197−206.

[27] P. Slepička, L. Peterková, S. Rimpelová, A. Pinkner, N.S. Kasálková, Z. Kolská, et al., Plasma activated perfluoroethylenepropylene for cytocompatibility enhancement, Polym. Degrad. Stab. 130 (2016) 277−287.

[28] L.F. Liu, S.C. Yu, L.G. Wu, C.J. Gao, Study on a novel polyamide-urea reverse osmosis composite membrane (ICIC-MPD): II. Analysis of membrane antifouling performance, J. Memb. Sci. 283 (1−2) (2006) 133−146.

[29] S. Pal, S.K. Ghatak, S. De, S. DasGupta, Evaluation of surface roughness of a plasma treated polymeric membrane by wavelet analysis and quantification of its enhanced performance, Appl. Surf. Sci. 255 (5) (2008) 2504–2511.

[30] D. Liu, P. Chen, M. Chen, Z. Liu, Surface modification of high performance PBO fibers using radio frequency argon plasma, Surf. Coat. Technol. 206 (16) (2012) 3534–3541.

[31] G.H. Michler, W. Lebek, Electron microscopy of polymers. polymer morphology: principles, characterization, and processing, pp. 37–53.

[32] B.D. Tompkins, J.M. Dennison, E.R. Fisher, H$_2$O plasma modification of track-etched polymer membranes for increased wettability and improved performance, J. Memb. Sci. 428 (2013) 576–588.

[33] A. Pegalajar-Jurado, M.N. Mann, M.R. Maynard, E.R. Fisher, Hydrophilic modification of polysulfone ultrafiltration membranes by low temperature water vapor plasma treatment to enhance performance, Plasma. Process. Polym. 13 (6) (2016) 598–610.

[34] P. Das, N. Ojah, R. Kandimalla, K. Mohan, D. Gogoi, S.K. Dolui, et al., Surface modification of electrospun PVA/chitosan nanofibers by dielectric barrier discharge plasma at atmospheric pressure and studies of their mechanical properties and biocompatibility, Int. J. Biol. Macromol. (2018). Available from: https://doi.org/10.1016/j.ijbiomac.2018.03.115.

[35] C.S. Yung, N.A. Tomlin, K. Heuerman, M.W. Keller, M.G. White, M. Stephens, et al., Plasma modification of vertically aligned carbon nanotubes: superhydrophobic surfaces with ultra-low reflectance, Carbon. N. Y. 127 (2018) 195–201. Available from: https://doi.org/10.1016/j.carbon.2017.10.093.

[36] M. Gheorghiu, F. Arefi, J. Amouroux, G. Placinta, G. Popa, M. Tatoulian, Surface cross linking and functionalization of poly (ethylene terephthalate) in a helium discharge, Plasma Sour. Sci. Technol. 6 (1) (1997) 8.

[37] R. Reis, L.F. Dumée, A. Merenda, J.D. Orbell, J.A. Schütz, M.C. Duke, Plasma-induced physicochemical effects on a poly (amide) thin-film composite membrane, Desalination 403 (2017) 3–11.

[38] M. Chen, D. Liu, P. Chen, Z. Liu, Z. Ding, The interfacial adhesion of poly-p-phenylene benzobisoxazole/bismaleimide composites improved by oxygen/argon plasma treatment and surface aging effects, Surf. Coatings Technol. 207 (2012) 221–226.

[39] P. Galář, J. Khun, D. Kopecký, V. Scholtz, M. Trchová, A. Fučíková, et al., Influence of non-thermal plasma on structural and electrical properties of globular and nanostructured conductive polymer polypyrrole in water suspension, Sci. Rep. 7 (1) (2017) 15068.

[40] P.K. Chu, J.Y. Chen, L.P. Wang, N. Huang, Plasma-surface modification of biomaterials, Mater. Sci. Eng. R Reports 36 (5–6) (2002) 143–206.

[41] O. Neděla, P. Slepička, V. Švorčík, Surface modification of polymer substrates for biomedical applications, Materials (Basel) 10 (10) (2017) 1115.

[42] S.S. Silva, S.M. Luna, M.E. Gomes, J. Benesch, I. Pashkuleva, J.F. Mano, et al., Plasma surface modification of chitosan membranes: characterization and preliminary cell response studies, Macromol. Biosci. 8 (6) (2008) 568–576.

[43] B.D. Ratner, A. Chilkoti, G.P. Lopez, Plasma Deposition, Treatment, and Etching of Polymers. Plasma Deposition, Treatment, and Etching of Polymers, 1990.

[44] A.C. Ritts, Q. Yu, H. Li, S.J. Lombardo, X. Han, Z. Xia, et al., Plasma treated multi-walled carbon nanotubes (MWCNTS) for epoxy nanocomposites, Polymers (Basel) 3 (4) (2011) 2142–2155. Available from: https://doi.org/10.3390/polym3042142.

[45] J.Y. Park, I. Lee, J. Ham, S. Gim, J.L. Lee, Simple and scalable growth of AgCl nanorods by plasma-assisted strain relaxation on flexible polymer substrates, Nat. Commun. (2017) 8. Available from: https://doi.org/10.1038/ncomms15650.

[46] C.-S. Park, D.H. Kim, B.J. Shin, H.-S. Tae, J. Phillips, Synthesis and characterization of nanofibrous polyaniline thin film prepared by novel atmospheric pressure plasma polymerization technique, Materials (Basel) 9 (39) (2016) 1–12. Available from: https://doi.org/10.3390/ma9010039.

CHAPTER 12

Spectroscopic and Mass Spectrometry Analyses of Plasma-Activated Polymeric Materials

Parvathy Nancy[1], Jomon Joy[2], Jemy James[1,3], Blessy Joseph[4], Sabu Thomas[2,4] and Nandakumar Kalarikkal[1,2]

[1]School of Pure and Applied Physics, Mahatma Gandhi University, Kottayam, Kerala, India
[2]School of Chemical Sciences, Mahatma Gandhi University, Kottayam, Kerala, India
[3]FRE CNRS 3744, IRDL, University of Southern Brittany, Lorient, France
[4]International and Inter University Centre for Nanoscience and Nanotechnology (IIUCNN), Mahatma Gandhi University, Kottayam, Kerala, India

12.1 INTRODUCTION

The past few decades have witnessed an extensive increase in the use of polymer-based materials in industrial and engineering applications. Superior mechanical and thermal properties make polymers more popular than other conventional materials. However, most of the polymer surfaces are chemically inert in nature and, hence, pose difficulties in adhering with other substrates when specific applications are concerned. Plasma technologies have revolutionized the field of polymeric materials making them adequate to meet the increasing demands of the modern industry. Synthesis of plasma–activated polymeric materials has opened doors to the generation of a novel class of materials having tenability in the structure − property relationships depending on its final application. Functionalization of polymeric surfaces with desired groups of interest via plasma treatment of polymers is a well-established procedure which can be applied in various domains, whereas organic plasma polymer films (PPFs) has received fast-growing attention in biomedicine and microelectronics. Among the various processing techniques employed to modify polymer surfaces, radiation processing using plasma has been found to be very advantageous. Exposure to an electrical discharge or plasma is a common practice to improve the surface chemistry of polymer materials. This practice has been used to

Non-Thermal Plasma Technology for Polymeric Materials
DOI: https://doi.org/10.1016/B978-0-12-813152-7.00012-3

improve the properties of adhesion, wetting, printability, and biocompatibility of materials. Plasma can bring about significant changes in the surface properties of the polymers and the related mechanisms have been carefully analyzed by Liston et al. [1]. Plasma treatment is considered as an eco-friendly approach whereby the surface properties can be enhanced while maintaining the bulk properties [2,3]. Low-pressure plasma treatments find use the in food industry [4]. Surface modification of polymers by means of plasma polymerization is used in the creation of selectively permeable membranes for gas separation [5,6] and improving biocompatibility of implant materials [7−10], etc.

The most-widely used technique for characterizing plasma-treated polymers is spectroscopy. Looking at some illustrations, atmospheric plasma-treated polypropylene was analyzed by large area ToF-SIMS imaging and NMF by Gustavo et al. [11]. The multivariate calibration of ToF-SIMS and XPS data of plasma-treated polypropylene thin films were studied by Awaja et al. [12]. Plasma polymerized C: H: N: O thin films for the modulated release of antibiotic substances was reported by Kratochvíl et al., in which the surface chemical composition of magnetron sputtered C: H: N: O films were detected using XPS [13]. Vogel et al. employed ATR-FT-IR spectroscopy and gas chromatography to designate the titanium plasma spray (TPS) coatings on poly-ether-ether-ketone (PEEK) materials used for medical implants [14]. The following sections detail a wide range of spectroscopic and mass spectrometry techniques used to characterize plasma activated polymeric materials.

12.2 SPECTROSCOPIC ANALYSIS

12.2.1 Fourier Transform Infrared Spectroscopy

Plasma surface modifications are limited only to a few nanometers below the surface. Fourier transform infrared spectroscopy (FTIR) spectroscopy in attenuated total reflection (ATR) mode is one of the methods used to reveal the finer surface information. This technique is very handy to obtain the chemical changes induced by the plasma treatment and also used to characterize the polymer surface. The spectra of the treated and the untreated samples are compared to observe the changes. The plasma treatment generates a slight increase on surface roughness which can have an effect on FTIR spectra, but the most important contribution is attributed to surface functionalization. It is important to note that plasma functionalization occurs by insertion of active species (oxygen containing species) in the free radicals produced during the plasma treatment and some functionalization can be achieved after plasma treatment; when plasma-treated samples are exposed to ambient atmosphere, oxygen from the air can rapidly react with some free radicals and this contributes to intensify the surface functionalization phenomenon. Plasma treatment produces chemical changes in the polymer surfaces, which led to the improvement in the hydrophilic nature of polymer surface [15−20].

Cold plasma treatment of PP and PET polymers shows distinct absorption bands at 1548 and 1697 cm^{-1}, corresponding to COO$^-$ and CO stretching vibration, respectively, and a weak and broad absorption is also observed between 3500 and 3700 cm^{-1}, corresponding to −OH. The FTIR analysis shows that the oxygen functional groups, such as −CO, −OH, and COO$^-$, are created or increased in the polymer surface after plasma treatment, and that these chemical changes in the polymer surface led to hydrophilic improvement [19,20] (Fig. 12.1).

Guruvenket et al. studied the effect of plasma surface modification of polystyrene and polyethylene. When the untreated polystyrene spectrum is compared with the argon-treated polystyrene spectrum it is seen that a broad band occurs around 3300 and at 1600 cm^{-1}, which correspond to the hydroxyl group (−OH). When compared with the untreated polyethylene, argon plasma treated PE shows peaks at 3300, 1737, 1646, and 1535 cm^{-1}, which shows a hydrophilic transformation of the polymer surfaces. After argon plasma treatment, the activated polymer surface adsorbs the moisture that is present in the environment which is indicated by the broad band at 3300 and 1600 cm^{-1} in case argon plasma treated polystyrene and at 3300 cm^{-1} in the case of polyethylene. In case of

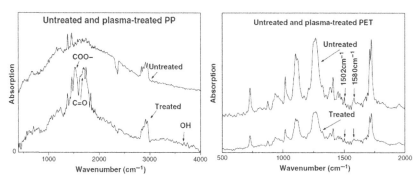

Figure 12.1 FTIR spectra of untreated and plasma-treated PP (left) and PET (right) fibers.

polystyrene treated with oxygen plasma, the peak at $1746 \, cm^{-1}$ corresponds to C = O stretching, which indicates that the surface is not only getting activated, but also the plasma reaction of oxygen with the surface to form a C = O group on the surface. When polyethylene is treated with oxygen, the band around $3500 \, cm^{-1}$ is due to the —OH group and the broadening could be due to its intermolecular or intramolecular reactions. The FTIR—ATR analysis manifests the absorption of moisture in the argon plasma polymer samples. Likewise, FTIR—ATR analysis of oxygen-treated polymer surface shows several oxygen-based functionalities at the surface (carbonyl, carboxyl, ether, and peroxide, among others). Plasma-treated samples at different exposure times clearly show the evolution of polar groups on the solid surface. These polar groups are comprised of mainly hydroxyl and carboxyl, carbonyl groups which are emanated from the interaction between the O_2 plasma and the LDPE surface [19—22]. The presence of these polar groups strongly contributes to an increase the hydrophilic nature of the LDPE surface. The presence of carbonyl groups is evident from the observation of the FTIR spectra since the C = O stretch absorbs in the range $1750—1600 \, cm^{-1}$. The presence of carboxyl groups is confirmed by the presence of a strong peak around $1650—1560 \, cm^{-1}$, which corresponds to antisymmetrical deformation of COO groups and a weak peak in the range of $1400—1310 \, cm^{-1}$ corresponding to the symmetrical deformation of those groups. Furthermore, the presence of a strong peak in the $680—580$ and $3700—3600 \, cm^{-1}$ ranges can be indicative of the presence of hydroxyl groups as a result of the O_2 plasma treatment [18,23,24]. The electron-impact dissociation of surface hydrogen atoms is due to the treatment of PE with air RF-plasma and pure oxygen plasma. The latter generates radicals on the polymer surface and unsaturated carbon—carbon bonds were formed. The strong peak at $1700—1730 \, cm^{-1}$ is an indication of the presence of the (C = O) group. A weak maximum at $1400 \, cm^{-1}$ was related to the presence of COO- groups. Strong peaks at 1255 and $1090 \, cm^{-1}$ were assigned to C-O-C group. The IR spectra at $900—950 \, cm^{-1}$ show the out-of-plane deformation of carboxyl OH as well as a small peak for alkyl peroxides at $870 \, cm^{-1}$.

Fig. 12.2 shows the FTIR spectra of LDPE powder; for the pristine polyethylene powder without plasma treatment (in black), and the LDPE plasma treated at different process gas conditions, that is, air, oxygen, and H^{2+} O_2 atmosphere. In particular, the strong bands at 2918 and $2848 \, cm^{-1}$ are assigned to asymmetric and symmetric C-H stretching

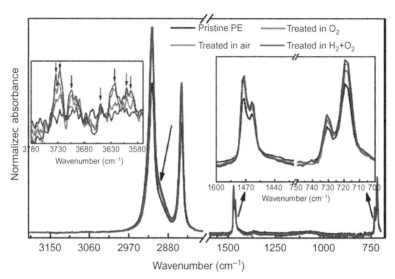

Figure 12.2 ATR-FTIR spectra of pristine and plasma treated polyethylene powder for 2 min at different working gas condition.

vibration in CH_2 respectively, while the strong bands at 1471 and 1463 cm^{-1} are assigned to the C—H bending deformation. The medium band at 731 and 719 cm^{-1} are the characteristic C—C rocking deformation. The weak band at 1376 cm^{-1} is assigned to the CH_3 symmetric deformation. The medium band at 1367 cm^{-1} is the wagging deformation. The weak band at 1306 cm^{-1} is the crumple deformation. FTIR spectra in Fig. 12.2 clearly indicates that the polar groups were formed on the LDPE powder surface. Polar groups mainly consist of carboxyl, carbonyl, and hydroxyl groups [25], which is due to the interaction of plasma with the LDPE surface. The hydrophilicity of LDPE powder surface increases due to the polar groups' contribution. The presence of a new band for plasma-treated LDPE around the region of 3750—3580 cm^{-1} may be an indication of the presence of hydroxyl groups as a result of the plasma treatment. The plasma treatments at atmospheric pressure affect only the outmost layer of the surface which is the why it is not an easy task to observe surface chemical modification by FTIR analysis. However, in the case of heavy surface oxidation of polymers their infrared spectra may present some new characteristics. The plasma treatment initiates a minimal increase in the surface roughness which can influence the FTIR spectra, but the major contribution is accredited to surface functionalization. It is important to note that plasma functionalization occurs by insertion of active species (oxygen containing species) in the free radicals generated during the plasma treatment and some functionalization can be achieved after plasma treatment; when plasma-treated samples are exposed to ambient atmosphere, oxygen from the air can rapidly react with some free radicals and this contributes to intensify surface functionalization phenomenon. So FTIR in a very good technique to identify the changes that happen after plasma modification.

12.2.2 X-Ray Photoelectron Spectroscopy

Surface-specific techniques are required to analyze polymers modified by plasma because only the top layers of the samples are modified by plasma. A variety of surface-specific techniques are available for the characterization of polymers. Among the most powerful and frequently used is X-ray photoelectron spectroscopy (XPS). XPS is an ideal technique for studying plasma-induced chemistry at polymer surfaces because of its surface sensitivity (1−10 nm), semi-quantitative nature, and the ability to obtain chemical bonding information. XPS analysis of plasma-modified samples became relatively familiar only in the 1990s with the increase in the commercially available XPS equipment. Indeed, it is useful to remember that XPS is still a relatively young method, as Siegbahn was awarded the Nobel Prize for it only in 1981. Not only have XPS instruments became easily commercially available, but the development of the technology was also a considerable factor. The development of new detectors (multiple channel tron or channel plates) led to a fabulous increase in the number of counts per second (cps) and, therefore, to a higher sensitivity and to a better signal/noise ratio. This favors the use of monochromators which are frequently necessary to unequivocally identify functionalities grafted onto a polymer surface [26−37]. A commercial XPS instrument is shown below (Kratos Axis Ultra DLD) in Fig. 12.3.

Most of the time, polymers are nonconductive. Due to the X-ray excitation mechanism, and to the ejection of the (photo) electrons from the sample, a positive charge spontaneously appears on the surface. Because of this surface charging, spectra are shifted of a few eV towards higher binding energies (lower kinetic energies). To compensate for this charging effect, some authors use an electron flood gun [38−45] or a magnetic lens [46]. If no flood gun is used, the spectra are shifted by the user with the software using internal references. Most of the time the internal reference used is the C1s peak of CC/CH bonds coming from

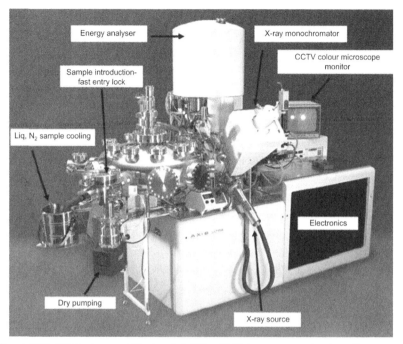

Figure 12.3 XPS instrument (Kratos Axis Ultra DLD) (www.researchgate.net).

the polymer itself or the CC signal coming from small contaminations by adventitious carbon. The C1s peak is usually the one which is strongly modified due to the change in chemical environment. On fluorinated samples some authors use the F1s signal [47−49] instead of the C1s one. The F1s peak is very symmetrical and can be fit with just one component, even for co-polymers having fluorine atoms with different neighborhoods [50]. The only difference between the different neighborhoods is the position of the F1s peak (Table 12.1).

As one of the major applications of the plasma treatment of polymers is to chemically modify their surface, the analysis of the high-resolution C1s peak (HRC1s) shape with peak fitting is a powerful tool to identify the functionalities grafted onto the sample. Depending on the chemical environment of the carbon atom, the C1s peak can present high chemical shifts making it relatively easy to identify its main components. The addition of O, N, and F substituent results in strong changes in the C1s envelope due to the various chemical environments of the carbon atoms (chemical shifts). The peak fitting can help to identify the grafted components.

Table 12.1 Binding energy of various carbon species

Chemical function	C1s B.E. (eV)
CF3−CF2	294.6
CF3−or CF2−O	294.1
−(CF2−CF2)−n	292.5
CF2 or CFO	292.2
−(CHF−CF2)−n	291.6
−(CH2−CF2)−n	290.9
CF	289.8
O−C O or HFC−CF2 or CF−CF2	289.2
CHF−CHF	288.4
C O or CF or N−C O	288.0
C O or CHF−CH2	287.9
C N	286.9
C−O or C−CF	286.5
C−O−C	286.45
C−N	286.3
(CH2)n	285.0
C	284.7
C−C	284.6

For some components, like C−N and C−O, the binding energies of the C1s peak are very close, making it almost impossible to identify the components unambiguously. In such cases, a chemical derivatization may help to identify specific species. The basic principle is to have a very specific reaction between the target functionalities to be studied and a special marker molecule. The marker molecule usually contains fluorine atoms because of the very large chemical shift it induces in the C1s binding energy. More generally, the marker must have at least one element not present in the original sample. Therefore, the target functionality can be detected unambiguously by the new element present in the spectra. Derivatization is a very powerful technique, but the overall analysis procedure is delicate, and not trivial [51,52]. Those derivatization reactions are of great interest for surfaces containing both nitrogen and oxygen. The most common groups to be derivatized are carboxyl [53−58], hydroxyl [59,60], carbonyl [60], and amine [61].

In-depth information in XPS is generally achieved by means of ion etching ("depth profiling"). On the other hand, this approach is mostly prescribed for polymers, as the information about the chemical bonds and the chemical environment will be lost by the sputtering [62].

Only the very first atomic layers of polymers will be chemically modified by the treatment with plasma. As during XPS measurements, the information is averaged out over a depth of $3-6$ nm, depending on the attenuation of the emergent electrons, the entire useful in-depth information can, therefore, be obtained by angle-resolved XPS (ARXPS). It allows discriminating between the compositions of the top layers (those modified by plasma) and the averaged out surface. XPS is well-known for quantifying the amount of polar species grafted and their nature. One of the most well-known and capable treatments to improve the surface energy of polymers and, therefore, their adhesion properties is the exposure to oxygen-containing plasma. Other gases are also used: CO_2, N_2, NH_3, air, H_2, and Ar treatment [35,57,61,63−72]. Most of them result in an increase in the oxygen content. This oxygen originates either introduced through postplasma reaction or reactions of the grafted species with oxygen or water vapor present in the atmosphere or directly from the plasma itself [73,74]. Sometimes long lifetime of radicals produced by the plasma but present at the surface or in the bulk of the polymer is still a concern for industrial applications. Unwanted posttreatment reactions resulting in a degradation of the final product due to the reaction of radicals with the environment.

XPS is also used to characterize plasma polymerized films (pp film). It is one of the few methods that provide access to the atomic composition of the films. In fact, one of the benefits of XPS (over SIMS for example) is the comparatively easy and strong atomic quantification. This can also provide information about the degradation of the monomer in the plasma and, therefore, provide information about the mechanism of reaction. Indeed, if the ratios of the peaks of the elements are the same in the pp-film and in the monomer one can conclude that there was not much degradation of the monomer functionalities by the plasma. The HRC1s peak could also be used to study fragmentation and/or damage of the monomer during the polymerization process. In the case of aromatic monomers, the Π-Π* shake-up can be used to study whether the aromatic structure of the monomer was preserved in the polymerized film. Some authors use the valence band (VB) analysis to study their plasma polymerized films [75,76]. Although usually, XPS deals with core electrons (C1s, N1s, O1s, etc.), there is also significant information at low-binding energies representative from electrons originating from the valence band of the sample. However, most of the time, XPS analysis alone is not adequate to achieve a correct chemical characterization of plasma-polymerized films. This requires a multitechnique approach [77]. A lot of fragmentation from the monomer could happen during the plasma polymerization, which result in very complex chemistry. Among the other techniques, TOF-SSIMS, infrared spectroscopy, and nuclear magnetic resonance are the most powerful and complementary to XPS.

XPS helps to detect the presence or the absence of protein on the surface of the sample. Usually the N1s signal coming from the protein is used to detect it. Some authors utilize plasma treatment to decline the adsorption of protein on membranes. The adsorption of proteins on the membrane causes a strong reduction of the permeate flux [78]. Wang et al. [79] used the N1s signal from the peptide bond of the protein to detect the adsorption of protein onto their plasma-treated membranes. The presence of the protein can also be seen on the HRC1s peak shape. Samples that do not resist the protein adsorption shows new components coming from the protein itself. Those new components are $C-C$ at 285.0 eV, $C-N$ or $C-O$ between 285.7 and 286.5 eV and $N-C$ O at 288.2 eV [80]. XPS is an essential tool to probe the physical and chemical properties of the plasma-treated polymers: adhesion, biocompatibility, and ageing are today studied using XPS in conjunction with other techniques. XPS is mostly used to probe for the identification of chemical groups grafted, through the compositional analysis and the peak fitting (mostly of the C1s peak). With the help of derivatization and ARXPS, a decent characterization of the plasma-treated surface can now be achieved. Modification of polytetrafluoroethylene (PTFE), polyvinylidene fluoride (PVDF), and polyvinyl fluoride (PVF) polymers after exposure to nitrogen and oxygen plasmas were reported by Vandencasteele et al. [81]. The XPS analysis revealed that fluorination occurred in all samples (Fig. 12.4). Except for PTFE surface treated with oxygen, grafting of multiple nitrogen or oxygen functionalities were observed.

XPS was found to be effective in studying plasma modification in polyurethane membranes treated with acrylic acid (AA) [82]. The studies clearly showed the incorporation of $C-O$ and COO groups after modification (Fig. 12.5).

12.2.3 Raman Spectroscopy

Raman spectroscopy is a nondestructive analytical technique based on the analysis of inelastically scattered light that contains information about the vibrational modes of a sample's constituent molecules [83]. Raman

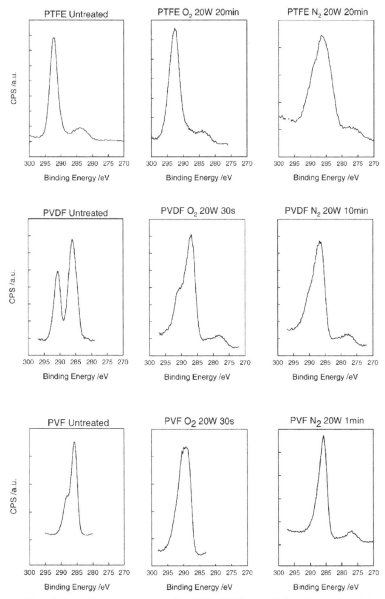

Figure 12.4 C 1 s photoelectron peak for PTFE, PVDF, and PVF untreated and treated by oxygen and nitrogen plasmas.

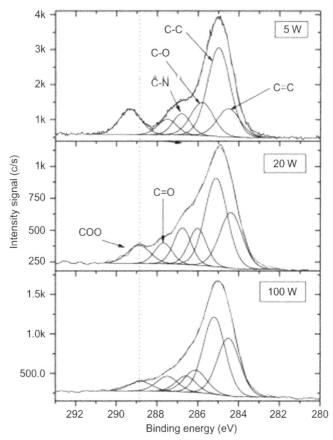

Figure 12.5 Effect of the plasma power on the high-resolution XPS spectra of the C 1 s peak of AA plasma treated PUs. Treatment time: 1 min.

spectroscopy has been effective in monitoring the variation of stress (or strain) of a short single fiber embedded in a matrix subjected to a deformation process [84,85]. The use of this technique in composite-related studies relies on the fact that when certain fibers are deformed, several Raman bands shift in a well-defined way. Raman spectroscopy studies of high-modulus (HM) carbon fibers in order to understand the effect of oxygen plasma on the structural properties of the fibers were done by Montes-Moran et al. [86]. Three untreated HM carbon fibers were studied namely: P120 and P100 pitch-based, and T50 PAN-based carbon fibers. Fig. 12.6 shows the Raman spectra of the treated and untreated T50 carbon fibers. It was evident that the spectra exhibited four well-resolved bands, namely D ($\sim 1330 \, \text{cm}^{-1}$), G ($\sim 1580 \, \text{cm}^{-1}$),

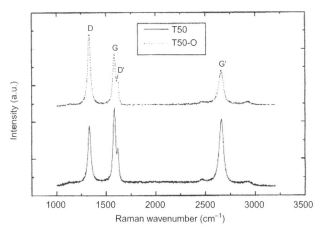

Figure 12.6 Raman spectra of the T50 fibers, untreated (top) and plasma treated (bottom).

D″ (1620 cm^{-1}), and G′ (2660 cm^{-1}) bands. Two weaker features were also identified in the second-order region of the spectra, one at around 2470 cm^{-1} and another at 2920 cm^{-1}. It was deduced that there was a considerable increase in the intensity ratio of the two first-order bands D and G of the treated fibers when compared to the neat ones. This could be due to the increase of the structural disorder, mostly through the presence of a higher proportion of edge planes in the surface of the treated fibers.

The influence of oxygen plasma treatment on the interfacial micro-mechanics of single poly (p-phenylene terephthalamide (PPTA) and poly (p-phenylene benzobisoxazole (PBO) fibers embedded in an epoxy resin was studied by Tamargo-Martínez et al. [87]. PPTA and PBO fibers exhibit poor adhesion to epoxy resins due to their inert chemical structure and smooth surface, which prevent chemical and mechanical bonding to various substrates [88,89]. Hence the aim of this work was to improve the interfacial adhesion between PPTA and PBO fibers and polymer matrices via plasma modification. The composite was subjected to a small axial tensile strain (3.5% for PPTA and 2.5% for PBO and the corresponding changes in the stress distributions along the fiber was studied using Raman spectroscopy.

It could be seen that the Raman band associated to the vibration mode of the backbone p-phenylene ring (1613 cm^{-1} for PPTA and 1618 cm^{-1} for PBO) shifted to a lower wavenumber and broadened under an axially applied stress for both types of plasma treated fibers referred as KO 4 for PPTA and ZO 4 for PBO, as shown in Fig. 12.7.

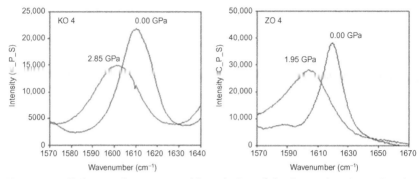

Figure 12.7 Shift in peak position and broadening of the Raman band associated to the *p*-phenylene vibration mode for an applied tensile stress specified on the graphs.

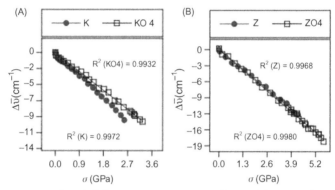

Figure 12.8 Raman band shift versus stress curves obtained for a single tested filament of PPTA (A) and PBO (B) fibers.

A lab-made stress rig, used to break single filament tabs of 50 mm in length was used to analyze the dependence of the peak position of the selected Raman bands on axial stress. The rate of stress-induced band shift at each tested length is shown in Fig. 12.8. It was concluded that under the applied stress, band shift varied linearly up to fracture, as shown in Fig. 12.8B. For KO 4 and ZO 4 of 50 mm. It has been shown that in the case of polymeric fibers having a highly oriented three-dimensional structure, the shift of the Raman band position ($\Delta\bar{\nu}$), for a given backbone vibrational mode, is only due to the chain stretching and is directly proportional to the applied stress (σ).

12.3 MASS SPECTROMETRY ANALYSIS

12.3.1 Traditional Gas Chromatography-Mass Spectrometry

Combination of gas chromatography-mass spectrometry (GC–MS) with Electron ionization (EI) or chemical ionization (CI) is the traditional and persisting method for analyzing polymers. GC–MS could be the best technique to probe the volatile, minute, and residual parts of the polymers including the residual monomers and contaminants if any. GC–MS is a bi-dimensional analysis utilizing the possibilities of gas chromatography and mass spectrometry. Almost all volatile compounds can be characterized and analyzed by GC–MS. The limitation of this technique is that it can best analyze the compounds which are volatile enough to wash from the chromatograph.

GC–MS is the most popular tool in polymer analysis to identify and characterize volatile elements and contaminants [90]. By using this application, one can identify the odor issues in industrial products. Maeno et al. characterized an odor issue of a wet poly-acrylate super absorbent polymer by using sniff port GC–MS [91,92]. Maeno et al. detected that presence of compounds with vinyl-ketone structure caused the odor issues and identified that 5-Methylhex-1-en-3-one (isobutyl vinyl ketone) was particularly mal-odorous. In case of a human observer as GC detector while doing this analysis, extra care must be taken to avoid any kind of exposure of the observer to hazardous components, may be present in the column effluent. GC–MS is widely used to characterize oligomeric compounds with low molecular weight [93]. These oligomers having low molecular mass have less potential to be actionized with metals and cannot effectively be analyzed by other mass spectrometry techniques such as matrix-assisted laser desorption/ionization (MALDI), ESI, fast atom bombardment (FAB) and secondary ion mass spectrometry (SIMS). Characterization of low molecular mass industrial ethoxylated surfonyl surfactant (2, 4, 7, 9-tetramethyl-5-decyne-4, 7-diol) named S 40, by means of GC–MS is an excellent example of this application. Another application of GC–MS is the characterization of degraded products of polymers. Pyrolysis, photolysis, and thermos chemolysis are the methods to degrade polymer materials. Pyrolysis will be discussed separately. Richer et al. characterized the photo degradation of poly (2, 6-dimethyl-1, 4-phenylene oxide) PPE polymers by using GC–MS and liquid chromatography (LC)-MS techniques. To improve the light stability of PPE, understanding of photo degradation process is necessary.

The combination of GC-MS with solid-phase micro extraction (SPME) can also be used for characterization of degraded products in polymers [94]. Hakkarainen et al. describes that SPME followed by GC-MS is more effective than GC-MS to characterize degradation products from low-density polyethylene film [95]. They treated the LDPE films with ultra violet (UV) radiation for 100 h and then did mild thermal ageing at 80°C for 5 weeks. Silicon-based fibers coated with poly (dimethyl-siloxane) (PDMS) and polar carbowax (divinyl benzene) were used for SPME. Several degradation product species such as ketones, carboxylic acids, keto acids, and furanones were detected. Explicitly, the SPME experiments detected more degradation species than the GC-MS.

12.3.2 Time-of-Flight Secondary Ion Mass Spectrometry

Time-of-flight secondary ion mass spectrometry (ToF-SIMS) has been shown to be an acceptable technique for the characterization of plasma surface-treated polyolefins, with several classes of work describing the characteristic secondary ion fragments that a treated surface can yield [96]. It has conjointly been reported that energetic surface treatments can drive the surface segregation of additives and also the presence of such molecules at the surface of a polymer sample can have a detrimental impact on the required application performance. ToF-SIMS characterization of the treated surfaces is often assisted by multivariate analysis (MVA) strategies like principal component analysis to see the important characteristic peaks after the treatment or maybe the depth of oxidation employing a dual beam depth profiling set up [97,98]. XPS is widely used for the characterization of the reactivity zones of plasma-treated polymer surfaces. No similar investigation has been done by exploiting ToF-SIMS. The spatial distribution of the treatment is far larger than the standard raster size of a ToF-SIMS primary ion beam, but modern instruments offer stage raster modes that helps imaging analysis of such large areas. The present challenges lie on the acquisition of machine-driven large space datasets of robust insulators and the post process of the big quantity of data generated.

For an example, Tof-SIMS analysis of positively charged secondary ions can be conducted using a ToF-SIMS system using 25 keV Bi^{3+} ion beam operated within 0.3 pA, high current bunched mode with 100 µs cycle time, leading to a mass varying between 0 and 694 µ. Secondary ion maps were obtained by employing the stage raster mode. One patch represents the maximum beam raster area of $500 \times 500\,\mu m^2$. The total $2 \times 2\,cm^2$ analyzed area contains 2000×2000 pixels which comprise of 40×40 patches sized 50×50 pixels each. The complete area can be scanned once with one shot per pixel, guaranteeing static conditions. A low-energy electron flood cab be used to neutralize charge build-up on the sample surface. Generally, we can perform ToF–SIMS characterization straight after surface treatment of polymer samples. After that the data can be analyzed by the MVA. Fig. 12.9 represents the NMF results with four components for the samples treated with the plasma torch at 4 mm (first row of plots) and 8 mm (second row of plots) from the surface.

The most relevant ToF–SIMS peaks found for nontreated (hydrocarbons) and treated (oxygen containing) polypropylene surfaces are depicted in Table 12.2.

Both the plasma-treated samples exhibit very similar results even though their datasets have undertaken individual factorizations. Results

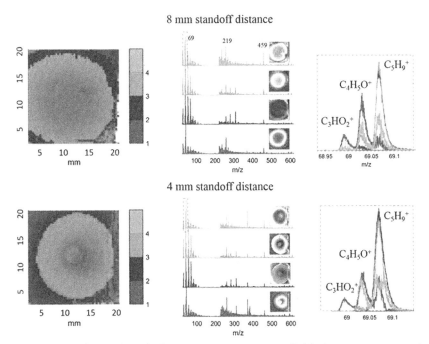

Figure 12.9 Left: overlay of all components intensities (field-of-view: 2 cm × 2 cm). Each component is represented by a different color with opacity proportional to their intensities. Right: characteristic spectra of components. The inset images are the individual maps of each component.

Table 12.2 ToF-SIMS peaks found for nontreated (hydrocarbons) and treated (oxygen containing) polypropylene surface

Fragment	Mass
CHO^+	29.001
$C_2H_5^+$	29.038
CH_3O^+	31.017
CH_5O^+	33.034
$C_2H_2O^+$	42.008
$C_3H_6^+$	42.044
$C_2H_3O^+$	43.0189
$C_3H_7^+$	43.0559
CHO_2^+	44.9958
$C_2H_5O^+$	45.0329
$C_3H_3O^+$	55.0228
$C_4H_7^+$	55.0583
$C_3H_5O^+$	57.0373
$C_4H_9^+$	57.0768
$C_2H_3O_2^+$	59.014
$C_3H_7O^+$	59.0511
$C_3HO_2^+$	68.9968
$C_4H_5O^+$	69.035
$C_5H_9^+$	69.0754

show the datasets from the two samples had very different channel to mass calibrations (due to charging effects) and different peak shapes. Hence ToF-SIMS can be chosen for the surface and in–depth analysis of plasma-modified polymer surfaces because it is the only surface technique possessing extremely good selectivity and sensitivity.

12.3.3 Static Secondary-Ion Mass Spectrometry (Static SIMS)

Static secondary ion mass spectrometry is the technique which comprises of several succeeding steps such as primary ion bombardment, energy transfer, particle desorption, particle mass analysis, and detection. In SIMS technique, the plasma-treated polymer is annihilated with a primary ion beam, usually Xe, Ar, Ga, or Cs in the range of 5−25 kV. The annihilation of primary ions will trigger the process collision cascade in which energy and momentum is transferred into the sample. Thus, the collision cascade culminates in the desorption of secondary electrons, ions, and neutral species. These secondary products are mass analyzed and detected. The points proximal to the primary ion impact site where the energy level will be maximum and desorbs ions and electrons; distal from the impact site molecules are desorbed. There are two types of SSIMS which are being used to analyze mass spectra of polymers in terms of the samples used, such as thin layer samples and thick layer samples. In both these experiments there is no need of a solution or matrix to interpret the sample. In this method, solid samples can be analyzed. Charge compensation is required to obtain proper mass spectra of solid samples, since most of the solid polymers are insulators. By using SSIMS molecular weight and chemical structure of the plasma-activated polymer can be revealed. A significant volume of information regarding SSIMS is available in the literature and in review articles by Van Vaeck, Adriaens, and Gijbels, Adriaens, VanVaeck, and Adams, Wien, Bertrand, and Weng [99−102]. Also matrix enhanced secondary ion mass spectrometry (MESIMS) is a version of SSMIS, in which the samples are prepared the same as that of MALDI. The matrix effects are rationalized to enhance the secondary ion signal. This method is widely used in probing low-molecular weight polymers and MALDI samples. It is noted that the sample surface is greatly dependent on the composition of matrix and polymer. We can find the relative solubility of different MALDI matrices by using MESIMS while probing the sample surfaces. This is a time-efficient method to analyze plasma modified polymer surfaces.

12.4 CONCLUSION

Plasma modification has been found to be an effective strategy for imparting surface functionalities to polymers. The modified polymer exhibits enhanced surface properties when compared to the pristine polymer. This chapter provides insight into the various spectroscopic techniques used for characterizing the plasma-activated polymers. Varying the plasma parameters leads to varied functionalities and changes in the treated polymers. Hence a combination of various spectroscopic approaches are essential in order to investigate the plasma processes. The surface sensitivity of XPS makes it the preferred choice for understanding the surface chemistry of plasma-irradiated polymers in combination with other techniques. Other techniques that have proved effectiveness to probe the plasma−surface interaction are ToF-SIMS, Static SIMS, and ME-SIMS. GC-MS has been used for the characterization of volatile compounds. The limitation with this technique is that the material to be characterized should be sufficiently volatile. FTIR, Raman spectroscopy, etc., have all been extensively used for studying the functional enhancement of plasma-activated polymers. Briefly, plasma treatment modifies the surface properties of the polymer without affecting the quality of the material as such.

REFERENCES

[1] E.M. Liston, L. Martinu, M.R. Wertheimer, J. Adhesion, Sci. Technol. 7 (10) (1993) 1091−1127.
[2] Y. Kusano, J. Adhesion 90 (9) (2014) 755−777.
[3] J. Cognard, Compt. Rend. Chim. 9 (1) (2006) 13−24.
[4] Ulbin-Figlewicz, et al., Ann. Microbiol. 65 (3) (2015) 1537−1546.
[5] Stancell, A.F., Spencer, A.T., Separating Fluids with Selective Membranes. US Patent 3,657,113, 1972.
[6] Anand, M., Costello, C.A., Campbell, K.D., Plasma-Assisted Polymerization of Monomers onto Polymers and Gas Separation Membranes Produced Thereby. US Patent 5,013,338, 1991.
[7] Narayanan, P.V., Rowland, S.M., Stanley, K.D., Treatment of Metallic Surfaces Using Radiofrequency Plasma Deposition and Chemical Attachment of Bioactive Agents. US Patent 5,336,518, 1994.
[8] Marchant, R.E., Process for Producing Hydroxylated Plasma-Polymerized Films and the Use of the Films for Enhancing the Compatiblity of Biomedical Implants. US Patent 5,112,457, 1992.
[9] Hoffman, A.S., Patel, A.S., Polyethylene Oxide Coated Intraocular Lens. US Patent 5,618,316, 1997.
[10] Okamura, M., Nakajima, I., Intraocular Implant Having Coating Layer. US Patent 5,007,928, 1991.
[11] Gustavo, et al., Surf. Interface Anal. (2018) 1−7.
[12] Awaja, et al., Plasma Process. Polym. (2014) 745−754.
[13] Kratochvíl, et al., Plasma Process Polym. (2018).
[14] Vogel, et al., J. Mech. Behav. Biomed. Mater. 77 (2018) 600−608.
[15] K. Tanaka, et al., Thin Solid Films 386 (2001) 217−221.
[16] Fridman, Surf. Coat. Technol. 131 (2000) 528−541.
[17] O.J. Kwon, et al., Surf. Coat. Technol. 192 (2005).
[18] C. Cheng, et al., Coat. Technol. 200 (2006) 6659−6665.
[19] S. Guruvenket, et al., Appl. Surf. Sci. 236 (2004) 278−284.
[20] R. Hansen, et al., Polym. Sci. Part 4 (1966) 203−209.
[21] I.H. Coopes, et al., Sci. Part A - Chem. 17 (1982) 217−226.
[22] S. Guruvenket, et al., J. Appl. Polym. Sci. 90 (2003) 1618−1623.
[23] S.K. Øiseth, et al., Appl. Surf. Sci. 202 (2002) 92−103.
[24] N.V. Bhat, et al., J. Appl. Polym. Sci. 86 (2002) 925−936.
[25] N. Patra, et al., Polym. Degrad. Stab. 98 (2013) 1489−1494.
[26] H. Yasuda, et al., J. Polym. Sci. 15 (1977) 991−1019.
[27] J.M. Burkstrand, J. Vac. Sci. Technol. 1 (5) (1978) 223−226.
[28] Dilks, J. Polym. Sci. 19 (1981) 1319−1327.
[29] D.T. Clark, et al., J. Polym. Sci. 21 (1983) 837−853.
[30] Dilks et al, in: K. L. Mittal (Ed.), Physiocochemical Aspects of Polymer Surface, (vol. 2, pp. 749−772). Plenum Press, New York.
[31] R.G. Nuzzo, et al., Macromolecules 17 (1984) 1013−1019.
[32] J.F. Evans, et al., Anal. Chem. 319 (1984) 841−849.
[33] L.J. Gerenser, J. Adhesion Sci. Technol. 1 (1987) 303−318.

[34] T.G. Vargo, et al., J. Polym. Sci. A., Polym. Chem. 27 (1989) 1267—1286.
[35] J.E. Klemberg-Sapieha, et al., J. Vac. Sci. Technol. 9 (1991) 2975—2981.
[36] R. Foerch, et al., J. Adhesion Sci. Technol. 5 (1991) 549.
[37] R. Foerch, et al., Surface Interface Anal. 17 (1991) 847—854.
[38] Larrieu, et al., Surf. Coat. Technol. 200 (2005) 2310—2316.
[39] M. Atreya, et al., Polym. Degrad. Stab. 65 (1999) 287—296.
[40] M. Dhayal, et al., Appl. Surf. Sci. 252 (2006) 7957—7963.
[41] L.J. Gerenser, et al., Surf. Interf. Analysis 29 (2000) 12—22.
[42] S. Vasquez Borucki, et al., Surf. Coat. Technol. 138 (2001) 256—263.
[43] D. Wolany, et al., Surf. Interf. Anal. 27 (1999) 609—617.
[44] K.J. Kim, et al., Thin Solid Films 398—399 (2001) 657—662.
[45] C.E. Moffitt, et al., Plasma Polym. 6 (2001) 193—209.
[46] Walker, et al., Charged Particle Energy Analyzers, Kratos Analytical Ltd., UK, 1994.
[47] J. Carpentier, et al., Surf. Coat. Technol. 192 (2005) 189—198.
[48] Sugimoto, et al., Polymer 41 (2000) 511—522.
[49] V. Yanev, et al., Surf. Sci. 566—568 (2004) 1229—1233.
[50] A.M. Ferraria, et al., Polymer 44 (2003) 241—7249.
[51] Chilkoti, et al., Surf. Interf. Anal. 17 (1991) 567—574.
[52] Chilkoti, et al., Chem. Mater. 3 (1991) 51—61.
[53] L.A. Langley, et al., Chem. Mater. 18 (2006) 169—178.
[54] R.P. Popat, et al., J. Mater. Chem. 5 (1995) 713—717.
[55] Sutherland, et al., J. Mater. Chem. 4 (1994) 683—687.
[56] M.R. Alexander, P.V. Wright, B.D. Ratner, Surf. Interf. Anal. 24 (1996). 217—.
[57] N. Inagaki, et al., J. Appl. Polym. Sci. 46 (1990) 399—413.
[58] M.R. Alexander, T.M. Duc, Polymer 40 (1999) 5479—5488.
[59] J.F. Friedrich, et al., Surf. Coat. Technol. 74—75 (1995) 664—669.
[60] G. Kühn, et al., Surf. Coat. Technol. 142—144 (2001) 494—500.
[61] F. Poncin-Epaillard, et al., Macromol. Chem. 192 (1991) 1589—1599.
[62] S. Lucas, et al., Nucl. Instr. Methods Phys. Res. B 266 (2008) 2494.
[63] K. Ozeki, et al., Appl. Surf. Sci. 254 (2008) 1614—1621.
[64] Banik, et al., Polymer 44 (2003) 1163—1170.
[65] R. Foerch, Les vide, les couches minces 246 (1989) 213—215.
[66] M. Bryjak, et al., Europ. Polym. J. 38 (2002) 717—726.
[67] P. Favia, et al., Plasma Polym. 1 (1996) 91—112.
[68] N.-Y. Cui, et al., Appl. Surf. Sci. 189 (2002) 31—38.
[69] C. Riccardi, et al., Appl. Surf. Sci. 211 (2003) 386—397.
[70] D.T. Clark, et al., J. Polym. Sci. 16 (1978) 911—936.
[71] V. Stelmashuk, et al., Vacuum 77 (2005) 131—137.
[72] J.C. Caro, et al., Europ. Polym. J. 35 (1999) 1149—1152.
[73] S. O'Kell, et al., Surf. Interf. Anal. 23 (1995) 319—327.
[74] A.J. Wagner, et al., Plasmas Polym. 8 (2003) 119—134.
[75] G. Maggioni, et al., Surf. Coat. Technol. 200 (2005) 481—485.
[76] J.F. Friedrich, et al., Surf. Coat. Technol. 142—144 (2001) 460—467.
[77] I. Retzko, et al., J. Electr. Spectr. Rel. Phen. 121 (2001) 111—129.
[78] K.S. Kim, et al., J. Mem. Sci. 199 (2002) 135—145.
[79] P. Wang, et al., J. Mem. Sci. 195 (2002) 103—114.
[80] C.D. Tidwell, et al., Surf. Interf. Anal. 31 (2001) 724—733.
[81] Vandencasteele, et al., XPS and contact angle study of N2 and O2 plasma-modified PTFE, PVDF and PVF surfaces, Surf. Interf. Anal. 38 (4) (2006) 526—530.
[82] Weibel, et al., Surface modification of polyurethane membranes using acrylic acid vapour plasma and its effects on the pervaporation processes, J. Membr. Sci. 293 (1—2) (2007) 124—132.

[83] Atkins, et al., Appl. Spectr. 71 (5) (2017) 767—793.
[84] Young, R.J., Evaluation of composite interfaces using Raman spectroscopy. In *Key Engineering Materials* (Vol. 116, pp. 173—192).
[85] Bennett, et al., Plast. Rubber Compos. 32 (5) (2003) 199—205.
[86] M.A. Montes-Moran, R.J. Young, Carbon 40 (6) (2002) 845—855.
[87] Tamargo Martinez, et al., Compos. Sci. Technol. 71 (6) (2011) 784—790.
[88] Park, et al., J. Colloid Interf. Sci. 264 (2) (2003) 431—445.
[89] Luo, et al., J. Macromol. Sci. Part B 45 (4) (2006) 631—637.
[90] M.J. Stadler, Am. Lab. (2000, Jan) 7—8.
[91] Dravineks, et al., J. Agric. Food Chem. 19 (1971) 1049—1056.
[92] Maeno, et al., J. Chromatogr 849 (1999) 217—224.
[93] Parees, et al., J. Am. Soc. Mass Spectrom. 9 (1998) 282—291.
[94] Arthur, et al., J. Anal. Chem. 62 (1990) 2145.
[95] Hakkarainen, et al., Polym. Degrad. 5 (1997) 67—73.
[96] Williams, et al., Int. J. Adhesion Adhesives (2015) 1—23.
[97] Awaja, et al., Surf. Interf. Anal. (2008) 1454—1462.
[98] Cristaudo, et al., Plasma Process. Polym (2016) 1104—1117.
[99] Van Vaeck, et al., Mass Spectrom. Rev. 18 (1999) 1—47.
[100] Adriaens, et al., Mass Spectrom. Rev. 18 (1999) 48—81.
[101] Wien, et al., Methods Phys. Res. B 131 (1997) 38—54.
[102] Bertrand, et al., Mikrochim. Acta 13 (1996) 167—182.

FURTHER READING

Baudewyn L. Baudewyn, Dielectric barrier discharge for surface treatment, 2008.
Cumpson, 1995 P.J. Cumpson, J. Electr. Spectr. Rel. Phen. 73 (1995) 25—52.
Holländer 2008 Holländer, S. Kröpke, F. Pippig, Surf. Interf. Anal. 40 (2008) 379—385.
Jeong et al., 1998 J.Y. Jeong, et al., Plasma Sour. Sci. Technol. 7 (1998) 282—285.
Kim et al., 2007 J. Kim, et al., Appl. Surf. Sci. 253 (2007) 4112—4118.
Koinuma et al., 1992 H. Koinuma, et al., Appl. Phys. Lett. 60 (1992) 816—817.
Paynter, 2002 R.W. Paynter, Surf. Interf. Anal. 33 (2002) 14—22.
Tsai et al., 1997 P.P. Tsai, et al., Text. Res. J. 67 (1997) 359—369.

CHAPTER 13

Plasma Treatment of High-Performance Fibrous Materials

Marija Gorjanc
Faculty of Natural Sciences and Engineering, University of Ljubljana, Ljubljana, Slovenia

13.1 INTRODUCTION

In the second half of the twentieth century, with the advances in polymer chemistry and textile engineering, new fibrous materials have emerged that perform chemically and physically much better than natural ones. These high-performance fibrous polymers currently prevail as components in many products that need to meet critical and sophisticated performance requirements. That is why high-performance fibrous materials are used in biomedicine (artificial organs, prosthetic devices, antimicrobial sutures), transportation (lightweight aircrafts, space shuttle shields), protective clothing (bulletproof, fireproof, cut-resistant, protection from nerve, and toxic gases), geotextiles (road-bed liners, soil erosion protection, pond liners for water preservation), architectural and construction engineering (transparent rooftops, concrete reinforcement), high-performance sporting goods, filters (for air, water, gas, and chemical), and separators in batteries. However, due to almost perfect crystallinity and orientation, these fibrous materials have extremely low adhesion properties and are chemically inert. These properties are a drawback when used in a matrix with other materials because of decreased interfacial interactions. Moreover, the absence of certain functional groups and high hydrophobicity of high-performance fibrous materials limit their use in regenerative medicine since cell attachment and growth cannot be ensured. Currently, one of the most promising and environmentally benign methods to overcome the drawbacks of high-performance fibrous materials is plasma technology. By using a gaseous plasma at nonequilibrium conditions, we can tailor the polymer surface properties without affecting the bulk properties. This chapter will provide information on the plasma treatment of high-performance fibrous

Non-Thermal Plasma Technology for Polymeric Materials
DOI: https://doi.org/10.1016/B978-0-12-813152-7.00013-5

341

materials, such as glass, aramid, carbon, polyethylene, and poly(p-pheny-lene benzobisoxazole).

13.2 INTERACTION BETWEEN A PLASMA AND HIGH-PERFORMANCE FIBROUS MATERIALS

The main purpose of treating high-performance fibrous materials with a plasma is to change their properties, especially wetting behavior, structure, and chemical composition [1]. These changes are crucial for the improvement of the adhesion properties of high-performance fibrous materials as they are often used for reinforcement in the composites. Plasma modification promotes chemical surface changes which can be affected by further treatment. The longer exposure of fibrous material to a plasma leads to an etching effect. In Fig. 13.1, a schematic illustration of plasma modification mechanisms is presented. The chemical and morphological changes of high-performance fibers after plasma modification are analyzed with Fourier transform infrared spectroscopy (FTIR), X-ray photoelectron spectroscopy (XPS), Time-of-flight secondary ion mass spectrometry (ToF–SIMS), scanning electron microscopy (SEM), and atomic force microscopy (AFM). In some cases, the surface changes are analyzed measuring the wettability of the samples by measuring dynamic and static contact angles, surface free energy, and performing thermogravimetric

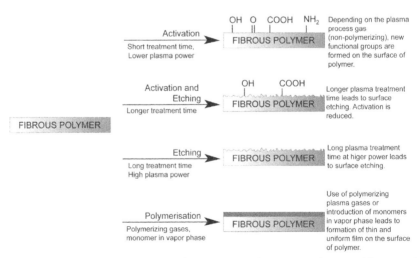

Figure 13.1 Schematic illustration of mechanism of plasma surface modification of fibrous polymers.

analysis. Since the majority of research on the plasma treatment of high-performance fibers has been performed for increasing the adhesion of fibrous materials in the matrix, the adhesion properties of composites were tested by analyzing the interlaminar shear strength (ILSS) and interfacial shear strength (IFSS).

13.2.1 Plasma Treatment of Glass Fibers

Glass is the oldest and most familiar performance fiber. Fibers have been manufactured from glass since the 1930s. Although early versions of glass fibers were strong, they were relatively inflexible and not suitable for many applications. Today's glass fibers offer a much wider range of properties and can be found in a wide range of end uses such as insulation batting, fire-resistant fabrics, and reinforcing materials for plastic composites [2]. Items such as bathtub enclosures and boats, often referred to as "fiberglass" are in reality plastics (often crosslinked polyesters) with glass fiber reinforcement. The chemical composition of glass determines the physico-chemical properties of glass fibers, such as the chemical resistance and electrical and optical properties [3]. The production of glass fibers involves the high-temperature conversion of various raw materials (predominantly borosilicates) into a homogeneous melt, followed by the fabrication of this melt into glass fibers. There are several types of known glass fibers [4]: type A (neutral) is made from "waste" glass and has a low chemical resistance and low strength; type E (electric insulating) has a high thermal and moisture resistance and is used in manufacturing structural, insulating, electrical materials; type C (chemical resistant) has a high chemical resistance to acids and alkalis; type D (low dielectric permittivity) has very low dielectric parameters, low strength, and chemical resistance and it is used as reinforcing material in electronic boards and radar housing; and type S (high-strength, high-modulus) has high strength and elasticity moduli parameters necessary to develop composites with high mechanical properties. The fiber-reinforced composites are strongly influenced by the affinity between the fiber and matrix polymer. Better adhesion of glass fibers in the polymer matrix is achieved by wet-chemical procedures, which sometimes lower the strength of the fibers, and the technique requires the treatment of waste chemicals. An alternative to wet-chemical treatments is plasma modification.

Only a few studies were published on treating glass fibers with a plasma. Iriyama et al. [5] treated glass fabrics in a bell-jar reactor with a 40 kHz frequency and a power of 70 W. Four nonpolymer-forming gases, such as argon (Ar), oxygen (O_2), air, and ammonia (NH_3) were used for the modification of glass fabrics at different pressures (from 24 to 93 Pa). Each treatment was carried out for 60 min. It was found that at the lower treatment pressure, the carbon content on the fibers was lower, which suggested that chemical etching occurred at a high energy density. No greater differences in carbon to silicon (C/Si) or oxygen to silicon (O/Si) ratios were noticed between samples treated with O_2, air, and Ar. The samples treated with an NH_3 plasma exhibited a higher C/Si ratio and the presence of nitrogen (N) functional groups. As the plasma energy density increased, the greater the amount of N was introduced. The SEM analysis showed no morphological changes on the glass fibers regardless of plasma conditions. The adhesion of plasma-treated glass fabrics to epoxy resin increased for all used gases; however, samples treated with a NH_3 plasma showed constantly high tensile strengths in all pressure ranges. The oxygen (O) and nitrogen (N) rich functional groups can be introduced on the surface of fibrous-glass by an air atmospheric dielectric barrier discharge plasma (DBD) [6]. The reason is that the oxygen and nitrogen from air are excited and react with the surface of the glass fibers during the discharge process. The highest concentration of O and N on the glass fibers was found for samples treated for 180 s. Lim et al. [7] studied the dependence of discharge power (20−140 W) and treatment time (1−10 min) with a radiofrequency (RF) oxygen plasma on the modification of glass fibers. The optimal conditions, where the water contact angle on the fibers decreased from 26 to 12 degrees, were at plasma processing parameters of 100 W, 0.1 Torr, and 3 min. These samples also exhibited a lower dielectric constant and a higher tensile strength. The modification of glass fibers was also conducted using a plasma polymerization process [8,9] because the technology enables the formation of thin, uniform films on the surface of the substrate. Sever et al. [8] performed a series of depositions of the coupling agent γ-glycidoxypropyltrimethoxysilane (γ-GPS) on glass fibers using a low-frequency plasma. Afterwards, an epoxy resin and a hardener were applied onto the samples. The organic/inorganic character of the glass fibers was expressed as the C/Si ratio, which was highest for samples treated at the highest power (90 W). These samples also exhibited the highest ILSS, which means that the interfacial adhesion between the glass fiber and epoxy matrix increased after plasma polymerization. In a study by Çökeliler et al. [9], the flexural strength of the acrylic denture material increased by 36% when it was reinforced with plasma-polymerized ethylenediamine (EDA) glass fibers and by 10%

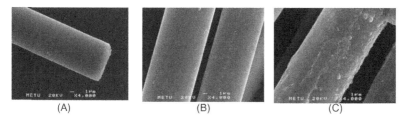

(A)   (B)   (C)

Figure 13.2 SEM images of untreated and plasma-treated glass fibers. (A) Untreated glass fiber, (B) Modified with HEMA, (C) Modified with EDA [9].

when it was reinforced with plasma-polymerized 2-hydroxyethyl methacrylate (HEMA) glass fibers. The most pronounced changes in the composition were found on EDA plasma-treated glass fiber surfaces. Additionally, the concentration of nitrogen was found to be as high as 15.5%. Surface morphological changes after plasma-polymerization are visible in Fig. 13.2. The use of monomer EDA increases the surface roughness of the glass fibers much more than monomer HEMA.

Glass fiber-reinforced polyester (GFRP) materials exhibit high strength-to-weight ratios and corrosion resistance and are mainly used in civil engineering. GFRP materials are often joined with other materials, and due to their smooth surface, they must be prepared by a chemically or a mechanically abrasive technique. Because the treatment of GFRP plates with a DBD helium plasma did not change either the surface roughness or surface chemistry [10], the samples were treated using a plasma-polymerization process with the deposition of hexamethyldisiloxane (HDMSO) and vinyltriethoxysilane (VTES) [11]. Before the plasma-polymerization of organosilicon monomers, the GFRP samples were treated with an Ar and O_2 gas plasma. The results revealed a high concentration of oxygen with respect to silicon on the VTES films. The adhesion of these samples increased with increasing effective power and was higher than those of the HMDSO samples. It was also found that treating GFRP materials with an Ar plasma increases the ILSS of the composite. The same group performed the modification of GFRP using an Ar plasma for 5 min and compared the results to samples polymerized with HMDS and VTES [12]. The IFSS increased for the Ar plasma-treated sample up to 52%, while for HMDS and VTES treated samples the shear strength increased by 27% and 25%, respectively, [13] treated the GFRP samples for 2−60 s with a DBD atmospheric plasma using a mixture of O_2/N_2 gases. The water contact angle dropped from 84 to 22 degrees after a 2 s treatment and decreased to 0 degree when the plates were treated for more than 30 s. The concentration of aluminum and oxygen increased as well as the surface roughness with the plasma treatment. The adhesive strength of plasma-treated samples was comparable or slightly higher than that achieved by conventional sand blasting.

13.2.2 Plasma Treatment of Aramid Fibers

According to the US Federal Trade Commission, aramid is a manufactured fiber in which the fiber-forming substance is a long-chain synthetic polyamide, where at least 85% of the amide ($-CONH-$) linkages are attached directly to two aromatic rings [14]. The first commercialized aramid fiber was Nomex (DuPont), a meta-aramid fiber with excellent thermal, chemical, and radiation resistance properties, which opened new uses for the fibrous material in many industrial applications, for example, aerospace, electrical, and electronic industries to produce protective apparel, hot gas filtration, automotive hoses, electrical insulation, aircraft parts, and sporting goods. A decade later, DuPont developed and commercialized Kevlar, a high-tenancy and modulus para-aramid fiber. Twaron is a trade-name of the para-aramid fiber produced by Teijin, which was commercialized in the 1980s, but was developed in 1970s. This material has similar properties as Kevlar. Another commercial aramid fiber that is produced in Russia is Armos. This material's main features are long lifetime under load (more than 15 years) and high modulus and is produced without any sizing. Armos fibers are used in fire-resistant materials, in radial tires, and as a composite in the aerospace industry.

The research on the plasma treatment of aramid polymers was performed on meta-aramid and para-aramid fibrous polymers. While meta-aramid has a high resistance to flames and chemicals, para-aramid has a high strength, modulus, and toughness and exhibits thermal stability. Nomex and Kevlar were treated with an atmospheric pressure plasma source of a diffuse coplanar surface barrier discharge, where air and nitrogen were used as the working gases [15]. The polymers were treated for 30 s, 2 and 5 min at a discharge power of 300 W and a frequency of 15 kHz. No significant morphological changes were found on Kevlar-treated polymer, while Nomex had a rougher surface after 5 min of a nitrogen-plasma treatment. The concentration of oxygen increased significantly on both polymers after plasma treatment and was higher on Nomex than Kevlar. The concentration of oxygen was higher after the air-plasma treatment than after the nitrogen–plasma treatment, whereas the concentration of nitrogen-containing groups was higher after the nitrogen–plasma treatment. Electron spin resonance (ESR) was used to investigate the free radicals trapped in aramid (Kevlar49) fibers after atmospheric pressure plasma jet (APPJ) treatments [16]. Different plasma operational conditions such as output power, gas flow rate (helium), oxygen content, treatment time, and jet-to-substrate distance were used to study the fiber moisture content. The results indicate that no significant differences in the ESR spectra line shapes were observed under different conditions; however, the amplitudes of spectra were higher in the lower fiber moisture region. The free radical concentration in the sample in the 0.5% moisture region was twice as much as that of the wet sample (4.8% moisture region). The influence of the aramid fiber moisture region on plasma treatment effectiveness and aging was further evaluated by contact angle measurements, XPS, and measurements of the bonding strength to the epoxy [17]. The results demonstrate that the hydrophobic recovery period (30 days) after plasma treatment was faster for fibers with a lower moisture region than for fibers with a higher moisture region. The reason attributed to this phenomenon is the reorientation and migration of the highly mobile polar groups introduced by the treatments. The concentration of oxygen and nitrogen increased for all samples after the plasma treatment. The highest concentrations of functional groups, even after aging, were found on fibers with a higher moisture region. The IFSS of the aramid/ epoxy composites were significantly improved after plasma treatment from 26.9 to 48.3 MPa and 55.7 MPa. Although aging caused the IFSS values to decrease, the sample with the highest moisture region still exhibited a very high IFSS value (45.8 MPa). The reason that the APPJ treatment performs better on aramid fibers with a higher moisture region is that the presence of moisture (water) facilitates the oxidation process during plasma treatment and, thus, oxidizes the fiber surfaces [18]. Investigated the APPJ penetration depth in a 3D aramid (Kevlar129) woven structure and the mechanical properties of the resultant composites. The thickness of fabric was 6 mm and had five layers. The samples were treated with APPJ on one side and both sides of fabric for the three laps. The results showed that the fibers from the outer-most layers of the plasma-treated fabric exhibited a significant improvement in their surface roughness, chemical bonding, wettability, and adhesion properties. The effect of the APPJ treatment gradually diminished for the inner-layered fibers. By now, the third layer had the same properties as the untreated layer. The composites of aramid/polyester resin were prepared with the two-sided plasma-treated samples. The flexural strength and modulus increased to 11% and 12%, respectively. For the purpose of using para-aramid cord (Twaron) in a rubber application, plasma and wet-chemical treatments were compared [19]. Since untreated aramid has almost no interaction with the resorcinol-formaldehyde latex, it has to be dipped in the epoxy-based dipping system. As an alternative method to a wet-chemical treatment, an air and nitrogen APPJ was used. A yarn speed of 15 m/min was used and the power setting varied between 250 and 425 W. The plasma treatments increased the concentration of oxygen and nitrogen on Twaron; however, the values were lower than those of the wet-chemically treated materials. The plasma treatment increased the surface roughness due to plasma surface etching and the nitrogen atmosphere led to a higher surface roughness than the air atmosphere. The plasma-treated aramid fibers showed a higher adhesion to natural rubber than that of an untreated surface, but the adhesion level of chemically treated fibers was not reached. The adhesion properties of the aramid polymer were also increased using a DBD. The samples of Kevlar29 fibers were treated in an argon DBD under different experimental conditions: treating power, treating time and gas flux [20]. The IFSS to epoxy resin increased 28% for a plasma treatment at 300 W, 60 s, and 2 L/min. At these conditions the concentrations of oxygen and nitrogen on the aramid surface were highest (O/C = 30.62; N/C = 11.48) compared to the concentrations of the untreated sample (O/C = 16.49; N/C = 4.49), which lowered the water contact angles from 65 to 49 degrees. The surface roughness of argon plasma-treated aramid increased from 197.56 to 271.30 nm (Fig. 13.3).

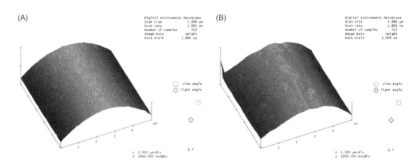

Figure 13.3 AFM images of Kevlar fibers: (A) untreated, (B) plasma-treated [20].

Different processing parameters of the air DBD plasma were also studied on the aramid (III) polymer [21]. The aramid samples were treated at different power densities (from 0 to 33.33 W/cm^3) and treatment times (0−21.6 s). The optimum plasma conditions, which increased the O/C ratio, surface roughness and wettability of polymer, were at a higher power density and a longer treatment time. However, the O/C ratio decreased slowly at too high a power density or too long a treatment time. The reason could be due to an etching of the fiber surface. Jia et al. [22] studied the effect of an air DBD plasma on the original (treated with oily finish) and acetone-washed Twaron fibers and the fiber performance in the composite with thermoplastic poly(phthalazinone ether sulfone ketone) (PPESK). The aramid samples were treated in an air atmosphere for 12 s. The volume of the discharge area was 5.2 cm^3, and the discharge power was 143.5 W. The XPS analysis revealed that the concentration of oxygen was the highest on the original fibers and the lowest on the acetone-cleaned fibers. Plasma treatment of the original fibers lowered the concentration of oxygen and increased the concentration of nitrogen, which is a result of plasma etching the oily finish on the fiber. The etching effect was confirmed by SEM. The concentration of oxygen increased on the acetone-cleaned fiber after plasma treatment. The surface free energy was the highest on the plasma-treated acetone-cleaned aramid due to the incorporation of oxygen-containing active groups and roughening of the fibers after plasma treatment. The results of the IFSS test show that the strength of the composites made with plasma-treated aramid increased up to 35.3%. Moreover, the highest IFSS value was obtained for the plasma-treated acetone-washed sample. In the following study, Jia et al. [23] treated acetone-washed aramid at different air DBD discharge powers (from 150 to 450 W). The adhesive properties of the plasma-treated fibers were improved in the composite with PPESK. The best plasma treatment of fibers was at a discharge power of 300 W. For the resulting composite, the IFSS value increased by 34.1%. The use of low-pressure systems enables the use of various plasma-forming gases. The synthesized meta-aramid nanofiber mats were treated with an argon low-pressure plasma at different treatment times (5−20 s) [24]. After plasma treatment, the O/C ratio increased by 25%, and the contact angle decreased two-fold. As the time of the plasma treatment increased, the nanofiber diameter decreased and so did its tensile stress. The IFSS values of the epoxy composites increased by 10% for nanofibers treated with plasma for 5 s, which was the optimum treatment time. Su et al. [25] studied the influence of oxygen inductively coupled radiofrequency (ICRF) plasma at different treatment powers (30−120 W) and times (3−10 min) on acetone-washed Kevlar fabrics and their adhesion in a bismaleimide (BMI) composite. The optimum plasma working parameters were 5 min of treatment at 70 W plasma power. The plasma-treated fibers had the highest O/C ratio and highest roughness (Fig. 13.4), and the fiber/BMI composites had the highest IFSS values.

Another possible method to incorporate polar groups on the polymer is the use of an ammonia plasma [26]. Acetone-washed aramid III fibers were treated in an ICRF plasma system under an ammonia atmosphere at different experimental conditions. The treating powers were 50−200 W; the treating times ranged from 5 to 20 min and the gas pressures changed from 20 to 50 Pa. The water contact angle of untreated fibers was 71.4 degrees, which decreased rapidly with an increase in treatment time, power, and discharge pressure of the ammonia plasma. The minimum contact angle (47.8 degrees) was achieved on fibers treated at a power of 100 W for 15 min and at a discharge pressure of 30 Pa. The concentration of carbon and oxygen decreased with the ammonia plasma treatment, but

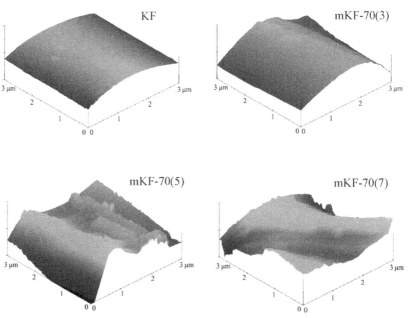

Figure 13.4 AFM images of untreated (KF) and treated Kevlar fibers at 70 W for different lengths of time [25].

the concentration of nitrogen-containing groups increased to almost 10%. The fiber surface roughness increased from 119 to 158 nm. The resulting chemical and physical changes after the plasma increased the IFSS values of the fiber/epoxy composite by 42%. It is well-known that the chemical plasma modification declines with extended storage time. Supposedly, the aging effect can be overcome with a modification of the polymers with a coating from an organophosphorus monomer through plasma-induced graft-polymerization (PIGP) or plasma-enhanced chemical vapor deposition (PECVD). Acetone-washed Kevlar29 fibers were first treated with different working gases (N_2, O_2, Ar) at a low pressure for different treatment times (5−20 min) and different output powers (100−400 W) [27]. Second, the fibers were treated with acrylic acid by a PIGP method for 50 min. The most effective working parameters were with argon as the working gas for 15 min under 300 W power. The value of IFSS increased from 25 to 45 MPa. Here, the adhesion properties after aging the modified fibers were not analyzed. The stability of the PECVD treatment of the para-aramid fabrics, which were applied for designing soft ballistic inserts in body protectors was evaluated in an investigation by Struszczyk et al. [28]. The fabrics were modified with monomer tetradecafluorohexane in the PECVD system. The properties of the aged samples were evaluated after accelerated aging conditions at a temperature of 70°C and relative humidities of 15% and 65%. The aging had a similar effect on both the untreated and PECVD-treated samples. However, the initial improved mechanical properties of the PECVD-treated samples remained much higher than those of the untreated samples. The exception was the abrasion resistance that decreased drastically after accelerated aging, almost to the same value as the untreated samples.

13.2.3 Plasma Treatment of Carbon Fibers

Ninety percent of carbon fibers are made from polyacrylonitrile (PAN) fibers since PAN-based polymers are the optimum precursors for carbon fibers owing to a combination of tensile and compressive properties as well as the carbon yield. The remaining 10% of C fibers are made from petroleum feedstock (pitch-based carbon fibers) or are vapor-grown by a catalytic process from carbon-containing gases. While pitch-based carbon fibers are used for general products, PAN-based carbon fibers are used for high-performance products [3]. The typical manufacturing steps involved in the production of PAN-based carbon fibers are polymerization of

PAN-based precursors, spinning of fibers, and thermal stabilization (oxidation) which can take up to several hours of carbonization and graphitization under an inert atmosphere of gases such as nitrogen or argon; surface treatment (liquid and gaseous oxidation treatments) and washing; and finally drying, sizing (lubrication) and winding [29]. Carbon fibers have excellent tensile properties, low densities, high thermal and chemical stabilities in the absence of oxidizing agents, good thermal and electrical conductivities, and excellent creep resistance [29,30]. In recent years, the carbon fiber industry has been growing steadily to meet the demands arising from different applications such as aerospace (aircraft and space systems), military, turbine blades, construction, lightweight cylinders and pressure vessels, medical, automobile, and sporting goods.

The ILSS, flexural strength, and elastic modulus of carbon fibers/phenolphthalein polyaryletherketone composites were improved with the plasma ICP oxygen plasma treatment of the carbon fibers [31]. The optimum mechanical properties of composites were obtained when the carbon fibers were treated for 5 min at a power of 300 W. The excessive plasma treatment negatively affected the mechanical properties of the single fibers and the composites. With an increasing power of the plasma treatment, the ratio of oxygen to carbon atoms (O/C ratio) decreased again and the etching effect was more pronounced. The extended treatment time of the carbon fibers with an atmospheric oxygen plasma also reduced the mechanical properties of the fibers/epoxy composites [32]. The concentrations of the carboxyl/ester groups and hydroxyl groups increased until the plasma treatment time was 4 min. Longer treatment times slowly decreased the concentration of the abovementioned functional groups. The pH value of the untreated fibers was slightly basic (pH = 7.2), while plasma-treated fibers had a slight acidic character (pH = 6.94). The maximum strength values of composites were exhibited under a plasma treatment time of 4 min, which was a consequence of improving the acid-base intermolecular interaction of the composites. A very high concentration of oxygen-containing groups on the surface of carbon fibers was found after modification with atmospheric He plasma for 1 min and afterwards with ethanol pyrolysis for 5 min [33,34]. The O/C ratio was 0.25 for treated fibers and 0.02 for untreated fibers [33]. In the two-stage modification of carbon fibers, the pyrolytic carbon (PyC) coating was deposited uniformly and the surface roughness and surface energy increased significantly [34]. The surfaces of the fibers were covered with numerous particles with a size of less than 10 nm and some

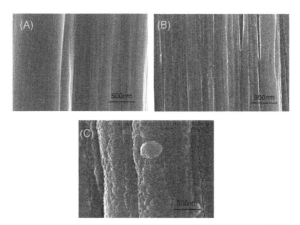

Figure 13.5 SEM images of surface morphology of (A) untreated, (B) plasma treated, (C) PyC coated [33].

of the particles agglomerated (Fig. 13.5). The particles provided more contact points and increased the contact area between the fiber surface and the matrix, which was beneficial for the enhancement of the single fiber tensile strength (increased by 15.7%) and the interfacial adhesion of the composite (from 65.3 to 83.5 MPa).

Carbon fibers were also treated with active screen plasma, which is an advanced plasma technology developed in early 2000s [35]. The active screen plasma is related to the exposure of treated surfaces to active species in the plasma with minimum or no damage by ion bombardment. Carbon fibers were exposed to the 25% N_2/75% H_2 plasma from 5 to 40 min. The surface of fibers was enriched with nitrogen functional groups and linked to the NH and N^* excited species in the plasma. Even under severe treatment conditions, the surfaces of the carbon fibers remained relatively smooth. The decrease in contact angle from 80 to 10 degrees was remarkable for the treated samples. However, the mean lifetime of the newly formed functional groups and the wettability of fibers lasted only 3 days. The treatment of the carbon fibers with hydrogen PECVD for 1 h increased the mechanical properties of the composites with polyetherimide by 30% [36]. This improvement was the result of the combined effect of the increased density of sp^3 bonds and increased surface roughness of the fibers. The hydrogen bonding between the plasma-treated carbon fibers and epoxy was also the reason for the higher IFSS in the research performed by Feih et al [37]. Carbon fibers were exposed to the acetylene and oxygen low-pressure RF plasma at different gas mixtures. Acetylene polymerizes completely during the plasma treatment and the oxygen molecules split up into free radicals. The mixture of both gases during the treatment leads to a copolymerization process involving oxygen. A significant improvement in the IFSS was obtained for fibers exposed to the plasma for 1 min at a gas ratio of 2:1 (acetylene/oxygen). On the surface of the carbon fibers treated with air microwave plasma, new carboxyl and hydroxyl groups were formed, which increased the ILSS and thermal resistance of the carbon fiber/polyimide composite [38]. Under the same load, the plasma-treated composite showed the lowest friction coefficient, which was connected to the high wear resistance. Moosberger–Will et al. [39] performed a small industrial scale continuous plasma modification of a carbon fibers tow (1200 m long 50,000 single fibers) as a promising treatment in composite production. In this research, the atmospheric plasma jet treatment of the combined plasma activation and plasma polymerization was used to improve adhesion to the epoxy matrix. Methyltrimetoxysilane was used as the precursor gas to deposit a silicon organic layer. The concentration of silicon on the treated fibers was 2.5%. The results also showed a homogeneous plasma polymerized layer of a thickness of only a few nanometers along and across the carbon tow. Three different DBD plasma systems (coaxial cylinder-type, parallel plated-type, and needle plate-type reactors) using different gases (air, oxygen, nitrogen, and ammonia) were used for the treatment of carbon fibers to evaluate their catalytic performance for the removal of carbon disulfide (CS_2) [40]. The results indicate that the catalytic hydrolysis of CS_2 was improved to a great extent when the carbon fibers were treated with a nitrogen plasma in a coaxial cylinder-type reactor for 5 min with a 7 kV output voltage and at a 7.5 mm distance from the discharge gap. The main reason for this phenomenon was attributed to the moderate amount of oxygen and nitrogen functional groups that formed on the surface of the carbon fibers after plasma treatment. Additionally, a higher volume of nanopores was produced, which provided more reaction sites for the catalytic hydrolysis of CS_2. To reduce the cost of preparing carbon fiber composites, the use of recycled carbon fibers was proposed. However, recycling causes the loss of the sized layer on the fibers, which causes the fibers to form bundles, which consequently affect the mechanical properties of the resulting composite. Lee at al. [41,42] studied the effects of an atmospheric plasma treatment on recycled carbon fibers. The fibers were plasma-treated using dry air (20.9% oxygen, 79.1% nitrogen, relative humidity > 3 ppm) and CO_2 (99% purity) [41]. The fibers were treated for five irradiation periods up to 1.67 s. There was no noticeable difference in surface morphology between untreated and plasma-treated fibers, except on the sample treated for 1.67 s. The latter had a slight scratch-type damage on its surface. The O/C ratios on the fibers increased with increasing plasma treatment time and were much higher on samples treated with a dry air plasma than with a CO_2 plasma. Accordingly, the mechanical properties increased for air plasma-treated samples, whereas no distinguishable difference in the untreated samples was noticed for the CO_2 plasma-treated samples. Increasing the plasma treatment time (up to 10 s) with the dry air plasma deepened the surface grooves on the carbon fibers [41]. Consequently, the IFSS decreased slightly. For samples treated for shorter times, the flexural strength increased by 17% and the mechanical properties were close to those of fresh carbon fibers.

13.2.4 Plasma Treatment of Polyethylene Fibers

Polyethylene was accidently synthesized in 1898 by the German chemist Hans von Pechmann, who was at that the time investigating diazomethane. However, it was not commercialized until the discovery of Ziegler-Natta catalysts in the 1950s. Polyethylene fibrous polymers (including high-density PE (HDPE) and ultrahigh molecular weight PE (UHMWPE)) have specific strength and fracture toughness characteristics, high elastic moduli, high tensile strengths, good biocompatibility, and high chemical resistances. PE is used for various products, from food packaging to implants that support material for cell growth, battery separators, bullet-protective gear, and pipes, etc. The negative characteristics that are attributed to PE are the smooth and hydrophobic surface properties that cause an inadequate interfacial adherence with a matrix.

The PE membrane separator was treated with an atmospheric pressure air plasma jet [43]. The surface of PE membrane was found to be highly hydrophilic as the power input of the jet was increased. The hydrophilic character of PE membrane increased due to the incorporation of oxygen-rich functional groups, which were found in the highest concentration on the PE treated with the highest power of plasma jet (125 W). The surface morphology also changed after plasma treatment but there were no distinguishable differences among the samples treated at different jet powers. The gas type (i.e., He, He/O_2, CO_2) influenced the plasma-polymer interactions and prevalence of the oxygen-rich groups on the

Figure 13.6 Photographs of water contact angles of untreated and plasma-treated PE separators with different oxygen plasma treatment time [46].

polymer [44,45]. The PE film treated with a He plasma had a higher oxygen content but clearly a lower percentage of $-COO$ than the one treated with a He/O_2 plasma [44]. The treatment of PE with a CO_2 plasma increased the carboxylic acid, ketone/aldehyde, and hydroxyl/epoxide groups [45]. The incorporation of oxygen-rich groups on the surface of PE membrane separator was also achieved using an oxygen RF plasma [46]. Longer treatment times with the oxygen plasma (10 min) increased the concentration of oxygen-rich groups on the polymer surface, so the wettability increased drastically (the decrease in contact angle was from 114 to 8.1 degrees) (Fig. 13.6). The effect induced by the plasma decreased the tensile strength of the PE separator but increased the electrochemical performance of the lithium-ion batteries.

The performance of PE fibers for prosthodontics applications was also enhanced after Ar or O_2 low-pressure plasma treatment [47]. The O_2 plasma was more successful at incorporating oxygen-rich groups on the surface of the polymer and the etching effect was more pronounced than after Ar plasma modification. Both treatments increased the flexural strength and deflection of the fiber-reinforced dental composite. The oxygen plasma was proven to be effective in improving the wettability of HDPE as well increasing its microhardness [48]. An increased negative surface charge of the plasma-treated HDPE was a consequence of the oxygen-introduced groups, such as carbonyl, carboxyl, ether, and peroxide. Simultaneously, a vigorous increase in the surface roughness was found to be a result of successful plasma etching. The latter effects were strongly dependent on the flow rate of the applied plasma gas. The equilibrium values of the effects were found for samples treated for 20 min at lower flow rates. The Ar plasma was used for the creation of nanostructured UHMWPE to enhance the adhesion of mouse embryonic fibroblasts (L929 cells) [49]. The samples were treated in the direct glow diode Ar^+ plasma device and subsequently sputtered with a gold coating. The surface roughness increased for plasma-treated samples. The difference in the morphology of the rough surface was apparent in the case of samples

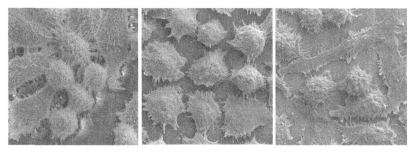

Figure 13.7 SEM images of L929 cells cultivated for 72 h on a pristine PE (left), PE treated with plasma for 60 s (middle) and PE treated with plasma for 60 s and sputtered with gold for 30 s (right) [49b].

treated with the plasma for different time periods. For the samples treated by the plasma for 60 s, sharper hills and protrusions were formed, whereas the samples treated with plasma for 240 s had more rounded surface roughness. Sputtering decreased the surface roughness by filling the protrusions. The results showed that the surface of PE should be rougher and without an excessive amount of gold clusters to have a positive effect on cell adhesion and cell proliferation (Fig. 13.7).

It was found that treating the HDMWPE fibers with a combination of a DBD plasma and a chitosan coating was successful in increasing the wettability, adhesion, and dyeability with a reactive dye [50]. The combined treatment was successful only when the treatment time was 100 s, while shorter or longer treatments did not provoke satisfactory surface chemistry or morphology. HDMWPE is also used in composites for ballistic applications. However, the friction between the HDMWPE fibers is too small, which represents a weakness. The friction between HDMWPE fibers was increased by PECVD at atmospheric pressure [51]. The carrier gas was nitrogen and the chemical subject to plasma action was dimethyl-dichlorosilane. With an increase in treatment time, the surface of the fibers became rougher with a higher number of particles and striation structures. The surface of fibers also became richer with carboxyl, carbonyl, hydroxyl, and amine groups and compounds containing silicon. The friction of the plasma-treated fibers increased compared to that of the untreated sample. In the case where hexametyldisiloxane (HMDSO) and tetradecafluorohexane (TDFH) were applied on UHMWPE by the PECVD technique, it was found that a modification with HMDSO leads to a reduction in tensile strength by 8%, while a modification with TDFH leads to an increase in tensile strength by 22% [52]. However, both treatments lead to an increased resistance to water penetration by 25% for the TDFH-treated sample and by 39% for the HDMSO-treated sample. Low-density polyethylene (LDPE) is another PE polymer that needs increased reactivity for better bonding, wettability, and adhesion, etc. The performance of LDPE was analyzed after modification with an atmospheric plasma jet at different speeds and treatment distances [53]. At a lower speed and distance, the wettability of polymer was the highest and so was the adhesion of joints. The reason was mainly due to the higher exposure time to the plasma that caused chemical and morphological surface changes, [54] modified the LDPE using the Ar/D_2O postdischarge of the atmospheric plasma jet at different distances and treatment times. The oxidative functionalization was analyzed using ToF/SIMS. Two different oxidation trends were observed: type I for distances from 3 to 7 mm where the functionalization decreased with increasing time, and type II for 10 mm distance where oxidation behaved inversely. The highest reactivity of the D_2O vapors with LDPE was achieved for times of a few tens of seconds at the distance of 5 mm from the plasma source. The oxidation of the LDPE was also achieved by an oxygen low-pressure RF plasma [55]. The samples were treated for $1-30$ min at a working pressure of $31-32$ Pa. The contact angles were reduced after 1 min of plasma modification and longer treatments did not cause additional significant changes. A slight etching of LDPE was also observed. The tests on durability of the plasma treatment showed that the incorporation of polar groups is not permanent and almost all functionality disappears in a few days.

13.2.5 Plasma Treatment of the Poly(p-Phenylene Benzobisoxazole) Fibrous Polymer

To date, the PBO fibers are the strongest known synthetic polymer fibers with excellent mechanical properties paired with extreme thermal stability. The PBO polymer is a rigid-rod lyotropic liquid crystal polymer derived from the reaction of terephthalic acid and 4,6-diaminoresorcinol [56]. Fibers are commercially produced by Toyobo, under the name Zylon. The high-performance PBO fibers are excellent candidates for use in the high-performance composites in space and aviation, military and industrial applications, and other advanced domains. Although PBO has excellent properties, its surface smoothness and lack of polar functional groups in its polymer repeat units, hinders the efficient bonding between the PBO fibers and matrix resin [57]. Therefore, a substantial improvement in the adhesion strength would open the broad field of composites for applications including PBO fibers.

An atmospheric corona plasma at different power treatments (50–300 W) was used for the modification of PBO fibers that were used in a composite with epoxy resin [58]. By increasing the plasma treatment power, the content of oxygen-rich groups and roughness increased as well as the ILSS. The ILSS evaluates the interfacial adhesion of fibrous composites and reflects the fiber surface adhesive properties. Although the maximum achieved ILSS value of composite was measured for the PBO fibers treated at 300 W, it was not much higher than of the composite with untreated PBO fibers (33.3 MPa). The large increase (up to 130%) in interfacial adhesion of the PBO/epoxy composite was achieved by treating the PBO fibers with an atmospheric pressure plasma jet (APPJ) [59,70]. In this study, the PBO fibers were treated with the He/O$_2$ APPJ for 30 s at a power of 30 W. The results were connected to an increase in oxygen-rich groups and the etching of PBO after plasma modification. The etching effect is visible from SEM images (Fig. 13.8).

The influence of the DBD plasma system on the modification of PBO fibers and their interface properties in composites with poly(phthalazinone ether sulfone ketone) (PPESK) or bismaleimide (BMI) was studied by several groups. In all studies, the air or oxygen DBD plasma increased the content of oxygen functional groups and etching on the surface of the PBO fibers. The ILSS values of the PBO/PPESK composite increased by 34.5% when the PBO fibers were treated by the air DBD system at discharge density of 41.4 W/m^3 [60]. The surface activity and surface free energy of PBO fibers improved after the air DBD plasma treatment at varying plasma powers (75–300 E); however, only to one extent [61].

Figure 13.8 SEM images of untreated (left) and plasma-treated (right) PBO fiber [59].

The surface free energy of the untreated samples was $44.62 \, mJ/m^2$ and increased to $71.17 \, mJ/m^2$ when the plasma power was 225 W. The higher plasma power (300 W) caused a decrease in surface free energy of the PBO fibers ($57.77 \, mJ/m^2$) and increased the fiber surface roughness. Thermogravimetric analysis and SEM confirmed that there was more PPESK resin adhering to the PBO fibers with a higher wettability (those with higher free surface energy). The same research group confirmed that the surface wettability of PBO fibers plays an important role in increasing the adhesion to PPESK [62]. In the following study by Jia et al. [63], the PBO/PPESK composites were prepared by treating the PBO fibers with an air DBD plasma and then impregnating the material with resins or first impregnating the sample with resins and then treating with the DBD plasma. The results show that the ILSS values increased regardless of the plasma modification order. Nevertheless, the highest ILSS values were measured for the sample that was first impregnated with resin and then treated with the plasma. It is possible that the DBD plasma initiated grafting polymerization of the resin on the fiber. Liu et al. [64] compared the effect of an air and oxygen DBD plasma on the performance of PBO fibers in BMI resin composites. The treatment time was set to 12 s with a plasma power density of $30 \, W/cm^3$. The ILSS values of the PBO/BMI composites increased after plasma modification of the PBO fibers from 43.9 to 51.2 MPa and 57.1 MPa for the oxygen-DBD plasma and the air-DBD plasma, respectively. Since the physical changes in the PBO fiber surface after plasma modification were same for both plasma gases, it is presumed that the main reason for the different ILSS values of composites were the chemical changes. The air-DBD plasma introduced oxygen and nitrogen atoms onto the PBO surface, while the oxygen-DBD plasma introduced only oxygen atoms. However, in another research study [65,66], an increase in the treatment time (to 24 s) with the same plasma power density ($30 \, W/cm^3$) demonstrated that the ILSS of the composite increased for the oxygen DBD plasma-treated PBO fiber (to 62 MPa). Nevertheless, the ILSS of the PBO/BMI composites decreased to 47 MPa, when they were prepared with 7-day-aged plasma-treated PBO fibers [65]. In another study [67], the aging effect of the oxygen DBD treatment of PBO fibers was already noticed after 5 days of aging and remained the same for next 30 days. A low-pressure plasma was also used in the experiments for increasing the ILSS values of composites with PBO fibers. Mainly oxygen, argon and their mixture were used as the working gases. Liu et al. [68] found that treating the PBO fibers for

7 min with the Ar low-pressure plasma at high input power induced a very small change in the O/C ratio (from 0.25 to 0.29). At higher power conditions, more effective ablation and sputtering effects drastically reduced the tensile strength, surface oxidation and roughness of the PBO fibers. The ILSS values of the PBO/BMI composites increased with plasma treatment, and the best results (an ILSS value higher by 39.7%) were obtained at power of 200 W. However, the ILSS values decreased rapidly (by 26.5%) in the first 3 days of aging and then slowly in next 30 days [69]. The ILSS of the PBO/BMI composite increased by 38.1% when the PBO fibers were treated by an O_2/Ar low-pressure plasma [70] but the brittleness of the composite increased. The results indicate that the different plasma conditions influence the efficiency of the O_2 low-pressure plasma on the ILSS of the PBO/PPESK composite [71]. The order of the working parameter's influence is the plasma power $>$ treatment time $>$ treatment pressure.

13.3 CONCLUSIONS

The major problem in using high-performance fibrous polymers is their poor adhesive properties in the composites. However, from the extensive review of the performed studies, it can be concluded that plasma technology plays an important role in increasing the interfacial strength of the composites with high-performance fibrous materials by changing their surface from inert to reactive through new functional groups and increased roughness. Various plasma systems were used to change the reactivity of high-performance fibrous polymers working at atmospheric or low pressure. Fibrous materials were modified using a DBD, an active screen plasma, a corona, a jet plasma, an inductively coupled radiofrequency plasma, and PECVD of polymerizing organosilicon monomers. Depending on the process gas (Ar, O_2, NH_3, air, He, CO_2, H_2, and mixtures N_2/H_2, He/O_2, and Ar/D_2O) the surface of the fibers was functionalized either with oxygen-rich or nitrogen-rich groups (carbonyl, carboxyl, hydroxyl, or amine groups). However, the lifetime of newly formed functional groups and the wettability of the fibers lasts a short time (i.e., 3 days). The aging effect can be increased by formation of a thin film on the surface of the substrate using the plasma-polymerization process. The modification of high-performance fibrous materials is also influenced by other plasma working parameters, and the order of their influence is as follows: plasma power $>$ treatment time $>$ distance of substrate from the plasma source > pressure. The overall results in the modification of high-performance fibrous materials increased the interfacial adhesion between the fibers and the epoxy matrix. However, it must be stressed that the plasma working parameters should be carefully selected since excessive treatment can cause negative mechanical properties of the single fibers and composites.

REFERENCES

[1] M. Gorjanc, M. Mozetič, Modification of Fibrous Polymers by Gaseous Plasma: Principles, Techniques and Applications, LAP Lambert Academic Publishing, Saarbrücken, 2014.
[2] F.R. Jones, Glass fibers, in: J.W.S. Hearle (Ed.), High-Performance Fibres, Woodhead Publishing Limited, Boca Raton, 2001, pp. 191—235.
[3] M.S. Aslanova, in: A. Kelly, Y.N. Rabotnov (Eds.), Handbook of Composites: Strong Fibers, Elsevier Science Publishers B.V, Amsterdam, 1985.
[4] Y.I. Kolesov, M.Y. Kudryavtsev, N.Y. Mikhailenko, Types and compositions of glass for production of continuous glass fiber (review), Glass Ceram. 58 (5) (2001) 197—202. Available from: https://doi.org/10.1023/a:1012386814248.
[5] Y. Iriyama, M. Taira, T. Ihara, T. Matsuoka, Effect of plasma treatment of glass fabrics on the mechanical strength of epoxy composites, J. Photopolym. Sci. Technol. 10 (1) (1997) 139—142.
[6] G. Yang, Y. Feng, Y. Yang, D. Wang, C. Kou, A study of surface modification of e-glass fiber by low temperature plasma treatment, 2016 IEEE International Conference on High Voltage Engineering and Application (ICHVE) (2016). Available from: https://doi.org/10.1109/ICHVE.2016.7800739. 0—2.
[7] K.-B. Lim, D.-C. Lee, Surface modification of glass and glass fibres by plasma surface treatment, Surf. Interf. Anal. 36 (3) (2004) 254—258. Available from: https://doi. org/10.1002/sia.1682.
[8] K. Sever, M. Sarikanat, Y. Seki, M. Mutlu, I. Hakk, Improvement of interfacial adhesion of glass fiber/epoxy composite by using plasma polymerized glass fibers, J. Adhesion 83 (9) (2010) 915—938. Available from: https://doi.org/10.1080/00218464.2010.506160.
[9] D. Çökeliler, S. Erkut, J. Zemek, H. Biederman, M. Mutlu, Modification of glass fibers to improve reinforcement: a plasma polymerization technique, Dental Mater. 23 (3) (2007) 335—342. Available from: https://doi.org/10.1016/j.dental.2006.01.023.
[10] Y. Kusano, K. Norrman, S.V. Singh, F. Leipold, P. Morgen, A. Bardenshtein, et al., Ultrasound enhanced 50 Hz plasma treatment of glass-fiber-reinforced polyester at atmospheric pressure, J. Adhesion Sci. Technol. 27 (7) (2013) 825—833. Available from: https://doi.org/10.1080/01694243.2012.727156.
[11] V. Cech, R. Prikryl, R. Balkova, J. Vanek, A. Grycova, The influence of surface modifications of glass on glass fiber/polester interphase properties, J. Adhesion Sci. Technol. 17 (10) (2003) 1299—1320. Available from: https://doi.org/10.1163/156856103769172751.
[12] V. Cech, R. Prikryl, R. Balkova, A. Grycova, J. Vanek, Plasma surface treatment and modification of glass fibers, Compos. Part A-Appl. Sci. Manuf. 33 (10) (2002) 1367—1372. Available from: https://doi.org/10.1016/S1359-835X(02)00149-5.

[13] Y. Kusano, H. Mortensen, B. Stenum, P. Kingshott, T.L. Anderson, P. Brøndsred, et al., Atmospheric pressure plasma treatment of glass fibre composite for adhesion improvement, Plasma Process. Polym. 4 ((Suppl.1) (2007) 455—459. Available from: https://doi.org/10.1002/ppap.200731206.

[14] S. Rebouillat, Aramids, in: J.W.S. Hearle (Ed.), High-Performance Fibres, Woodhead Publishing Limited, Cambridge, 2001, pp. 23—58.

[15] M. Stepankova, J. Šašková, J. Grégr, J. Wiener, Using of DSCBD plasma for treatment of Kevlar and Nomex fibers, Chemicke Listy 102 (16 SPEC. ISS.) (2008) 1515—1518.

[16] J. Sun, L. Yao, S. Sun, Y. Qiu, ESR study of atmospheric pressure plasma jet irradiated aramid fibers, Surf. Coat. Technol. 205 (23—24) (2011) 5312—5317. Available from: https://doi.org/10.1016/j.surfcoat.2011.05.045.

[17] Y. Ren, C. Wang, Y. Qiu, Influence of aramid fiber moisture regain during atmospheric plasma treatment on aging of treatment effects on surface wettability and bonding strength to epoxy, Appl. Surf. Sci. 253 (23) (2007) 9283—9289. Available from: https://doi.org/10.1016/j.apsusc.2007.05.054.

[18] X. Chen, L. Yao, J. Xue, D. Zhao, Y. Lan, X. Qian, et al., Plasma penetration depth and mechanical properties of atmospheric plasma-treated 3D aramid woven composites, Appl. Surf. Sci. 255 (5 PART 2) (2008) 2864—2868. Available from: https://doi.org/10.1016/j.apsusc.2008.08.028.

[19] P.J. de Lange, P.G. Akker, Adhesion activation of Twaron aramid fibers for application in rubber: plasma versus chemical treatment, J. Adhesion Sci. Technol. 26 (6) (2012) 827—839. Available from: https://doi.org/10.1163/016942411X580036.

[20] R. Gu, J. Yu, C. Hu, L. Chen, J. Zhu, Z. Hu, Surface treatment of para-aramid fiber by argon dielectric barrier discharge plasma at atmospheric pressure, Appl. Surf. Sci. 258 (24) (2012) 10168—10174. Available from: https://doi.org/10.1016/j.apsusc.2012.06.100.

[21] M. Xi, Y.L. Li, S. Shang, D.H. Yong, Li, Y.X. Yin, X.Y. Dai, Surface modification of aramid fiber by air DBD plasma at atmospheric pressure with continuous on-line processing, Surf. Coat. Technol. 202 (24) (2008) 6029—6033. Available from: https://doi.org/10.1016/j.surfcoat.2008.06.181.

[22] C. Jia, P. Chen, B. Li, Q. Wang, C. Lu, Q. Yu, Effects of Twaron fiber surface treatment by air dielectric barrier discharge plasma on the interfacial adhesion in fiber reinforced composites, Surf. Coat. Technol. 204 (21—22) (2010) 3668—3675. Available from: https://doi.org/10.1016/j.surfcoat.2010.04.049.

[23] C. Jia, P. Chen, Q. Wang, J. Wang, X. Xiong, K. Ma, The effect of atmospheric-pressure air plasma discharge power on adhesive properties of aramid fibers, Polym. Compos. 37 (2016) 620—626. Available from: https://doi.org/10.1002/pc.23219.

[24] S.Y. On, M.S. Kim, S.S. Kim, Effects of post-treatment of meta-aramid nanofiber mats on the adhesion strength of epoxy adhesive joints, Compos. Struct. 159 (2017) 636—645. Available from: https://doi.org/10.1016/j.compstruct.2016.10.016.

[25] M. Su, A. Gu, G. Liang, L. Yuan, The effect of oxygen-plasma treatment on Kevlar fibers and the properties of Kevlar fibers/bismaleimide composites, Appl. Surf. Sci. 257 (8) (2011) 3158—3167. Available from: https://doi.org/10.1016/j.apsusc.2010.10.133.

[26] S. Li, K. Han, H. Rong, X. Li, M. Yu, Surface modification of aramid fibers via ammonia-plasma treatment, J. Appl. Polym. Sci. 131 (10) (2014) 1—6. Available from: https://doi.org/10.1002/app.40250.

[27] C.X. Wang, M. Du, J.C. Lv, Q.Q. Zhou, Y. Ren, G.L. Liu, et al., Surface modification of aramid fiber by plasma induced vapor phase graft polymerization of acrylic acid. I. Influence of plasma conditions, Appl. Surf. Sci. 349 (2015) 333—342. Available from: https://doi.org/10.1016/j.apsusc.2015.05.036.

[28] M.H. Struszczyk, A.K. Puszkarz, M. Miklas, B. Wilbik-Hałgas, M. Cichecka, W. Urbaniak-Domagała, et al., Performance stability of ballistic para-aramid woven fabrics modified by plasma-assisted chemical vapour deposition (PACVD), Fibres Textiles Eastern Europe 24 (4) (2016) 92−97. Available from: https://doi.org/10.5604/12303666.1201137.

[29] S. J. Park, G. Y. Heo, Precursors and manufacturing of carbon fibers, Carbon Fibers. Springer Series in Materials Science vol. 210 (2015) 31−66. Available from: https://doi.org/10.1007/978-94-017-9478-7_2.

[30] B.A. Newcomb, Processing, structure, and properties of carbon fibers, Compos. Part A: Appl. Sci. Manuf. 91 (2016) 262−282. Available from: https://doi.org/10.1016/j.compositesa.2016.10.018.

[31] W. Li, S.Y. Yao, K.M. Ma, P. Chen, Effect of plasma modification on the mechanical properties of carbon fiber/phenolphthalein polyaryletherketone composites, Polym. Compos. 34 (3) (2013) 368−375. Available from: https://doi.org/10.1002/pc.22385.

[32] S. Park, J. Oh, Effect of atmospheric plasma treatment of carbon fibers on crack resistance of carbon fibers-reinforced epoxy composites, Carbon Sci. 6 (2) (2005) 106−110.

[33] J. Sun, F. Zhao, Y. Yao, Z. Jin, X. Liu, Y. Huang, High efficient and continuous surface modification of carbon fibers with improved tensile strength and interfacial adhesion, Appl. Surf. Sci. 412 (2017) 424−435. Available from: https://doi.org/10.1016/j.apsusc.2017.03.279.

[34] J. Sun, F. Zhao, Y. Yao, X. Liu, Z. Jin, Y. Huang, A two-step method for high efficient and continuous carbon fiber treatment with enhanced fiber strength and interfacial adhesion, Mater. Lett. 196 (2017) 46−49. Available from: https://doi.org/10.1016/j.matlet.2017.03.007.

[35] C.S. Gallo, C. Charitidis, H. Dong, Surface functionalization of carbon fibers with active screen plasma, J. Vacuum Sci. Technol. A 35 (2) (2017) 21404−1−10. Available from: https://doi.org/10.1116/1.4974913.

[36] E.S. Lee, C.H. Lee, Y.S. Chun, C.J. Han, D.S. Lim, Effect of hydrogen plasma-mediated surface modification of carbon fibers on the mechanical properties of carbon-fiber-reinforced polyetherimide composites, Compos. Part B: Eng. 116 (2017) 451−458. Available from: https://doi.org/10.1016/j.compositesb.2016.10.088.

[37] S. Feih, P. Schwartz, Modification of the carbon fiber/matrix interface using gas plasma treatment with acetylene and oxygen, J. Adhesion Sci. Technol. 12 (5) (1998) 523−539. Available from: https://doi.org/10.1163/156856198X00209.

[38] W. Wenxia, Plasma treatment of carbon fiber on the tribological property of polyimide composite, Surf. Interf. Anal. 49 (7) (2017) 682−686. Available from: https://doi.org/10.1002/sia.6208.

[39] J. Moosburger-Will, M. Bauer, F. Schubert, C. Kunzmann, E. Lachner, H. Zeininger, et al., Methyltrimethoxysilane plasma polymerization coating of carbon fiber surfaces, Surf. Coat. Technol. 311 (2017) 223−230. Available from: https://doi.org/10.1016/j.surfcoat.2017.01.017.

[40] K. Li, P. Ning, K. Li, C. Wang, X. Sun, L. Tang, et al., Low temperature catalytic hydrolysis of carbon disulfide on activated carbon fibers modified by non-thermal plasma, Plasma Chem. Plasma Process. 37 (4) (2017) 1175−1191. Available from: https://doi.org/10.1007/s11090-017-9813-y.

[41] H. Lee, I. Ohsawa, J. Takahashi, Effect of plasma surface treatment of recycled carbon fiber on carbon fiber-reinforced plastics (CFRP) interfacial properties, Appl. Surf. Sci. 328 (2015) 241−246. Available from: https://doi.org/10.1016/j.apsusc.2014.12.012.

[42] H. Lee, H. Wei, J. Takahashi, The influence of plasma in various atmospheres on the adhesion properties of recycled carbon fiber, Macromol. Res. 23 (11) (2015) 1026−1033. Available from: https://doi.org/10.1007/s13233-015-3141-y.

[43] Y.C. Tseng, H.L. Li, C. Huang, Atmospheric-pressure plasma activation and surface characterization on polyethylene membrane separator, Jap. J. Appl. Phys. 56 (1) (2017) 01AF03. Available from: https://doi.org/10.7567/JJAP.56.01AF03.

[44] J. Sun, Y. Qiu, The effects of gas composition on the atmospheric pressure plasma jet modification of polyethylene films, Plasma Sci. Technol. 17 (5) (2015) 402−408. Available from: https://doi.org/10.1088/1009 0630/17/5/07.

[45] J.G.A. Terlingen, H.F.C. Gerritsen, A.S. Hoffman, J. Feijen, Introduction of functional groups on polyethylene surfaces by a carbon dioxide plasma treatment, J. Appl. Polym. Sci. 57 (8) (1995) 969−982. Available from: https://doi.org/10.1002/app.1995.070570809.

[46] S.Y. Jin, J. Manuel, X. Zhao, W.H. Park, J.H. Ahn, Surface-modified polyethylene separator via oxygen plasma treatment for lithium ion battery, J. Ind. Eng. Chem. 45 (2017) 15−21. Available from: https://doi.org/10.1016/j.jiec.2016.08.021.

[47] S.M.M. Spyrides, M. Prado, J.R. de Araujo, R.A. de Simão, F.L. Bastian, Effects of plasma on polyethylene fiber surface for prosthodontic application, J. Appl. Oral Sci. 23 (6) (2015) 614−622. Available from: https://doi.org/10.1590/1678-775720150260.

[48] M. Lehocký, H. Drnovská, B. Lapčíková, A.M. Barros-Timmons, T. Trindade, M. Zembala, et al., Plasma surface modification of polyethylene, Colloids Surf. A: Physicochem. Eng. Aspects 222 (1−3) (2003) 125−131. Available from: https://doi.org/10.1016/S0927-7757(03)00242-5.

[49] Z. Novotná, S. Rimpelová, P. Juřík, M. Veselý, Z. Kolská, T. Hubáček, et al., The interplay of plasma treatment and gold coating and ultra-high molecular weight polyethylene: on the cytocompatibility, Mater. Sci. Eng. C 71 (2017) 125−131. Available from: https://doi.org/10.1016/j.msec.2016.09.057.

[50] Y. Ren, Z. Ding, C. Wang, C. Zang, Y. Zhang, L. Xu, Influence of DBD plasma pretreatment on the deposition of chitosan onto UHMWPE fiber surfaces for improvement of adhesion and dyeing properties, Appl. Surf. Sci. 396 (2017) 1571−1579. Available from: https://doi.org/10.1016/j.apsusc.2016.11.215.

[51] Y. Chu, X. Chen, L. Tian, Modifying friction between ultra-high molecular weight polyethylene (UHMWPE) yarns with plasma enhanced chemical vapour deposition (PCVD), Appl. Surf. Sci. 406 (2017) 77−83. Available from: https://doi.org/10.1016/j.apsusc.2017.02.109.

[52] M.H. Struszczyk, A.K. Puszkarz, B. Wilbik-Halgas, M. Cichecka, P. Litwa, W. Urbaniak-Domagala, et al., The surface modification of ballistic textiles using plasma-assisted chemical vapor deposition (PACVD), Textile Res. J. 84 (19) (2014) 2085−2093. Available from: https://doi.org/10.1177/0040517514528559.

[53] V. Fombuena, R. Balart, O. Fenollar, Atmospheric plasma treatment of polyethylene substrates for improved mechanical performance of adhesion joints, Bull. Trans. Univ. Brașov Series I 4 (1) (2011) 43−50.

[54] V. Cristaudo, S. Collette, N. Tuccitto, C. Poleunis, L.C. Melchiorre, A. Licciardello, et al., Molecular surface analysis and depth-profiling of polyethylene modified by an atmospheric Ar-D2O post-discharge, Plasma Process. Polym. 13 (11) (2016) 1104−1117. Available from: https://doi.org/10.1002/ppap.201600061.

[55] M.R. Sanchis, V. Blanes, M. Blanes, D. Garcia, R. Balart, Surface modification of low density polyethylene (LDPE) film by low pressure O_2 plasma treatment, Eur. Polym. J. 42 (7) (2006) 1558−1568. Available from: https://doi.org/10.1016/j.eurpolymj.2006.02.001.

[56] C.J. McHargue, J.B. Darby Jr, M.J. Yacamán, J. Gasga (Eds.), Synthesis and Properties of Advanced Materials, Springer, 1996.

[57] E. Mäder, S. Melcher, J.W. Liu, S.L. Gao, A.D. Bianchi, S. Zherlitsyn, et al., Adhesion of PBO fiber in epoxy composites, J. Mater. Sci. 42 (19) (2007) 8047−8052. Available from: https://doi.org/10.1007/s10853-006-1311-1.

[58] C. Zhang, C. Li, B. Wang, R. Wang, H. Li, Effects of atmospheric air plasma treatment on interfacial properties of PBO fiber reinforced composites, Appl. Surf. Sci. 276 (2013) 190−197. Available from: https://doi.org/10.1016/j.apsusc.2013.03.064.

[59] R. Zhang, X. Pan, M. Jiang, S. Peng, Y. Qiu, Influence of atmospheric pressure plasma treatment on surface properties of PBO fiber, Appl. Surf. Sci. 261 (2012) 147−154. Available from: https://doi.org/10.1016/j.apsusc.2012.07.123.

[60] Q. Wang, P. Chen, C. Jia, M. Chen, B. Li, Improvement of PBO fiber surface and PBO/PPESK composite interface properties with air DBD plasma treatment, Surf. Interf. Anal. 44 (5) (2012) 548−553. Available from: https://doi.org/10.1002/sia.3846.

[61] C. Jia, Q. Wang, P. Chen, S. Lu, R. Ren, Wettability assessment of plasma-treated PBO fibers based on thermogravimetric analysis, Int. J. Adhesion Adhesives 74 (2017) 123−130. Available from: https://doi.org/10.1016/j.ijadhadh.2017.01.010. October 2016.

[62] Q. Wang, P. Chen, C. Jia, M. Chen, B. Li, Effects of air dielectric barrier discharge plasma treatment time on surface properties of PBO fiber, Appl. Surf. Sci. 258 (1) (2011) 513−520. Available from: https://doi.org/10.1016/j.apsusc.2011.08.078.

[63] C.X. Jia, Q. Wang, P. Chen, Y.W. Pu, Surface adhesive properties of continuous PBO fiber after air-plasma-grafting-epoxy treatment, J. Central South Univ. 23 (9) (2016) 2165−2172. Available from: https://doi.org/10.1007/s11771-016-3273-z.

[64] Z. Liu, Q. Zeng, P. Chen, Q. Yu, Z. Ding, Comparison of effects on PBO fiber by air and oxygen dielectric barrier discharge plasma, Vacuum 121 (2015) 152−158. Available from: https://doi.org/10.1016/j.vacuum.2015.08.012.

[65] Z. Liu, P. Chen, D. Han, F. Lu, Q. Yu, Z. Ding, Atmospheric air plasma treated PBO fibers: wettability, adhesion and aging behaviors, Vacuum 92 (2013) 13−19. Available from: https://doi.org/10.1016/j.vacuum.2012.11.002.

[66] Z. Liu, P. Chen, X. Zhang, Q. Yu, K. Ma, Z. Ding, Effects of surface modification by atmospheric oxygen dielectric barrier discharge plasma on PBO fibers and its composites, Appl. Surf. Sci. 283 (2013) 38−45. Available from: https://doi.org/10.1016/j.apsusc.2013.05.137.

[67] D. Liu, P. Chen, Q. Yu, K. Ma, Z. Ding, Improved mechanical performance of PBO fiber-reinforced bismaleimide composite using mixed O2/Ar plasma, Appl. Surf. Sci. 305 (2014) 630−637. Available from: https://doi.org/10.1016/j.apsusc.2014.03.150.

[68] D. Liu, P. Chen, M. Chen, Z. Liu, Surface modification of high performance PBO fibers using radio frequency argon plasma, Surf. Coat. Technol. 206 (16) (2012) 3534−3541. Available from: https://doi.org/10.1016/j.surfcoat.2012.02.033.

[69] D. Liu, P. Chen, M. Chen, Q. Yu, C. Lu, Effects of argon plasma treatment on the interfacial adhesion of PBO fiber/bismaleimide composite and aging behaviors, Appl. Surf. Sci. 257 (23) (2011) 10239−10245. Available from: https://doi.org/10.1016/j.apsusc.2011.07.029.

[70] Z. Liu, P. Chen, X. Zhang, Q. Yu, Degradation of plasma-treated poly(p-phenylene benzobisoxazole) fiber and its adhesion with bismaleimide resin, RSC Adv. 4 (8) (2014) 3893−3899. Available from: https://doi.org/10.1039/c3ra46192k.

[71] X. Zhang, P. Chen, X. Kang, M. Chen, Q. Wang, Improvement of the interfacial adhesion between PBO fibers and PPESK matrices using plasma-induced coating, J. Appl. Polym. Sci. 123 (2012) 2945−2951. Available from: https://doi.org/10.1002/app.34917.

CHAPTER 14

Plasma Modified Polymeric Materials for Implant Applications

Ladislav Cvrček and Marta Horáková
Department of Materials Engineering, Faculty of Mechanical Engineering, Czech Technical University in Prague, Prague, Czech Republic

14.1 INTRODUCTION

Polymeric materials are widely used in orthopedics and traumatology where they have an irreplaceable role in the construction of implants. Unlike metals and ceramics, polymers have a low elastic modulus close to that of cortical bone. This can be successfully used in the suppression of the stress shielding that is in contact between the implant and the bone [1,2]. The easy manufacturing allowing for the production of different shapes (films, fibers, sheets, special implant shapes, etc.) provides a base to use the polymeric materials as high-quality biomaterials. Polymeric biomaterials have to be biocompatible, sterilizable, they require adequate mechanical, physical, and surface properties (e.g., required roughness, hydrophilicity/hydrophobicity, adhesive properties, etc.) [3].

Implant applications are in direct contact with the human living body and blood. Thus, the surface, chemical, and biological properties play a crucial role with regards to the biomedical polymers [4]. However, due to the variability of polymeric implant applications, the large number of different polymer types, the variability of their properties and chemistry and, last but not least, many possible methods of their treatment to reach or improve the demanded properties, the polymers used in implant applications are a complex and complicated topic.

The surface and biological properties are of high importance, but the preservation of the required mechanical properties is required at the same time. Many possible treatment methods exist to reach this goal. Most of them are based on chemical procedures, which can be harmful

Non-Thermal Plasma Technology for Polymeric Materials
DOI: https://doi.org/10.1016/B978-0-12-813152-7.00014-7

to the environment [5−7]. Plasma surface treatment methods provide independent modification possibilities to the surface and biological properties without any alteration or degradation to the bulk (mechanical) properties. This chapter will propose an overview of plasma-treated polymeric materials for implant applications. Firstly, the required biological, chemical, and surface properties of polymer-based implants will be briefly described. Afterwards, we will focus on plasma modification for various implant applications.

14.2 POLYMERS FOR IMPLANTS

The main reason for the introduction of polymers into implant production is their excellent elasticity, chemical stability and, in most cases, their nontoxicity. Polymers suitable for implant production must meet the basic criteria that are standardised. These are a biocompatibility, especially nontoxicity or mutagenicity, and chemical resistance in the aggressive environment of the human body. Depending on the time when the implant is to be used, they are tested either in the short or long term. However, the required properties of the polymers strongly vary on their biomedical application. For applications where a bioactive surface is required, for example, for bone cell adhesion, polymers have a disadvantage in their inert surface. It is necessary to modify the surface due to this reason. The change can only be in the surface morphology or surface chemical composition. Therefore, it is necessary to test the properties of the polymers as a set of complex parameters.

14.2.1 Polymer Properties

The chemical and biological polymer properties include the chemical composition of the bulk and the surface, biocompatibility or biodegradation properties. Physical properties comprise density, surface topology, porosity, texture, roughness, form and geometry, the coefficient of thermal expansion, electrical conductivity, refractive index, opacity, or translucency. The following will focus primarily on the chemical, biological, and surface properties, because most of these can be modified through plasma treatment.

14.2.1.1 Biocompatibility

Biocompatibility is a complicated process depending on various factors. This process includes bio-functionality, bio-inertia, bioactivity, and bio-stability. Biocompatibility leads to the surrounding tissue and the human body accepting the synthetic implants without any undesirable immunity response, allergic reactions, inflammatory or chronical problems and, moreover, biocompatible materials are not carcinogenic. Biocompatibility strongly depends on the type of the application. The basic factors that influence biocompatibility are [3,5,8]:

1. Interaction with the surroundings—influence of cytotoxins, toxicological or allergic reactions, carcinogenic or mutagenic reactions, inflammatory processes, degree and quality of the biodegradation, contact with human blood.
2. Period of the implant application—long-term or short-term implant applications.
3. Surface biocompatibility—suitability of the implanted surface for the host tissue (chemical, biological and morphological).
4. Structural biocompatibility—optimal adaptability of the implant's mechanical properties to the mechanical properties of the host tissue.
5. Function—optimal friction coefficient, mechanical properties demanded by the application.
6. Proportion—size and shape.
7. Material—aggressiveness of the synthetic material to the host tissue and vice versa.

14.2.1.2 Biodegradability

Two main groups of polymeric materials (1) nondegradable and (2) degradable polymers are used in implantology based on their applications. Nondegradable polymers frequently used for implants are based on polyolefins, polyethers, polycarbonates, polyurethanes, poly-siloxanes, poly-sulphones, and polyamides. Degradable polymers used in biomedical application can be divided into two main groups: The first group is classified as biodegradable polymers which degrade under physiological conditions. The degradation of the biodegradable polymers is a biological process induced by enzymes (human or microbial) or hydrolysis (hydrogels). The second group of degradable polymers is characterized by the degradation process when the chemistry of the polymers is not fundamentally changed, but the physical state changes from a solid state to a solubilised structure. Biodegradable synthetic polymers include poly(glycolic acid) (PGA), poly(lactic acid) (PLA) and their copolymers—poly (lactic-co-glycolide) (PLGA) or poly(L-lactic acid) (PLLA), polydioxanone (PDO), poly(caprolactone) (PCL)—and the copolymers of trimethylene carbonate and glycolide [9−11].

Polymers like PLA or PGA degrade by nonspecific hydrolytic scission of ester bonds. A product of the PLA hydrolysis is lactic acid, which belongs to the group of secondary products of the human body's metabolism. Biodegradation of the PGA occurs due to the combination of enzymatic processes and hydrolysis [12–14].

The biodegradability of the polymers is also a complicated factor in biomedical applications. Biodegradable PLA stents are widely used in vascular intervention because the biodegradation period is sufficient enough to support the vessel walls after the stent implantation. However, the autocatalysation of aliphatic polyesters such as PLA accelerate ester hydrolysis by carboxyl end groups (RCOOH) generated from degradation, somewhat acidify the surrounding area. Moreover, residues consisting of crystalline molecules originating from spherulites may cause local inflammation following the degradation [15].

PDO degrades hydrolytically in two stages, where the degradation of the amorphous phase is induced more easily than the crystalline phase. Scission of PDO chains in the amorphous regions leads to the higher mobility of the chains which results in an increase in the crystallinity. PDO degradation products are in accordance with the human body's metabolites and they are excreted by respiration or the digestive tract [16].

14.2.1.3 Chemical Composition

Generally, the chemical composition of the polymers is based on the carbon which is bound between them creating long carbon-based chains. The carbon-based chemistry of the polymers makes them closer to the human tissue than inorganic materials. Different polymers contain other atoms like hydrogen, fluorine, oxygen, silicone, sulfur, or aromatic rings. However, it is not only the chemical composition which influences polymer properties. The strong variability of their properties is caused by various chain lengths, crystallinity, side groups, and copolymers (Fig. 14.1).

The surface chemical composition of the polymers plays a special role in biomedical applications. The polymer surface is in the direct contact with living tissue, thus, it must provide nontoxicity together with the preservation of the required mechanical properties. The chemical composition is also a highly significant factor influencing processes of surface treatment of polymeric implants. The chemical composition of the polymer surface can influence the morphology of the subsequently deposited nanolayers. For example, Slepička et al. described changing the content of O_2 in first 10 atomic layers from the surface towards the bulk of the

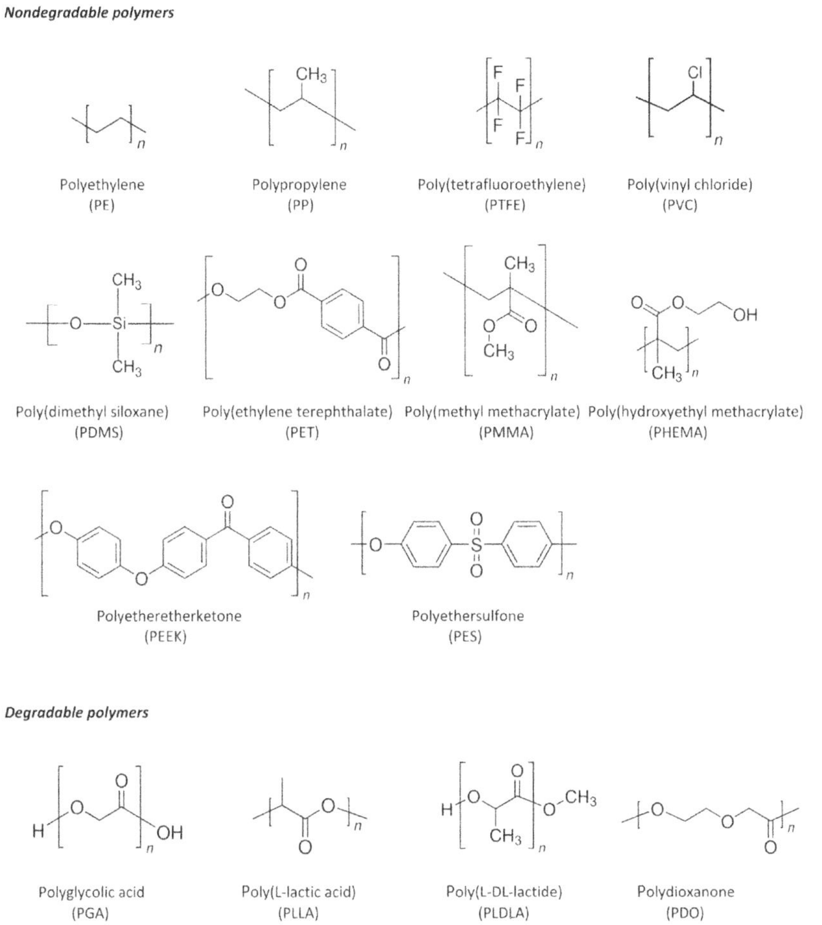

Nondegradable polymers

| Polyethylene (PE) | Polypropylene (PP) | Poly(tetrafluoroethylene) (PTFE) | Poly(vinyl chloride) (PVC) |

Poly(dimethyl siloxane) (PDMS)   Poly(ethylene terephthalate) (PET)   Poly(methyl methacrylate) (PMMA)   Poly(hydroxyethyl methacrylate) (PHEMA)

Polyetheretherketone (PEEK)   Polyethersulfone (PES)

Degradable polymers

Polyglycolic acid (PGA)   Poly(L-lactic acid) (PLLA)   Poly(L-DL-lactide) (PLDLA)   Polydioxanone (PDO)

Figure 14.1 Examples of the chemical structures of the polymers used in biomedical applications [17].

polyethylene terephthalate (PET). This phenomenon was explained by Kwang et al. by the reorientation of the polar oxygen groups in the surface layer to the bulk of the PET [18,19].

It is well-known that the degradation processes of the polymers start on the polymer surface and they spread out to the bulk of the polymeric components and major degradation processes of the polymers are based on chemical reactions. These reactions mainly have an oxidation, enzymatic, or hydrolytic character. As well as presence of the different types of the chemical elements or chemical groups on the polymer surface (O_2, H_2, N, F, S, Cl, CH_3, NH_2, OH, etc.) it is a chemical bond. The covalent bond dissociative energy between two carbon atoms (-C-C-) is high, approximately 348 kj/mol and the chemical bond -Si-O-, the basic component of silicone, has a similar dissociative energy. The chemical bond C-F reaches a dissociative energy of up to 544 kj/mol which causes the exceptional stability and resistance of polytetrafluoroethylene (PTFE). In addition, side polymer chains as well as their orientation and intermolecular interaction (hydrogen bridges, van der Waals interactions, hydrophobic interactions, etc.) strongly influence the polymers' stability. The surface properties related to the surface chemistry, which can be successfully influenced by plasma treatment, are corrosion resistance, oxidation and catalytic properties, or electrochemistry [20].

14.2.1.4 Other Surface Properties

Many publications are concerned with the biological properties, like biocompatibility or biodegradability, which are described above. Although, they play a crucial role in implantology, other surface properties are of a high importance also. Chu et al. summarized the surface properties of biomaterials modified using plasma treatment methods [20]. The surface energy is related to the chemical composition of the surface and is frequently modified by plasma, not only for biomedical applications, but also in industrial or high-tech applications. The surface energy influences hydrophobicity, hydrophilicity, reactivity, or inertness. Among other things, these properties affect antifouling properties, antibacterial properties, or cell adhesion. For long-term applications, some part of the implanted material has to couple with the living tissue and good cell adhesion is needed. On the other hand, cardiovascular components have to sustain adequate blood flow without a thrombotic response and ophthalmic implants or implants used in ureteral treatment should have the necessary antifouling properties, thus, harmful biological sediment residue and good cell adhesion is undesirable in these cases.

Many medical devices are used in single-phase homogenous and isotropic materials including metals, ceramics, and polymers. But polymeric materials are also used in implant applications as composite biomaterials or in combination with other polymers, metals, ceramics as well as carbon fibers, glass fibers, Kevlar fibers, bio-glass, or parts of the living body like bone particles, cells, etc. [14,21].

Thus, implant materials interact with the living tissue, but some particular implant components can oftentimes interact among themselves under specific conditions. For example, joint prostheses or mechanical heart-valves are strained by friction and wear, and they can suffer from fatigue fractures. In this case, polymers interact not only with blood or living tissue, but also with the components made from metals or ceramics. Tribological properties (friction, wear, lubricity) and morphology and texture (smoothness, roughness, patterns, or specific surface area) can be altered by plasma modification. Plasma deposition of functional thin films or plasma etching are used to achieve the required tribological properties and morphology. Mechanical properties, namely hardness, fatigue, or cracking properties can be influenced by plasma methods that influence crosslinking. For example, in the case of the polymeric implant applications, their porosity and the pore size (specific surface area) is very important for tissue ingrowth. A larger pore size can stimulate tissue ingrowth resulting in a good implant integration with the surrounding tissue. Well-integrated implants avoid additional use of screws or suture fixations. Porous high-density polyethylene (HDPE) has been used for craniofacial applications, for example, the chin, malar area, nasal reconstruction, ear reconstruction, orbital reconstruction, and the correction of craniofacial contour deformities [22].

14.2.2 Polymer Materials

Polymer implants have to meet high requirements for in vivo applications. First of all, they should not induce any harmful effects, for example, any inflammatory processes or the formation of unusual tissue. Polymer materials for implants need to meet the demands on bulk properties to function properly, as well as to reach the best surface properties. An overview of polymeric materials used for both short- and long-term implants are provided in Table 14.1 [17].

The group of the polymers utilized in implantology is wide-ranging. It involves polymers for ophthalmology, vascular intervention or temporary in vivo applications. For example, there are used silicones, polybutylene terephthalate (PBT), polydioxanone (PDO), 2-hydroxyethyl methacrylate (HEMA) and 6-hydroxyhexyl methacrylate (HEXMA), with a crosslinking agent ethylene glycol dimethacrylate (EGDMA), among many others.

14.3 MODIFICATION OF THE POLYMER SURFACES

The surfaces of the polymers are temperature-sensitive and can be treated by low-temperature methods only. In the case of polymeric implants, such surface treatments include a change in the roughness or chemical composition without affecting the bulk properties of the material [23].

Table 14.1 Polymers for implants [17]

Polymer	Implants
Polymethylmetacrylate (PMMA)	arch wires, brackets, dental bridges, boneplates, screws, bone cement
Polysulfone (PSU)	spine cage, plate, rods, screws and discs, total hip replacement, boneplates, bone replacement materials
Polypropylene (PP)	boneplates screws, arch wires, brackets
Polyethylene (PE)	tendon/ligament, total hip replacement, boneplates, screws, bone replacement material
Ultra-high-molecular weight polyethylene (UHMWPE)	total knee replacement, total hip replacement bone cement, finger joint
Poly(L-DL-lactide) (PLDLA)	boneplates, screws
Poly(L-lactic acid) (PLLA)	boneplates, screws, tendon/ligament, bone replacement materials
Polyglycolic acid (PGA)	boneplates, screws
Polycarbonate (PC)	arch wires, brackets
Polyethetretheketone (PEEK)	boneplates, screws, intramedullary nails, total hips replacement, spine cage, plate, rods, screws and discs
Polymethacrylate (PMA)	tendon/ligament
Polyurethane (PU)	vascular graft, cartilage replacement, abdominal walls prosthesis, spine cage, plate, rods, screws and discs, bone replacement materials
Polytetrafluorethylene (PTFE)	bone replacement material, vascular graft, total hip replacement, tendon/ligament, cartilage replacement
Polyethyleneterephthalate (PET)	cartilage replacement, tendon/ligament, finger joint, abdominal walls prosthesis, vascular graft, spine cage, plate, rods, screws and discs, bone replacement materials
Block co-polymer of lactic acid and polyethylene glycol (PELA)	vascular graft
Polyethyleneglycol (PEG)	bone replacement material
Polyhydroxybutyrate (PHB)	bone replacement material
Poly(20hydroxyethylmethacrylate) (PHEMA)	tendon/ligament
Liquid crystalline polymer (LCP)	intramedullary nails

Method principles, based on the generation of nonthermal plasma, are suitable for this surface modification because the electron temperature is kept higher than that of heavy particles, typically ions, and the average temperature can be close to the ambient temperature. Whereas for thermal plasma, the electrons' temperature is kept in equilibrium with that of the ions, which is between a hundred and a thousand degrees higher than the ambient temperature.

Nonthermal plasma can be generated at low pressure ($10^{-1}-10^3$ Pa) by electrical discharges such as direct current (DC) or pulsed glow discharges, radiofrequency (RF) and microwave discharges or ion beam assisted deposition (IBAD). At atmospheric pressure, the plasma is generated by a dielectric barrier discharge (DBD) or an atmospheric pressure plasma jet. In this case, the nonthermal character is achieved by a low-power density or a pulsed power supply, where the short pulse duration prevents the equilibrium state from being established [24,25]. Nonthermal plasma can be used in direct contact with the surface of the polymer or to serve as a source of charged particles that are blown or accelerated onto the surface of the polymer (Table 14.2). Generally, there are described selected methods that can be suitable for industrial use or to achieve unique material properties [26,27].

Table 14.2 Examples of plasma methods for biomedical applications [12]

Biomedical applications	Plasma modification
Modification of blood compatibility	plasma fluoropolymer deposition
	plasma siloxane polymer deposition
	plasma treatment using radio frequency glow discharge
Influencing cell adhesion and growth	surface treatment using NH_3 plasma
	plasma deposition of the acetone or methanol thin films
	plasma deposition of fluoropolymer
Controlling protein adsorption	plasma deposition of TiO_2 thin films
	plasma treatment to create covalent bonds in the fibronectin
Improving wear resistance, lubricity	plasma functionalization (surface addition of functional groups OH, COOH)
	plasma etching
Altering transport properties or modifying electrical characteristic	plasma deposition (methane, fluoropolymer, siloxane)
	plasma deposition of insulation or conductive layer
Drug releasing	plasma surface functionalization

14.3.1 Surface Modification

The polymers used for implant production have, in most cases, an inert surface. For implants or parts that provide moving joints, the inert surface is advantageous, and much more importantly is the case in machining and the final surface operations, such as surface polishing. On the contrary, it is necessary to change the inert surface to a bioactive one without affecting the bulk properties of the polymer for implants which will be integrated into the bone [28].

Modification of the polymer surface properties may involve a change in the chemical composition, roughness, and wettability, or a combination therein. There are a number of conventional methods, such as wet-chemical methods [5]. But it is essential to use nonthermal plasma-based methods in some applications, for example, the controlled change of surface roughness on the nanoscale or bonding of the defined chemical groups to the surface. Plasma methods can be combined with conventional methods also [29].

14.3.1.1 Dielectric Barrier Discharge

The dielectric barrier discharge (DBD) is based on an alternating current (AC) discharge which generates thermodynamic nonequilibrium plasma at atmospheric pressure. The typically working frequency is in the range from 1 kHz to 10 MHz and the working pressure is in the range from 10 to 500 kPa [30,31]. The created low temperature plasma is suitable for treatment of temperature sensitive or dielectric materials, such as polymers for example.

The basic arrangement consists of two electrodes, one is covered with a dielectric layer and the second electrode is conductive (Fig. 14.2). The electrodes are powered by time varying voltages due to the presence of capacitive coupling. DBD is created from many tiny microdischarges which last for nanoseconds or microseconds. For specific parameters, a

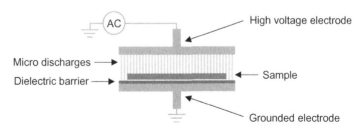

Figure 14.2 Basic arrangement of a dielectric barrier discharge.

homogeneous glow discharge can be obtained. It is usually necessary to add gas, for example helium allows to stabilize glow discharges at atmospheric pressure much easier than other gases [32].

The method was tested, for example, on the surface of carbon fiber-reinforced PEEK in the ambient air. It was shown that DBD plasma treatment led to a significant decrease in the water contact angle, an increase in roughness and the creation of new oxygen functional groups on the surface [33]. Further studies are focused on the deposition of coatings by DBD.

The method is still developing; many geometric configurations exist and have been tested with new pulsed power supplies [34]. An advantage of this method is the ability to work with low power in tens of Watts at atmospheric pressure. For industrial expansion, this is a crucial condition from an economic point of view.

14.3.1.2 Atmospheric Pressure Plasma Jet

The atmospheric pressure plasma jet (APPJ), see Fig. 14.3, consists of two coaxial electrodes between which a feed gas (usually helium, oxygen, argon, or other gases) flows at a high rate. The outer electrode is grounded and the inner electrode is powered by a RF power of 13.56 MHz. The plasma is created between the electrodes and exits the discharge volume through a nozzle onto the treatment material. The ions and electrons recombine, but the fast-flowing gas still contains neutral metastable species and radicals which can chemically modify the surface of the polymers. The plasma jet generates the thermal plasma in a central zone and the nonthermal plasma in a peripheral zone, where the heavy particle temperature is much lower than electrons temperature. The working temperature can by controlled by the distance of the plasma jet from the substrate [24,35].

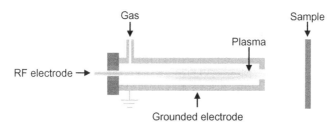

Figure 14.3 Basic arrangement of an atmospheric pressure plasma jet.

The APPJ method is primarily used for the treatment of polymeric surfaces to increase their wettability and adhesion to metal coatings. For example, the surface structuring of PEEK was tested this way. The plasma afterglow, which was used, contained activated neutral and metastable species without ions. The effects on the surface were mainly chemical in nature and the temperature at the nozzle exit did not exceed $200°C$ [36].

14.3.2 Coating Deposition

The coatings can change the biologically inert surface of the polymers to become bioactive or to improve their surface mechanical properties, primarily hardness or wear resistance. Unlike the modification of the surface, the coating can change the chemical composition of the surface completely and achieve unique properties in combination with the elastic polymer. A disadvantage of this solution may be the failure of the adhesion of the applied coating to the base material. Additionally, the coating must also be adapted to the elastic base material to avoid the delamination of the coating under load.

The most commonly used methods for a plastic coating is magnetron sputtering and its HIPIMS (high power impulse magnetron sputtering) modification which was developed more recently. Other methods can be used with similar principles also, such as electron beam scattering, which was tested for the deposition of a titanium coating on PEEK [37]. These are low-pressure methods, but polymers can be coated at atmospheric pressure also [38]. For example, the DBD method was tested for the deposition of the nanocrystalline anatase TiO_2 [39]. The principles of the atmospheric plasma methods are described in Section 3.1. Section 3.2 will be focused on the methods which are used in industry or the methods with future industrial potential.

14.3.2.1 Magnetron Sputtering

The basic layout is based on the glow discharge, where a high voltage potential is applied between two electrodes. Behind the cathode (Fig. 14.4), an asymmetric magnetic field, with permanent magnets, is placed. On the cathode surface, the magnetic tunnel is formed. It traps the electrons, which accelerate away from the cathode by the electric field. The electrons move around the magnetic field lines and they travel a much longer path-length in the plasma in comparison with conventional glow discharges. Ionization collisions are more intense and, consequently, higher ion fluxes are formed. The ions have a higher mass and

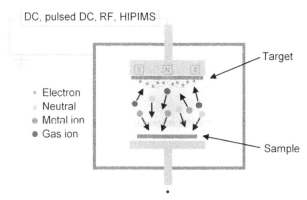

Figure 14.4 Magnetron sputtering method with various power supplies (DC, pulsed DC, RF, or HIPIMS) for coating of nonconductive materials.

are not influenced by the magnetic field lines as much as the electrons, but they bombard the cathode and cause a secondary electron emission. The enhanced ionization and the emission of the secondary electrons allow the magnetron to generally operate at higher currents (up to 0.1 A cm^2), lower voltages (300–700 V) and a lower pressure (0.1–1 Pa) than conventional glow discharges. Such an arrangement is referred to as a balanced magnetron and is advantageous for temperature sensitive materials where a low-current density is necessary for the deposition. On the contrary, an unbalanced magnetron does not have magnetic field lines closed on the cathode surface, but it leads the electrons and ions to the substrate [40].

Magnetron sputtering can be operated in DC, pulse DC, AC, or RF mode. For example, using RF magnetron sputtering, a hydroxyapatite (HA) coating can be deposited onto the polymeric materials. In general, the sputtered HA shows low crystallinity, which results in increase the dissolution of the HA layer in the body's fluids. To decrease dissolution, hydrothermal treatment was tested. The HA coating with a titanium intermediate layer was deposited onto PEEK and recrystallized by the hydrothermal treatment. The crystallinity increased, but the titanium adhesion to PEEK decreased with an increasing thickness of the titanium interlayer. The reason was the difference between the thermal expansion coefficients of titanium and PEEK [41].

Also, antibacterial properties of nano-silver coated PEEK [42] or tribological properties of DLC (diamond-like carbon) coated PEEK by a combination of magnetron sputtering and a PACVD (plasma assisted chemical vapor deposition) method [43] were tested.

14.3.2.2 High-Power Impulse Magnetron Sputtering

The method is based on the principle of magnetron sputtering (Fig. 14.4), but the power is applied in high energetic pulses with a low duty cycle (<0.1) with a frequency (<1.5 kHz), current density (0.2 A/cm^2) and power density (>0.5 kW/cm^2) on the target. The result is the creation of a very dense plasma that allows a high degree of ionization. Depending on the sputtering material, the ionization degree can reach up to 90 percent compared to the magnetron sputtering, where it does not exceed 10 percent.

The ions of the sputtered material can bombard the polymer substrate with a higher energy than the neutral particles. The enhanced mobility of the particles on the substrate surface and the coating structure results are denser without defects. The bombardment of the substrate with energetic ions can be used in ion implantation or the modification of surface morphology increasing the coating adhesion to the polymer substrate. Another advantage is the creation of hot electrons that are created in the initial phase of the pulse. They can cause further surface modifications, once they are absorbed by the substrate [44]. But generally, the temperature delivered to the substrate is lower for HIPIMS in the comparison with DC or pulsed DC magnetron sputtering [45].

It is not possible to apply the negative bias on the electrically nonconductive polymer materials. In this case, ions are partially accelerated by gradient between the plasma potential and the floating potential created on the polymer only. There are several possibilities how to ensure the accelerations of the positive ions.

The RF power can be applied to the sample holder (Fig. 14.4) on which a negative self-bias and a plasma are created with a stabilising effect on the HIPIMS discharge. The ion bombardment of the growing coating occurs during the time-off period [46].

Next option is a new approach called "hiPlus." The principle is based on the use of a reverse positive voltage pulse to the cathode after the primary plasma-ON sputtering pulse, thus accelerating the positive ions present in the plasma [47]. The additional acceleration of the ions can considerably improve the adhesion of the coating to the polymer and affect the growth mechanism of the coating. This new method has great potential for medical applications, coating PEEK with titanium for example.

The standard method of HIPIMS was used for the deposition of crystalline columnar TiO_2 coatings with anatase and rutile phases on a bioinert PEEK. The coatings with the phase exhibited greater osteoblast compatibility than the coatings with the anatase phase. The crystallinity HIPIMS coatings were higher than the coatings prepared by magnetron sputtering and the HIPIMS coating also had a stronger adhesion than the magnetron sputtering coating [48].

14.3.3 Ion Implantation

Ion implantation is one of the methods that can affect the mechanical or chemical properties of the surface layer of the polymeric material. The methods can be divided into three basic directions based on different configurations of the equipment: ion beam implantation, plasma immersion ion implantation, and cluster ion beam implantation [49].

14.3.3.1 Ion Beam Implantation

This method is referred to as *conventional beam line ion implantation*. The basic diagram of the ion implantation is provided in Fig. 14.5. The system consists of a plasma source, where the ions of the atoms or molecules are created in the plasma discharge using a plasma generator. Then the ions are extracted and accelerated to a high energy level and the ions bombard the solid material target. The current supplied to the implanter is typically in microamperes and the total amount of implanted material (dose), calculated from the integral over time of the ion current, is small.

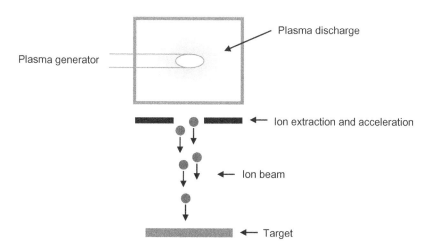

Figure 14.5 Basic arrangement of an ion beam implantation.

The ion beam implanter can work in two modes: in a continuous or pulsed one. A glow discharge, inductively coupled plasma, microwave induced plasma, or ECR ion source, etc., can be used as such an ion source. This is the reason why we can include ion implantation into the nonthermal plasma methods.

Depending on the implanted ion and implantation parameters (energy and dose), different effects can be obtained. Ion implantation of light elements (He, Ar, N, C) can increase the hardness due to a "crosslinking effect" of the polymeric chains induced by the ionization. On the contrary, ion implantation of heavier ions produces a "graphitisation" of the outer surface of the polymer, influencing, for example, the reduction of the friction coefficient or electrical conductivity [50,51].

The He^+ ion implantation was used for the treatment of the low abrasion resistant Bis-GMA/TEGDMA bio-compatible resin. The maximum surface hardening of the resin surfaces was achieved at He ion implantation energies of around 50 keV and fluences of $1 \times 10^{16} \, cm^{-2}$. The wear rate of the resin surface decreases by a factor of 2 with respect to the pristine resin [52].

UHMWPE was implanted by using H^+, He^+, Ar^+ and Xe^+ ion beams having a 300 keV energy level. The presented data referred to very thin graphite-like films on the polyethylene sheet, whose maximum thickness is 3.8 μm (for the 300 keV proton beams) [53].

14.3.3.2 Plasma Immersion Ion Implantation

The technology plasma immersion ion implantation (PIII) has been primarily developed for the surface treatment of conductive materials. The material is located directly on the active electrode, which is powered by a pulse bias with a high negative potential. A plasma sheet is formed around the whole electrode, including the treated material, in which the ions are accelerated and bombard the surface. The technology is still developing and allows for the modification of conductive materials as well as semiconductors or nonconductive materials, such as polymers (Fig. 14.6).

Changing the chemical composition of the bioinert surface is essential in the case of polymeric implants integrating with the bone. For example, the PEEK material, which has mechanical properties very close to that of bone, needs to be surface modified. One option is the integration of nitrogen-containing functional groups on the PEEK surface through nitrogen ion implantation. The modified PEEK has demonstrated enhanced biological activity [54].

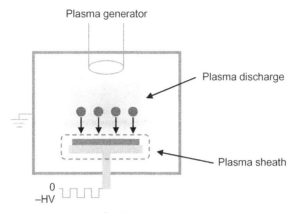

Figure 14.6 Basic arrangement of a plasma immersion ion implantation.

The method was also tested for increase surface hardness of UHMWPE and PEEK using of nitrogen, helium and oxygen gas mixtures. The results show that the ion implantation of combined elements has a significant effect on surface hardness of UHMWPE and only partial on PEEK, where surface hardness increase only using N-rich gas mixtures [55].

The method can also be used for coating, then it is called immersion ion implantation and deposition (PIII&D). Unlike the modification of the surface layer of the polymer, reactive gases are used for the deposition of the coatings. For example, the DLC film was deposited by the PIII&D method where the ion bombardment during the DLC deposition created a gradual transition between the DLC film and PEEK. The connection was almost isoelastic to cortical bone and the biological results showed increased osteoblast attachment, proliferation, and differentiation on the DLC/PEEK [56].

14.3.3.3 Cluster Ion Beam Implantation

This method is based on the implantation of cluster ions consisting of a few hundreds to many thousands of atoms. A cluster ion beam is created by expanding an appropriate gas through a small nozzle into a vacuum to form a stream of gas clusters with weakly-bonded atoms. The clusters are ionized by electron impact and accelerated using a high potential to the target material. Different cluster sources can be used. For example, ion sputtering source, magnetron sputtering source, or arc cluster ion source [57]. The schematic principle of the gas cluster ion beam (GCIB) technique is

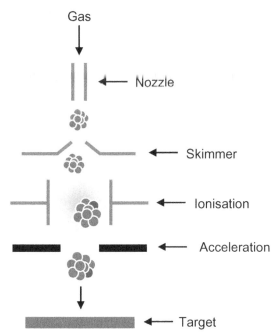

Figure 14.7 Basic arrangement of a cluster ion beam implantation.

provided in Fig. 14.7. The interaction of the cluster beams can by divided into two types: low energy interaction and energetic interaction.

The cluster ion beam implantation was modified for the surface treatment of polymers without surface chemical changes. A novel method of the accelerated neutral atom beam (ANAB) can modify the surface of PEEK to a shallow depth of no more than 5 nm [58]. The ANAB surface modification technique employs intense directed beams of neutral gas atoms having average energies which can be controlled over a range from a few eV per atom to beyond 100 eV per atom. These neutral atom beams are created by dissociating the energetic gas cluster ions produced by the GCIB technique [59].

14.4 PLASMA-MODIFIED POLYMER IMPLANTS

The in vivo applications and implants, where polymeric materials are used, are summarized by Maitz [17].

1. Temporary in vivo applications (vascular catheters, urinary catheters, and ureteral stents).

2. General surgical implants (suture materials, tissue adhesives and sealants, surgical meshes).
3. Orthopedic implants (joint prostheses, osteosynthesis materials, bone cements, scaffolds for ligament and tendon repair).
4. Vascular and cardio-vascular interventions (vascular stents, vascular grafts, polymeric heart valves.
5. Ophthalmology (intraocular lenses, cornea implants).
6. Dentistry (composites).
7. Neurosurgery (peripheral nerve guidance conduits).
8. Plastic, reconstructive and cosmetic surgery.

14.4.1 Temporary In Vivo Applications

Vascular catheters are used for both short and long durations and have to sustain low and high pressures. They have to meet vascular access needs to the body and reduce the complications connected with the insertion and placement in the desired position in the vascular tree. Different materials are used for vascular catheters to fulfil application requirements including silicone and polyurethanes. Different polyurethanes have been introduced to improve stability and vascular compatibility.

Implanted vascular catheters can induce two major problems: infections and clotting. Infections can occur in two ways: Firstly, infection is induced by a pathogenic bacteria or fungi that can migrate from the skin along the external catheter surface to cause a subcutaneous infection and the subsequent infection of the bloodstream. The second way is intraluminal bacterial colonization and migration to the bloodstream due to the contamination of the catheter hub. In order to reduce the incidence of infection, a topical application of chlorhexidine gluconate as a skin disinfectant for catheter site care is normally used. Another way is to treat vascular catheters with antimicrobial coatings [60−62].

Urinary catheters are used for both short or long duration also. Despite the large effort to avoid contamination and subsequent infections, catheters are still providers of microbial accumulation to form a single species biofilm which can even cover short-term catheters. More recently, materials such as gum-elastic, plastic (poly(vinylchloride), PVC), polyurethanes, silicone and latex rubbers have been used for their good malleability. Various processes to prevent infections related to urinary catheterisation are applied, such as the better handling of catheters, fabricating urinary catheter coatings, improving catheter design, and emphasizing short-term use. Urinary catheter coatings are designed to kill microbes. Clinically tested biocidal coatings are based on silver or antibiotics as the active agent. While these agents are predominant in the clinical field, many other agents (triclosan, chlorhexidine, nitric oxide, enzymes, peptides) or agent carriers (liposomes, polymers) are in the research stage [63,64].

Also, the application of nonvascular stents can cause some problems connected with their applications, such as patient discomfort (pain, urgency, frequency) and can become encrusted and infected [65].

Ways on how to avoid problems occurring during the use of the catheters or stents include changes in the device design, polymeric composition, drug-elution, and surface coatings. Surface coatings should prevent infection by inhibiting the bacteria as well as resisting urinary crystal formation and adherence. The application of an antiadhesive (modifying the surface charge, hydrophobicity, and roughness) and an antimicrobial (silver, antibiotics, detergents, or other) compounds are used to avoid these complications. Although plasma treatment of catheters or stents helps to achieve these various targets, most of the applications are still in the research stage. As it was mentioned above, during plasma treatment various types of the reactive species (electrons, radicals, ions, reactive molecules, UV irradiation, etc.) are generated, thus it is a very complicated process and many applications have not been clinically tested yet.

Polyurethane (PU) is often used as a material for catheters. A good example of the problems connected with the plasma treatment of catheters can be found in [66] Mrad et al., who studied plasma end electron beam sterilization of the catheters made of PU Pellethane 2363-80AE. Polyurethane cannot be sterilized by steam autoclaving, because this process may induce degradation and loss of shape. PU can suffer from different kinds of degradation processes. For example, the formation of quinoid structures in the polymer and photo Fries rearrangement with $C-N$ cleavage were reported [67−70]. Mrad et al. presume that low temperature plasma appears to be a good alternative to the existing sterilization processes. They used cold nitrogen atmospheric plasma (pulsed corona discharge) and their work focused on the following points:

1. Chain degradation, by evaluating branching, scission and crosslinking of the polymer chains, as these degradation processes might generate potential leachables, like oligomers, and change the mechanical properties of the catheter.
2. Hard/soft segment reorganization.

3. Degradation of the additives that could reduce polymer stability and create potential toxic and leachable compounds.
4. A preliminary study on surface modifications induced by plasma treatment.

The study shows that surface modification occurs on the depth of a few nanometers because no modification was observed by ATR on a micrometer scale. These results are very important, because the surface state has a significant impact on the biostability and the biocompatibility of the polymers. The results showed that the chemical composition changed after plasma treatment, with an increasing amount of the polar species (O and N). The topology was also modified due to a smoothing effect of the treatment. They found that plasma treatment did not damage the bulk of the polyurethane catheter nor affect the additives [66]. On the other hand, Lerouge et al. realised that different techniques of plasma treatment can lead to the degradation of the additive, slight changes in the hydrolytic stability, but not to the modification of the molecular weight [71].

APPJ was used by Weltmann et al. for the antimicrobial treatment of catheters for minimal invasive intracardial electrophysiological studies. RF-driven APPJ with Ar or Ar with small admixture of O_2 as working gases were used. The treatment device used was a movable plasma module consisting of two APPJs adjusted nearly against each other to ensure the complete treatment of the catheter. They proved to have a successful bactericidal effect of the plasma treatment on the Staphylococcus aureus bacteria [72].

All these results were observed for the external surface of the catheters only. Here, we can see another problem connected with the plasma surface treatment. Generally, it is the various shape of the treated implants. When surface modification should be homogenous and provide the required quality, all the treated surface has to be sufficiently exposed to the plasma. But when the treated implant has an intricate shape, a shadowing effect can occur during the plasma modification process resulting to inhomogeneous and inferior treatment. In the case of the catheters which have the shape of a long tube with a small diameter plasma cannot reach all the internal surface inside the catheter, thus, some part of the surface can remain untreated.

Among all the problems concerning the biological properties of stents and catheters, those made of the polymers could have worse printability due to the low-surface hydrophilicity. But marking the material used in healthcare is also important. For example, SurfaceTreat a.s. modified the wettability of PU catheters by APPJ using a glow discharge. The plasma treatment in the presence of air leads to the enrichment of the OH groups on the polymer surface providing an increase of hydrophilicity of the plasma modified surface.

Fig. 14.8 shows the increasing hydrophilicity of the modified catheters measured using a drop test where the contact angle decreased from 92 degrees before plasma modification to 40 degrees after plasma modification.

SurfaceTreat a.s. also studied the modification of nonvascular stents. Nonvascular stents designed using polydioxanone (PDO) were subsequently deposited by a copolymer of polybutylene terephthalate (PBT) and polyethylene glycol (PEG) (Fig. 14.9). Some of the stents showed poor adhesion of the PBT-PEG layer to the PDO surface, thus during

Figure 14.8 Plasma treatment of the catheters: (A) the drop test before modification, and (B) the drop test after modification.

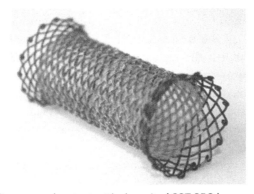

Figure 14.9 PDO nonvascular stent with deposited PBT-PEG layer.

mechanical stress, some disruption occurred. SurfaceTreat a.s. modified the surface of the PDO stent using low pressure RF plasma treatment with Ar and O_2 gases to increase hydrophilic properties, thus, the adhesion of the PBT-PEG layer to the PDO increased.

14.4.2 General Surgical Implants

Suture materials are usually not determined to be used for long-term applications. Nevertheless, suture materials have to meet some required properties. A surgical suture should have the needed tensile strength, tie easily, form secure knots, etc. They should not induce infection, should not have an adverse effect on the healing of the wound. Surgical sutures can be divided into main groups: (1) absorbable sutures, and (2) nonabsorbable sutures. Synthetic absorbable polymeric materials include aliphatic polyesters, polyglicolic acid (PGA), poly(glycolide-*co*-L-lactide) copolymer, polydioxanone (PDO), other glycolide-lactide copolymers, polylactide (PLA), polyhydroxyalkanoates (PHA). Synthetic nonabsorbable polymeric materials include polyesters, polyamides, polyolefins—polypropylene (PP), ultrahigh molecular weight polyethylene (UHMWPE), polytetrafluoroethylene (PTFE), and polyvinylidene fluoride (PDFD) [21,73–75].

Subsequently, some studies are concerned with plasma modification of suture materials. Saxena et al. used plasma-induced graft polymerization of acrylic acid followed by chitosan binding on the remaining carboxyl groups followed by immobilization of nanosilver and tetracycline hydrochloride to develop antimicrobial PP suture with drug release properties [76].

Serrano et al. investigated oxidative plasma treatment to nanopattern commercially available suturing threads (nonabsorbable monofilaments of PP, PET, and absorbable monofilament sutures of polydioxanone and modified glycolic acid). The possibility to prevent biofilm formation and to influence of the presence of nanosized topographies on a bacteria adherence was described [77].

14.4.3 Orthopedic and Spinal Implants

Orthopedic implants are a large group of surgical implants where polymers are used. It includes joint prostheses as a finger joint prosthesis, total hip, knee, ankle, or shoulder replacements where polymers are frequently used in combination with metals, ceramics, osteosynthesis materials, bone cements, scaffolds for ligament and tendon repair. Commonly used polymers for orthopedic applications are: acrylic, nylon, silicone, polyurethane, ultra-high molecular weight polyethylene (UHMWPE), and polypropylene (PP) [78–82]. The problems associated with the pathology of joint tissue replacement and regeneration can be considered under four headings:

1. Replacement of articular surfaces.
2. Replacement of nonarticulating intra-articular structures.
3. Intra-articular injectable agents.
4. Tissue engineering / regeneration [83].

Natural bone is a composite material that consists of a matrix based on collagen fibers hardened by hydroxyapatite crystals. The main requirements demanded for orthopedic implants are: high biocompatibility with no causation of inflammatory or toxicity for the human body, appropriate mechanical, tribological, and surface properties and economically feasible manufacturing and processing.

Generally, orthopedic implants have to fulfil the high demands on the required mechanical properties. The elastic modulus, tensile and compressive strengths are the mechanical properties investigated to ensure suitability of the biomaterial [78].

However, it is not only the mechanical properties that play a crucial role in the application of orthopedic implants. Surface and tribological properties are considered of high importance, for example, components used in bearing the joint prosthesis. Bearing surfaces should have low friction, should be resistant to wear and have good biocompatibility properties. Among other materials like metals or ceramics, a large proportion of bearing the joint prosthesis is designed as a metal-on-polymer. For example, high-density polyethylene used for the plastic acetabular component in most systems is hard wearing and has low friction. Polymer microparticles that are produced through wear may lead to lysis around the implants, resulting in aseptic loosening and the need for revision surgery, although this process is slow and predictable [84].

Ultra-high molecular weight polyethylene (UHMWPE) has an irreplaceable role in orthopedic implantology. UHMWPE is a semicrystalline polymer with roughly half its chains structurally organized in crystalline lamellae and the balance entangled in a random amorphous state [85]. UHMPWE is biologically inert and it has excellent wear resistance, high levels of strength, a low-friction coefficient, creep resistance, low-friction surface. Nevertheless, microparticles generated during the movement of the joint due to wearing can cause problems, such as loosening and osteolysis. To avoid these problems, UHMWPE is modified using several methods. For example, hydroxyapatite (HA) is used as a

(A)   (B)

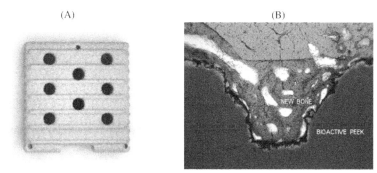

Figure 14.10 Spinal implant made of bioactive PEEK (LASAK Ltd.): (A) surface treatment includes a titanium coating deposited by magnetron sputtering and its chemical bioactive treatment, and (B) direct contact between the bone and implant (6 weeks after implantation, toluidine blue staining).

reinforcement agent which also helps to facilitate a biological fixation between the human cells and the implants [86].

In spinal implantology, PEEK is used for cervical intervertebral implants. Its mechanical properties are close to the properties of the bone, but the its bio-inert surface does not allow for the perfect integration with the bone. One way to change the bio-inert surface to a bioactive one is, for example, the application of a titanium coating and its chemical activation [87]. The company LASAK Ltd. along with the author's team tested a titanium coating deposited by magnetron sputtering. Fig. 14.10A shows a spinal implant with a bioactive surface treatment and Fig. 14.10B shows the direct contact between the bone and implant, 6 weeks after implantation.

14.4.4 Vascular and Cardio-Vascular Intervention

Nearly one-fourth of all deaths worldwide are attributable to cardiovascular diseases, including stroke and ischemic heart disease [88].

Vascular and cardio-vascular implants including vascular stents, vascular grafts, polymeric heart valves intervention is necessary to treat cardiovascular diseases. Vascular grafts are used to repair weakened blood vessels and to bypass blockages, polymeric artificial valves are used to replace damaged or diseased natural valves, and vascular stents are used to restore normal blood flow following coronary artery obstruction due to the presence of atherosclerosis [89,90].

Synthetic Dacron (PET) with a number of applied coatings including, for example, fluoropolymer, heparin-bound polymer, or silicone elastomer have been used for small diameter vascular grafts or Goretex (expanded PTFE), where the pore size can be varied from 30 to 100 μm, has also been used. Other materials used for vascular grafts are polyurethane, polyester-based urethane (PEU) or polyether urethane (PEEU) and poly-carbonate urethane (PCU).

Biodegradable polymers have been used in two ways: (1) as an implantable graft degrading while tissue remodeling occurs around it and inside it, and (2) as a scaffold for growing vascular tissue ex vivo, followed by implanting the cell-containing polymeric structure [91].

The main problem of implants used in vascular and cardio-vascular intervention is thrombogenicity. Vascular and cardio-vascular implants are in direct contact with the blood. But the surface of these implants is not recognized by the blood resulting in the deposition of plasma proteins. It leads to the formation of thrombosis on the implant's surface and adhesion of platelets. Thrombogenicity can be a problem especially in the case of the artificial grafts where the internal diameter is less than 5 mm, thus subsequent anticoagulation therapy is necessary to minimize thromboembolic complications [92,93].

In 1987 Hoffman et al. patented a plasma-treatment method (plasma polymerization) convenient for vascular graft materials to improve their biocompatibility and blood compatibility. The vascular graft material is exposed to a low-pressure plasma gas discharge in the presence of an inert gas following plasma treatment in an organic gas (halocarbon or halohydrocarbon) to form a thin biocompatible polymer film covalently bonded to the surface of the vascular graft material. Hoffman et al. proposed a plasma treatment for various materials such as polyethylene, polyacrylics, polypropylene, polyvinyl chloride, polyamides, polystyrene, polyfluorocarbons, polyesters, silicone rubber, hydrocarbon rubbers, polycarbonates, and other such synthetic resin materials and various vascular implants, such as a valve, pin, catheter, sleeve, vascular graft, and surgical tubing [94].

PTFE or e-PTFE is widely used in cardio-vascular intervention. The surface modification techniques of PTFE involve the removal of the F groups from the surface and the addition of special functional groups (such as OH, COOH or NH_3 groups). Exposure to plasma leads to a partial defluorination by $-C-F$ bond scission or polymer chain breakage. The $-C-F$ bond may arise from the ion interaction and it can react with other radicals on the surface of the treated polymer, air oxygen, thus $-C=C-$ bonds may be created on the plasma activated surface.

However, in comparison to other polymers, the PTFE surface exhibits a lower free radical count available for reaction with oxygen or nitrogen [26]. For example, a quick defluorination of fluoropolymers can be realised using RF hydrogen plasma [95]. Chandy et al. studied the surface modification of PET (Dacron) and PTFE (Teflon) vascular grafts. They showed that RF plasma surface grafting (in an Ar inert gas) of matrix components, such as collagen-type IV and laminin followed by the immobilization of the bioactive molecules (PGE, heparin or phosphatidyl choline) changed the surface conditioning of the vascular grafts and subsequently improved their biocompatibility [96].

The modification of small-diameter vascular grafts made of PCU was also investigated. Nanofibrous PCU grafts using low pressure RF plasma with O_2 and subsequently allylamine to create an ultra-thin film of stable primary and secondary functional groups of amines on the graft surface, which were subsequently employed to conjugate heparin via endpoint immobilization was investigated [97].

Many other studies have been made in the area of polymeric implants modification for cardio-vascular intervention. For example, Bilek et al. investigated and showed that an RF plasma enhanced chemical vapor deposition technique using a carbon-containing plasma can design polymers with the ability to covalently immobilize proteins in their bioactive form [98]. Guex et al. studied RF plasma surface functionalised electrospun poly(E-caprolactone) fibers as a matrix for bone-marrow-derived mesenchymal stem cell (MSC) cardiac implantation [99].

14.4.5 Ophthalmology

Biomedical applications of polymers in ophthalmology include vitreous replacement fluids, contact lenses, intraocular lenses, artificial orbital walls, artificial corneas, artificial lacrimal ducts, glaucoma drainage devices, viscoelastic replacements, drug delivery systems, sclera buckles, retinal tacks and adhesives, and ocular endotamponades [100].

Intraocular lenses (IOLs) are synthetic lenses implanted into the human eye to treat myopia and cataracts and they are used most frequently in optical surgeries. Various polymers are used for IOLs like rigid PMMA [101,102] with the satisfying biocompatibility of its surface, excellent light transmissibility, chemical inertness and possible incorporation of a UV absorber, and silicone-based IOLs poly(-dimethylsiloxane) poly(DMS), poly(diphenylsiloxane) poly(DPhS). According to the demanded surface properties, hydrophilic/hydrophobic IOLs are also based on acrylic polymers. Hydrophilic acrylic IOL polymers are designed from a network of hydrophilic chains able to absorb water. This network is based on 2-hydroxyethyl methacrylate (HEMA) and 6-hydroxyhexyl methacrylate (HEXMA), with a crosslinking agent (ethylene glycol dimethacrylate (EGDMA). Hydrophobic acrylic IOLs are based on esters of poly(meth)acrylic acid, mainly poly(2-phenethyl (meth) acrylate) [poly(PE(M)A], poly(ethyl(meth)acrylate) [poly(E(M)A)] and poly(2,2,2-trifluoroethyl methacrylate) [poly(TFEMA)]. They absorb less than 1 wt% of water [103,104].

Contact lenses are biomedical polymeric materials providing the correction of mild ametropia. As a material applied onto the human eye, it must be sufficiently hydrophilic, resistant to tear residue, and fulfil the required optical properties. Poly(hydroxyethylmethacrylate) (PHEMA) or hydrogels containing siloxane polydimethylsiloxane (PDMS) are also used [105].

Plasma modification of polymers for ophthalmic implants is realised in order to reach various goals to influence the biological response of the ophthalmic implants and to improve biocompatibility to improve the antibacterial properties by influencing surface properties such as wettability and roughness, to improve antithrombogenecity, to reduce ultraviolet transmittance or to provide drug release.

To improve anterior surface biocompatibility, Lin et al. modified acrylic IOLs made of a copolymer of butyl acrylate, ethyl methylacrylate, and methyl styrene with a crosslinking agent and ultraviolet-absorbent. Poly(ethylene glycol)s (PEGs) were immobilized by an atmospheric pressure glow discharge (APGD) treatment using argon as the discharge gas. PEG-grafted IOL showed nonageing increased hydrophilicity. The PEG-grafted surface also provided resistance to the platelets, macrophages, and lens epithelial cells attachment and the spread and growth of cells were suppressed [106].

Presently, many research activities are concerned with the plasma surface treatment of polymeric implants for ophthalmology. Thus, of course, various types of the bacteria are used to test the antibacterial properties of the plasma modified surface, most commonly *E. coli* and *Staphylococcus aureus*.

Rezaei et al. also studied cell adhesion on a modified PMMA surface. The PMMA used in the ophthalmic implants was modified by an RF low pressure plasma with O_2 as the working gas. It is well-known that the treatment of the polymeric surface using oxygen plasma discharge results in the increasing hydrophilicity and it was widely researched and it is already used in industrial applications. This phenomenon is caused by the formation of the functional groups (OH, COOH, etc.) on the treated surface changing the values of the surface energy, thus, the hydrophilicity and changes in the surface roughness caused by the etching process during plasma treatment. In this case, the bacterial adhesion of *E. coli* was successfully decreased. Razei et al. concluded that the hydrophilicity of the polymeric surface has a higher influence on the bacterial adhesion than the roughness [107].

However, not only PMMA is used in ophthalmology. Silicone elastomers (SE) are used for artificial cornea applications. For example, the surface modification of a silicone rubber membrane using an Ar plasma glow discharge was described in [20].

SE is functionalised utilizing atmospheric pressure plasma induced graft immobilization of poly(ethylene glycol) methyl ether methacrylate (PEGMA) [108] or poly(2-hydroxy ethyl methacrylate) (PHEMA) [20]. D'Sa et al. [108] used DBD modification to introduce reactive oxygen functionalities, and they tested cell adhesion on lens epithelial cells and bacterial adhesion of Staphylococcus aureus. Depending on the plasma treatment condition, the lens epithelial cells displayed evidence of shrinkage and were on the verge of detaching or no cell adhesion was detected. Contrary to this, no difference in the number of bacteria adhering to any of the tested surfaces was observed.

The other problem connected with ophthalmic implants can be thrombogenicity. To improve antithrombogenicity and reduce ultraviolet transmittance, polymethyl methacrylate intraocular lenses (PMMA IOLs) were pretreated with Ar plasma and combined with heparin (Hp), with polyglycol (PEG) and with both Hp and PEG in a plasma atmosphere in the work of Zhang et al. An RF vacuum plasma system was used for modification process. Platelet adhesion on the PMMA surface showed improvement of the antithrombogenicity and ultraviolet transmittance was reduced as well [109].

IOLs can be used as well for special applications as an alternative for extended, local drug delivery in the prevention of postoperative acute endophthalmitis. Modification of the surface of a hydrophilic acrylic material, used for manufacturing of IOLs, through plasma–assisted grafting copolymerization of 2-acrylamido-2-methylpropane sulfonic acid (AMPS) or [2-(methacryloyloxy)ethyl]dimethyl-(3-sulfopropyl)ammonium hydroxide (SBMA), to provide a controlled and effective drug release was realised. The modified material was loaded with moxifloxacin (MFX), a commonly used antibiotic for endophthalmitis prevention. This application is very promising because the swelling capacity, wettability, refractive index, and transmittance were not affected by the surface modification. The drug release of the modified IOLs was sufficient to be effective against Staphylococcus aureus and Staphylococcus epidermidis [110].

14.4.6 Dental Implants

The most frequently required tasks for implants in dentistry is to fill cavities occurring in the tooth or to substitute part or all of a missing tooth. Many different materials are used to treat dental problems such as cavity lining, cavity filling, fluting, endodontic, crown and bridge, prosthetic, preventive, orthodontic, and periodontal treatment of teeth [111,112]. The materials used in dental applications must resist high shear forces and load, forces occurring during thermal expansion and shrinkage and, at the same time, to provide esthetic quality, thus, optical properties. Dental resins are translucent with a refractive index matching that of the enamel and they are usually made from composites based on dimethacrylate or monomethacrylate monomers. Dental implants entering the jaw bone or fixed on its surface can be designed from metals or ceramics. However, various polymers are used as well. Mainly it is UHMPE, PMMA, PTFE, PS, PET, or in CF/C composites. But other utilization of polymers in dentistry like PEEK are studied [113,114]. When one tooth or more teeth are needed to be replaced, dental bridges are mainly based on composite materials from CF/PMMA, KF/PMMA, UHMWPE/PMMA, and GF/PMMA [111]. Some synthetic polymers are used as structural templates for stem cell differentiation and the proliferation of tissue scaffolds and to promote dental pulp tissue regeneration [115].

Plasma surface methods in dentistry is frequently applied to enhance osseointegration. The oxygen plasma modification of the PEEK surface by an RF glow discharge was tested. It has been confirmed that at the initial phase it improved osseointegration and implant stability in cancellous bones [116]. Another plasma treatment technique improving osseointegration is based on the coating methods such as deposition of plasma polymerized hexamethyldisiloxane, plasma polymerized allylamine, or plasma polymerized acrylic acid [117]. The adhesion of polymeric implant materials is connected with their surface wettability. PEEK, as many other polymers, may have level of the hydrophilicity which is unsatisfactory for good adhesion. The reactive magnetron sputtering for bioactive TiN-HA layer deposition on PEEK was tested by Boonyawan et al. They deposited clear films with high refractive index reducing the surface contact angle by more than 10 degrees. However, XPS results did not confirm the presence of TiN in the deposited films. In this case, water was released from the HA target and formed TiOH on the surface of the PEEK [118].

Several studies dealt with adhesion improvement. The good adhesion is necessary to create a stable bond/connection between composite resin and intact dentine. Zhang et al. investigated the positive influence of the plasma treatment using nonthermal argon plasma brush on the penetration of a model dental adhesive (based on BisGMA and HEMA) into demineralized dentine [119]. The influence of treatment by nonthermal atmospheric plasma brush on the surface wettability of different dental substrates (dentin, enamel, BisGMA, solirane-based) was investigated by Chen et al. In their work, they have successfully proven an increase in the dental surface hydrophilicity without altering the bulk properties or changing the surface morphology [120].

14.5 CONCLUSION

Polymeric materials provide a wide variability of properties allowing their utilization in different types of biomedical applications. However, they must meet the high requirements of the required functionalities and biomedical properties at the same time. Thus, some problems connected with a harmful effect due to the biological response of a body to an implanted material or insufficient functionality can occur.

Plasma treatment is a very promising method to modify and influence the required properties of polymeric materials used in implantology. Plasma processes involve a wide range of a various technologies which can be possibly used for the polymers' treatment such as surface modification/functionalisation, coating deposition or ion implantation. It involves different techniques operating under various conditions (low or atmospheric pressure, different working gases, various plasma sources, etc.). Nevertheless, the strong variability and complicated chemistry of the plasma processes together with the variability of the polymeric materials used in implantology result in the fact that plasma modification of polymeric in vivo applications and implants is a complex and difficult topic.

For example, plasma processes of polymers' modification were mostly applied on two-dimensional substrates so far. However, plasma treatment of three-dimensional implants (small-diameter implants, implants with an intrinsic shape or porous implants) is still in progress. The other problem is the price of the plasma treatment processes and potential complicated control of the processes, especially in case of the techniques carried out under low pressure using special chemicals. Thus, most of the applications of the plasma treatment of polymeric implants are still in the research phase and their utilization as a common technique is frequently not yet possible.

ACRONYMS

AC	alternating current
AMPS	2-acrylamido-2-methylpropane sulfonic acid
ANAB	accelerated neutral atom beam
APGD	atmospheric pressure glow discharge
APPJ	atmospheric pressure plasma jet
ATR	attenuated total reflectance
Bis-GMA/	2,2-bis[4-(2-hydroxy-3-methacryloxyl-oxypropoxy)
TEGDMA	phenyl]propane / triethylene glycol dimethacrylate
CF	carbon fibers
DBD	dielectric barrier discharge
DC	direct current
DLC	diamond like carbon
DMS	dimethylsiloxane
DPhS	diphenylsiloxane
ECR	electron cyclotron resonance
EGDMA	ethylene glycol dimethacrylate
(E(M)A	ethyl(meth)acrylate
e-PTFE	expanded polytetrafluoroethylene
GCIB	gas cluster ion beam
GF	glass fibers
HA	hydroxyapatite
HDPE	high-density polyethylene
HEMA	2-hydroxyethyl methacrylate
HEXMA	6-hydroxyhexyl methacrylate
HIPIMS	high power impulse magnetron sputtering
Hp	heparine
IBAD	ion beam assisted deposition
IOLs	intraocular lenses
KF	Kevlar fibers
LDPE	low-density polyethylene
PBT	polybutylene terephthalate
PACVD	plasma assisted chemical vapor deposition

PC	polycarbonate
PCL	poly(caprolactone)
PCU	polycarbonate urethane
PDFD	polyvinylidene fluoride
PDO	polydioxanone
PE	polyethylene
PEEK	polyether ether ketone
PEEU	polyether based urethane
PEGMA	poly(ethylene glycol) methyl ether methacrylate
PELA	block co-polymer of lactic acid and polyethylene glycol
PE(M)A	2-phenethyl (meth)acrylate
PET	polyethylene terephthalate
PEU	polyester-based urethane
PGA	poly(glycolic acid)
PGE	Prostaglandin E
PHA	polyhydroxyalkanoates
PHB	polyhydroxybutyrate
PHEMA	poly(2-hydroxyethylmethacrylate)
PIII	plasma immersion ion implantation
PIII & D	immersion ion implantation and deposition
PLA	poly(lactic acid)
PLDLA	poly(L-DL-lactide)
PLGA	poly (lactic-co-glycolide)
PLLA	poly(L-lactic acid)
PMA	polymetacrylate
PMMA	polymethylmetacrylate
PP	polypropylene
PS	polysulfone
PTFE	polytetrafluoroethylene
PU	polyurethane
PVC	poly(vinylchloride)
RF	radio frequency
SBMA	2-(methacryloyloxy)ethyl]dimethyl-(3-sulfopropyl)ammonium hydroxide
TFEMA	2,2,2-trifluoroethyl methacrylate
UHMWPE	ultra-high molecular weight polyethylene

REFERENCES

[1] D. Sumner, Long-term implant fixation and stress-shielding in total hip replacement, J. Biomech. 48 (2015) 797—800. Available from: https://doi.org/10.1016/j.jbiomech.2014.12.021.

[2] W. Lee, J. Koak, Y. Lim, S. Kim, H. Kwon, M. Kim, Stress shielding and fatigue limits of poly-ether-ether-ketone dental implants, J. Biomed. Mater. Res. Part B Appl. Biomater. 100B (2012) 1044—1052. Available from: https://doi.org/10.1002/jbm.b.32669.

[3] J. Wong, J. Bronzino, Biomaterials, CRC Press, Boca Raton, 2007.

[4] W. Schmoelz, U. Vieweg, F. Grochulla (Eds.), Manual of Spine Surgery: Implant Materials in Spinal Surgery, Springer, New York, 2012.

[5] R. Ma, T. Tang, Current strategies to improve the bioactivity of PEEK, Int. J. Mol. Sci. 15 (2014) 5426−5445. Available from: https://doi.org/10.3390/ijms15045426.

[6] L. Hallmann, A. Mehl, N. Sereno, C. Hämmerle, The improvement of adhesive properties of PEEK through different pre-treatments, Appl. Surf. Sci. 258 (2012) 7213−7218. Available from: https://doi.org/10.1016/j.apsusc.2012.04.040.

[7] Y. Zhao, H. Wong, W. Wang, T. Li, F. Ma, P. Chung, et al., Cytocompatibility, osseointegration, and bioactivity of three-dimensional porous and nanostructured network on polyetheretherketone, Biomaterials 34 (2013) 9264−9277. Available from: https://doi.org/10.1016/j.biomaterials.2013.08.071.

[8] M. Deng, S. Kumbar, C. Laurencin (Eds.), Natural and Synthetic Biomedical Polymers, first ed., Elsevier Science, 2014.

[9] L. Poole-Warren, P. Martens, R. Green (Eds.), Biosynthetic Polymers for Medical Applications, first ed., Elsevier Science, 2015.

[10] Z. Wang, N. Li, R. Li, Y. Li, L. Ruan, Biodegradable intestinal stents: a review, Progress Nat. Sci. Mater. Int. 24 (2014) 423−432. Available from: https://doi.org/10.1016/j.pnsc.2014.08.008.

[11] B. Ulery, L. Nair, C. Laurencin, Biomedical applications of biodegradable polymers, J. Polym. Sci. Part B: Polym. Phys. 49 (2011) 832−864. Available from: https://doi.org/10.1002/polb.22259.

[12] B. Ratner, Biomaterials Science: An Introduction to Materials in Medicine, third ed., Elsevier/Academic Press, Boston, 2013.

[13] Z. Sheikh, S. Najeeb, Z. Khurshid, V. Verma, H. Rashid, M. Glogauer, Biodegradable materials for bone repair and tissue engineering applications, Materials 8 (2015) 5744−5794. Available from: https://doi.org/10.3390/ma8095273.

[14] J. Middleton, A. Tipton, Synthetic biodegradable polymers as orthopedic devices, Biomaterials 21 (2000) 2335−2346. Available from: https://doi.org/10.1016/S0142-9612(00)00101-0.

[15] S. Im, Y. Jung, S. Kim, Current status and future direction of biodegradable metallic and polymeric vascular scaffolds for next-generation stents, Acta Biomater. 60 (2017) 3−22. Available from: https://doi.org/10.1016/j.actbio.2017.07.019.

[16] N. Goonoo, R. Jeetah, A. Bhaw-Luximon, D. Jhurry, Polydioxanone-based bio-materials for tissue engineering and drug/gene delivery applications, Eur. J. Pharm. Biopharm. 97 (2015) 371−391. Available from: https://doi.org/10.1016/j.ejpb.2015.05.024.

[17] M. Maitz, Applications of synthetic polymers in clinical medicine, Biosurf. Biotribol. 1 (2015) 161−176. Available from: https://doi.org/10.1016/j.bsbt.2015.08.002.

[18] P. Slepička, J. Siegel, O. Lyutakov, V. Švorčík, Nanostructuring of polymer surface stimulated by laser beam for electronics and tissue engineering, Chemické Listy 106 (2012) 875−883.

[19] K. Kim, C. Ryu, C. Park, G. Sur, C. Park, Investigation of crystallinity effects on the surface of oxygen plasma treated low density polyethylene using X-ray photo-electron spectroscopy, Polymer. (Guildf). 44 (2003) 6287−6295. Available from: https://doi.org/10.1016/S0032-3861(03)00674-8.

[20] P. Chu, Plasma-surface modification of biomaterials, Mater. Sci. Eng. R: Rep. 36 (2002) 143−206. Available from: https://doi.org/10.1016/S0927-796X(02)00004-9.

[21] M. King, B. Gupta, R. Guidoin, Biotextiles as Medical Implants, Woodhead Publishing, Philadelphia, 2013.

[22] D. Shi, Introduction to Biomaterials, World Scientific, Hackensack, NJ, 2006.

[23] S. Bauer, P. Schmuki, K. von der Mark, J. Park, Engineering biocompatible implant surfaces, Prog. Mater. Sci. 58 (2013) 261−326. Available from: https://doi.org/10.1016/j.pmatsci.2012.09.001.

[24] C. Tendero, C. Tixier, P. Tristant, J. Desmaison, P. Leprince, Atmospheric pressure plasmas: a review, Spectrochim. Acta Part. B. At. Spectrosc. 61 (2006) 2–30. Available from: https://doi.org/10.1016/j.sab.2005.10.003.

[25] M. Ruma, R. Habib, Basha, A survey of non-thermal plasma and their generation methods, Int. J. Renew. Energy Environ. Eng. 4 (2016) 6–12.

[26] H. Abourayana, D. Dowling, Plasma processing for tailoring the surface properties of polymers, Surface Energy, InTech, 2015, pp. 123–152. Available from: https://doi.org/10.5772/60927.

[27] L. Minati, C. Migliaresi, L. Lunelli, G. Viero, M. Dalla Serra, G. Speranza, Plasma assisted surface treatments of biomaterials, Biophys. Chem. 229 (2017) 151–164. Available from: https://doi.org/10.1016/j.bpc.2017.07.003.

[28] S. Yoshida, K. Hagiwara, T. Hasebe, A. Hotta, Surface modification of polymers by plasma treatments for the enhancement of biocompatibility and controlled drug release, Surf. Coat. Technol. 233 (2013) 99–107. Available from: https://doi.org/10.1016/j.surfcoat.2013.02.042.

[29] T. Yabutsuka, K. Fukushima, T. Hiruta, S. Takai, T. Yao, Effect of pores formation process and oxygen plasma treatment to hydroxyapatite formation on bioactive PEEK prepared by incorporation of precursor of apatite, Mater. Sci. Eng. C 81 (2017) 349–358. Available from: https://doi.org/10.1016/j.msec.2017.07.017.

[30] U. Kogelschatz, Dielectric-barrier discharges: their history, discharge physics, and industrial applications, Plasma Chem. Plasma Processing 23 (2003) 1–46.

[31] X. Deng, M. Kong, Frequency range of stable dielectric-barrier discharges in atmospheric He and N2, IEEE Trans. Plasma Sci. 32 (2004) 1709–1715. Available from: https://doi.org/10.1109/TPS.2004.831599.

[32] G. Borcia, A. Chiper, I. Rusu, Using a He + N 2 dielectric barrier discharge for the modification of polymer surface properties, Plasma Sour. Sci. Technol. 15 (2006) 849–857. Available from: https://doi.org/10.1088/0963-0252/15/4/031.

[33] H. Luo, G. Xiong, K. Ren, S. Raman, Z. Liu, Q. Li, et al., Air DBD plasma treatment on three-dimensional braided carbon fiber-reinforced PEEK composites for enhancement of in vitro bioactivity, Surf. Coat. Technol. 242 (2014) 1–7. Available from: https://doi.org/10.1016/j.surfcoat.2013.12.069.

[34] R. Brandenburg, Dielectric barrier discharges: progress on plasma sources and on the understanding of regimes and single filamentsPlasma Sour. Sci. Technol. 26 (2017) 053001. Available from: https://doi.org/10.1088/1361-6595/aa6426.

[35] F. Fanelli, F. Fracassi, Atmospheric pressure non-equilibrium plasma jet technology: general features, specificities and applications in surface processing of materials, Surf. Coat. Technol. 322 (2017) 174–201. Available from: https://doi.org/10.1016/j.surfcoat.2017.05.027.

[36] W. Knapp, D. Djomani, J. Coulon, R. Grunchec, Influence of structuring by laser and plasma torch on the adhesion of metallic films on thermoplastic substrates, Phys. Procedia 56 (2014) 791–800. Available from: https://doi.org/10.1016/j.phpro.2014.08.087.

[37] C. Han, E. Lee, H. Kim, Y. Koh, K. Kim, Y. Ha, et al., The electron beam deposition of titanium on polyetheretherketone (PEEK) and the resulting enhanced biological properties, Biomaterials. 31 (2010) 3465–3470. Available from: https://doi.org/10.1016/j.biomaterials.2009.12.030.

[38] D. Merche, N. Vandencasteele, F. Reniers, Atmospheric plasmas for thin film deposition: a critical review, Thin. Solid. Films. 520 (2012) 4219–4236. Available from: https://doi.org/10.1016/j.tsf.2012.01.026.

[39] L. Nie, C. Shi, Y. Xu, Q. Wu, A. Zhu, Atmospheric cold plasmas for synthesizing nanocrystalline anatase tio2 using dielectric barrier discharges, Plasma Process. Polym. 4 (2007) 574–582. Available from: https://doi.org/10.1002/ppap.200600212.

[40] A. Bogaerts, E. Neyts, R. Gijbels, J. van der Mullen, Gas discharge plasmas and their applications, Spectrochim. Acta Part. B. At. Spectrosc. 57 (2002) 609−658. Available from: https://doi.org/10.1016/S0584-8547(01)00406-2.

[41] K. Ozeki, T. Masuzawa, H. Aoki, Fabrication of hydroxyapatite thin films on poly-etheretherketone substrates using a sputtering technique, Mater. Sci. Eng. C 72 (2017) 1149—1157. Available from: https://doi.org/10.1016/j.msec.2016.11.111.

[42] X. Liu, K. Gan, H. Liu, X. Song, T. Chen, C. Liu, Antibacterial properties of nano-silver coated PEEK prepared through magnetron sputtering, Dent. Mater. 33 (2017) e348−e360. Available from: https://doi.org/10.1016/j.dental.2017.06.014.

[43] J. Dufils, F. Faverjon, C. Héau, C. Donnet, S. Benayoun, S. Valette, Evaluation of a variety of a-C: H coatings on PEEK for biomedical implants, Surf. Coat. Technol. 313 (2017) 96—106. Available from: https://doi.org/10.1016/j.surfcoat.2017.01.032.

[44] R. Bandorf, S. Waschke, F. Carreri, M. Vergöhl, G. Grundmeier, G. Bräuer, Direct metallization of PMMA with aluminum films using HIPIMS, Surf. Coat. Technol. 290 (2016) 77−81. Available from: https://doi.org/10.1016/j.surfcoat.2015.10.070.

[45] G. West, P. Kelly, P. Barker, A. Mishra, J. Bradley, Measurements of deposition rate and substrate heating in a HiPIMS discharge, Plasma Process. Polym. 6 (2009) S543−S547. Available from: https://doi.org/10.1002/ppap.200931202.

[46] S. Loquai, B. Baloukas, O. Zabeida, J. Klemberg-Sapieha, L. Martinu, HiPIMS-deposited thermochromic VO 2 films on polymeric substrates, Solar Energy Mater. Solar Cells 155 (2016) 60−69. Available from: https://doi.org/10.1016/j.solmat.2016.04.048.

[47] G. Eichenhofer, I. Fernandez, A. Wennberg, Industrial use of HiPIMS up to now and a glance into the future, a review by a manufacturer introduction of the hiP-V hiPlus technology, Univ. J. Phys. Applicat. 11 (2017) 73−79. Available from: https://doi.org/10.13189/ujpa.2017.110301.

[48] Y. Yang, H. Tsou, Y. Chen, C. Chung, J. He, Enhancement of bioactivity on medical polymer surface using high power impulse magnetron sputtered titanium dioxide film, Mater. Sci. Eng. C 57 (2015) 58−66. Available from: https://doi.org/10.1016/j.msec.2015.07.039.

[49] I. Yamada, Materials processing by cluster ion beams: history, technology, and applications, 1 ed., CRC Press, Boca Raton, n.d.

[50] R. Rodríguez, A. Medrano, J. García, G. Fuentes, R. Martínez, J. Puertolas, Improvement of surface mechanical properties of polymers by helium ion implantation, Surf. Coat. Technol. 201 (2007) 8146−8149. Available from: https://doi.org/10.1016/j.surfcoat.2006.12.030.

[51] J. García, R. Rodríguez, Ion implantation techniques for non-electronic applications, Vacuum 85 (2011) 1125−1129. Available from: https://doi.org/10.1016/j.vacuum.2010.12.024.

[52] G. Fuentes, J. Esparza, R. Rodríguez, M. Manso-Silván, J. Palomares, J. Juhasz, et al., Effects of He + ion implantation on surface properties of UV-cured Bis-GMA/TEGDMA bio-compatible resins, Nucl. Instr. Methods Phys. Res. Sect. B: Beam Inter. Mater. Atoms 269 (2011) 111−116. Available from: https://doi.org/10.1016/j.nimb.2010.11.026.

[53] A. Valenza, A. Visco, L. Torrisi, N. Campo, Characterization of ultra-high-molecular-weight polyethylene (UHMWPE) modified by ion implantation, Polymer. (Guildf). 45 (2004) 1707−1715. Available from: https://doi.org/10.1016/j.polymer.2003.12.056.

[54] K. Gan, H. Liu, L. Jiang, X. Liu, X. Song, D. Niu, et al., Bioactivity and antibacterial effect of nitrogen plasma immersion ion implantation on polyetheretherketone, Dent. Mater. 32 (2016) e263−e274. Available from: https://doi.org/10.1016/j.dental.2016.08.215.

[55] C. Díaz, G. Fuentes, Tribological studies comparison between UHMWPE and PEEK for prosthesis application, Surf. Coat. Technol. 325 (2017) 656−660. Available from: https://doi.org/10.1016/j.surfcoat.2017.07.007.

[56] H. Wang, M. Xu, W. Zhang, D. Kwok, J. Jiang, Z. Wu, et al., Mechanical and biological characteristics of diamond-like carbon coated poly aryl-ether-ether-ketone, Biomaterials. 31 (2010) 8181−8187. Available from: https://doi.org/10.1016/j.biomaterials.2010.07.054.

[57] V. Popok, E. Campbell, Beams of atomic clusters: effects on impact with solids, Rev. Adv. Mater. Sci, 11 (2006) 19−45.

[58] J. Khoury, S. Kirkpatrick, M. Maxwell, R. Cherian, A. Kirkpatrick, R. Svrluga, Neutral atom beam technique enhances bioactivity of PEEK, Nucl. Instr. Methods Phys. Res. Sect. B: Beam Inter. Mater. Atoms 307 (2013) 630−634. Available from: https://doi.org/10.1016/j.nimb.2012.11.087.

[59] A. Kirkpatrick, S. Kirkpatrick, M. Walsh, S. Chau, M. Mack, S. Harrison, et al., Investigation of accelerated neutral atom beams created from gas cluster ion beams, Nucl. Instr. Methods Phys. Res. Sect. B: Beam Inter. Mater. Atoms 307 (2013) 281−289. Available from: https://doi.org/10.1016/j.nimb.2012.11.084.

[60] G. Viola, J. Rosenblatt, I. Raad, Drug eluting antimicrobial vascular catheters: progress and promise, Adv. Drug Deliv. Rev. 112 (2017) 35−47. Available from: https://doi.org/10.1016/j.addr.2016.07.011.

[61] E. Brisbois, T. Major, M. Goudie, M. Meyerhoff, R. Bartlett, H. Handa, Attenuation of thrombosis and bacterial infection using dual function nitric oxide releasing central venous catheters in a 9 day rabbit model, Acta Biomater. 44 (2016) 304−312. Available from: https://doi.org/10.1016/j.actbio.2016.08.009.

[62] J. Sjollema, R. Dijkstra, C. Abeln, H. van der Mei, D. van Asseldonk, H. Busscher, On-demand antimicrobial release from a temperature-sensitive polymer — comparison with ad libitum release from central venous catheters, J. Control Release 188 (2014) 61−66. Available from: https://doi.org/10.1016/j.jconrel.2014.06.013.

[63] P. Singha, J. Locklin, H. Handa, A review of the recent advances in antimicrobial coatings for urinary catheters, Acta Biomater. 50 (2017) 20−40. Available from: https://doi.org/10.1016/j.actbio.2016.11.070.

[64] E. Dayyoub, M. Frant, S. Pinnapireddy, K. Liefeith, U. Bakowsky, Antibacterial and anti-encrustation biodegradable polymer coating for urinary catheter, Int. J. Pharm. 531 (2017) 205−214. Available from: https://doi.org/10.1016/j.ijpharm.2017.08.072.

[65] L. Yang, S. Whiteside, P. Cadieux, J. Denstedt, Ureteral stent technology: drug-eluting stents and stent coatings, Asian J. Urol. 2 (2015) 194−201. Available from: https://doi.org/10.1016/j.ajur.2015.08.006.

[66] O. Mrad, J. Saunier, C. Aymes Chodur, V. Rosilio, F. Agnely, P. Aubert, et al., A comparison of plasma and electron beam-sterilization of PU catheters, Radiat. Phys. Chem. 79 (2010) 93−103. Available from: https://doi.org/10.1016/j.radphyschem.2009.08.038.

[67] B. Ravat, M. Grivet, A. Chambaudet, Evolution of the degradation and oxidation of polyurethane versus the electron irradiation parameters: fluence, flux and temperature, Nucl. Instr. Methods Phys. Res. Sect. B: Beam Inter. Mater. Atoms 179 (2001) 243−248. Available from: https://doi.org/10.1016/S0168-583X(01)00571-7.

[68] B. Ravat, M. Grivet, Y. Grohens, A. Chambaudet, Electron irradiation of polyester-urethane: study of chemical and structural modifications using FTIR, UV spectroscopy and GPC, Radiat. Meas. 34 (2001) 31−36. Available from: https://doi.org/10.1016/S1350-4487(01)00116-0.

[69] B. Ravat, R. Gschwind, M. Grivet, E. Duverger, A. Chambaudet, L. Makovicka, Electron irradiation of polyurethane: some FTIR results and a comparison with a

EGS4 simulation, Nucl. Instr. Methods Phys. Res. Sect. B: Beam Inter. Mater. Atoms 160 (2000) 499–504. Available from: https://doi.org/10.1016/S0168-583X (99)00627-8.

[70] C. Wilhelm, A. Rivaton, J. Gardette, Infrared analysis of the photochemical behaviour of segmented polyurethanes: 3. Aromatic diisocyanate based polymers, Polymer. (Guildf). 39 (1998) 1323–1330.

[71] S. Lerouge, C. Guignot, M. Tabrizian, D. Ferrier, N. Yagoubi, L. Yahia, Plasma-based sterilization: effect on surface and bulk properties and hydrolytic stability of reprocessed polyurethane electrophysiology catheters, J. Biomed. Mater. Res. A. 52 (2000) 774–782.

[72] K. Weltmann, R. Brandenburg, T. von Woedtke, J. Ehlbeck, R. Foest, M. Stieber, et al., Antimicrobial treatment of heat sensitive products by miniaturized atmospheric pressure plasma jets (APPJs), J. Phys. D. Appl. Phys. 41 (2008). Available from: https://doi.org/10.1088/0022-3727/41/19/194008. 194008-.

[73] M. Kazemzadeh-Narbat, N. Annabi, A. Khademhosseini, Surgical sealants and high strength adhesives, Mater. Today 18 (2015) 176–177. Available from: https://doi.org/10.1016/j.mattod.2015.02.012.

[74] N. Annabi, A. Tamayol, S. Shin, A. Ghaemmaghami, N. Peppas, A. Khademhosseini, Surgical materials: current challenges and nano-enabled solutions, Nano Today 9 (2014) 574–589. Available from: https://doi.org/10.1016/j.nantod.2014.09.006.

[75] A. Duarte, J. Coelho, J. Bordado, M. Cidade, M. Gil, Surgical adhesives: systematic review of the main types and development forecast, Progress Polym. Sci. 37 (2012) 1031–1050. Available from: https://doi.org/10.1016/j.progpolymsci.2011.12.003.

[76] S. Saxena, A. Ray, A. Kapil, G. Pavon-Djavid, D. Letourneur, B. Gupta, et al., Development of a new polypropylene-based suture: plasma grafting, surface treatment, characterization, and biocompatibility studies, Macromol. Biosci. 11 (2011) 373–382. Available from: https://doi.org/10.1002/mabi.201000298.

[77] C. Serrano, L. García-Fernández, J. Fernández-Blázquez, M. Barbeck, S. Ghanaati, R. Unger, et al., Nanostructured medical sutures with antibacterial properties, Biomaterials. 52 (2015) 291–300. Available from: https://doi.org/10.1016/j.biomaterials.2015.02.039.

[78] K. Katti, Biomaterials in total joint replacement, Coll. Surf. B: Biointerf. 39 (2004) 133–142. Available from: https://doi.org/10.1016/j.colsurfb.2003.12.002.

[79] H. Mehboob, S. Chang, Application of composites to orthopedic prostheses for effective bone healing: a review, Compos. Struct. 118 (2014) 328–341. Available from: https://doi.org/10.1016/j.compstruct.2014.07.052.

[80] S. Affatato, A. Ruggiero, M. Merola, Advanced biomaterials in hip joint arthroplasty. A review on polymer and ceramics composites as alternative bearings, Compos. Part B: Eng. 83 (2015) 276–283. Available from: https://doi.org/10.1016/j.compositesb.2015.07.019.

[81] Applications of UHMWPE in Total Ankle Replacements, in: UHMWPE biomaterials handbook: ultra-high molecular weight polyethylene in total joint replacement and medical devices, 2nd ed., Academic, Amsterdam, 2009, pp. 153–167.

[82] S. Ghalme, A. Mankar, Y. Bhalerao, Biomaterials in hip joint replacement, Int. J. Mater. Sci. Eng. 4 (2016) 113–125.

[83] A. Freemont, The pathology of joint replacement and tissue engineering, Diagn. Histopathol. 23 (2017) 221–228.

[84] A. King, J. Phillips, Total hip and knee replacement surgery, Surgery (Oxford) 34 (2016) 468–474. Available from: https://doi.org/10.1016/j.mpsur.2016.06.005.

[85] F. Ansari, M. Ries, L. Pruitt, Effect of processing, sterilization and crosslinking on UHMWPE fatigue fracture and fatigue wear mechanisms in joint arthroplasty, J. Mech. Behav. Biomed. Mater. 53 (2016) 329–340. Available from: https://doi.org/10.1016/j.jmbbm.2015.08.026.

[86] D. Macuvele, J. Nones, J. Matsinhe, M. Lima, C. Soares, M. Fiori, et al., Advances in ultra high molecular weight polyethylene/hydroxyapatite composites for biomedical applications: a brief review, Mater. Sci. Eng. C 76 (2017) 1248−1262. Available from: https://doi.org/10.1016/j.msec.2017.02.070.

[87] M. Banghard, C. Freudigmann, K. Silmy, A. Stett, V. Bucher, Plasma treatment on novel carbon fiber reinforced PEEK cages to enhance bioactivity, Curr. Direct. Biomed. Eng. 2 (2016). Available from: https://doi.org/10.1515/cdbme-2016-0126.

[88] I. Cicha, R. Detsch, R. Singh, S. Reakasame, C. Alexiou, A. Boccaccini, Biofabrication of vessel grafts based on natural hydrogels, Curr. Opin. Biomed. Eng. 2 (2017) 83−89. Available from: https://doi.org/10.1016/j.cobme.2017.05.003.

[89] D. Xie, Y. Leng, F. Jing, N. Huang, A brief review of bio-tribology in cardiovascular devices, Biosurf. Biotribol. 1 (2015) 249−262. Available from: https://doi.org/10.1016/j.bsbt.2015.11.002.

[90] G. Catalano, A. Demir, V. Furlan, B. Previtali, Use of sheet material for rapid prototyping of cardiovascular stents, Proc. Eng. 183 (2017) 194−199. Available from: https://doi.org/10.1016/j.proeng.2017.04.019.

[91] S. Venkatraman, F. Boey, L. Lao, Implanted cardiovascular polymers: natural, synthetic and bio-inspired, Progress Polym. Sci. 33 (2008) 853−874. Available from: https://doi.org/10.1016/j.progpolymsci.2008.07.001.

[92] A. Trimukhe, K. Pandiyaraj, A. Tripathi, J. Melo, R. Deshmukh, Plasma Surface Modification of Biomaterials for Biomedical Applications, in: n.d.: p. 95. https://doi.org/10.1007/978-981-10-3328-5_3.

[93] A. Tripathi, J. Melo (Eds.), Advances in Biomaterials for Biomedical Applications, Springer Berlin Heidelberg, New York, NY, 2017.

[94] Plasma Gas Discharge Treatment for Improving the Biocompatibility of Biomaterials, 1987.

[95] D. Clark, D. Hutton, Surface modification by plasma techniques. I. The interactions of a hydrogen plasma with fluoropolymer surfaces, J. Polym. Sci. Part A: Polym. Chem. 25 (1987) 2643−2664. Available from: https://doi.org/10.1002/pola.1987.080251002.

[96] T. Chandy, G. Das, R. Wilson, G. Rao, Use of plasma glow for surface-engineering biomolecules to enhance bloodcompatibility of Dacron and PTFE vascular prosthesis, Biomaterials. 21 (2000) 699−712. Available from: https://doi.org/10.1016/S0142-9612(99)00231-8.

[97] X. Qiu, B. Lee, X. Ning, N. Murthy, N. Dong, S. Li, End-point immobilization of heparin on plasma-treated surface of electrospun polycarbonate-urethane vascular graft, Acta Biomater. 51 (2017) 138−147. Available from: https://doi.org/10.1016/j.actbio.2017.01.012.

[98] M. Santos, M. Bilek, S. Wise, Plasma-synthesised carbon-based coatings for cardiovascular applications, Biosurf. Biotribol. 1 (2015) 146−160. Available from: https://doi.org/10.1016/j.bsbt.2015.08.001.

[99] A. Guex, A. Frobert, J. Valentin, G. Fortunato, D. Hegemann, S. Cook, et al., Plasma-functionalized electrospun matrix for biograft development and cardiac function stabilization, Acta Biomater. 10 (2014) 2996−3006. Available from: https://doi.org/10.1016/j.actbio.2014.01.006.

[100] J. Calles, J. Bermúdez, E. Vallés, D. Allemandi, S. Palma, Polymers in ophthalmology, Advanced Polymers in Medicine, Springer International Publishing, Cham, 2015, p. 147. Available from: https://doi.org/10.1007/978-3-319-12478-0_6.

[101] T. Yasukawa, Y. Ogura, E. Sakurai, Y. Tabata, H. Kimura, Intraocular sustained drug delivery using implantable polymeric devices, Adv. Drug Deliv. Rev. 57 (2005) 2033−2046. Available from: https://doi.org/10.1016/j.addr.2005.09.005.

[102] B. Wang, Q. Lin, C. Shen, J. Tang, Y. Han, H. Chen, Hydrophobic modification of polymethyl methacrylate as intraocular lenses material to improve the cytocompatibility, J. Colloid. Interface. Sci. 431 (2014) 1—7. Available from: https://doi. org/10.1016/j.jcis.2014.05.056.

[103] D. Bozukova, C. Pagnoulle, R. Jérôme, C. Jérôme, Polymers in modern ophthalmic implants — Historical background and recent advances, Mater. Sci. Eng. R.¹ Rep. 69 (2010) 63—83. Available from: https://doi.org/10.1016/j. mser.2010.05.002.

[104] S. Kirchhof, A. Goepferich, F. Brandl, Hydrogels in ophthalmic applications, Eur. J. Pharm. Biopharm. 95 (2015) 227—238. Available from: https://doi.org/ 10.1016/j.ejpb.2015.05.016.

[105] A. Lloyd, R. Faragher, S. Denyer, Ocular biomaterials and implants, Biomaterials. 22 (2001) 769—785. Available from: https://doi.org/10.1016/S0142-9612(00) 00237-4.

[106] L. Lin, Y. Wang, X. Huang, Z. Xu, K. Yao, Modification of hydrophobic acrylic intraocular lens with poly(ethylene glycol) by atmospheric pressure glow discharge: a facile approach, Appl. Surf. Sci. 256 (2010) 7354—7364. Available from: https:// doi.org/10.1016/j.apsusc.2010.05.068.

[107] F. Rezaei, M. Abbasi-Firouzjah, B. Shokri, Investigation of antibacterial and wettability behaviours of plasma-modified PMMA films for application in ophthalmology, J. Phys. D. Appl. Phys. 47 (2014). Available from: https://doi.org/10.1088/ 0022-3727/47/8/085401.

[108] R. D'Sa, J. Raj, M. McMahon, D. McDowell, G. Burke, B. Meenan, Atmospheric pressure plasma induced grafting of poly(ethylene glycol) onto silicone elastomers for controlling biological response, J. Colloid. Interface. Sci. 375 (2012) 193—202. Available from: https://doi.org/10.1016/j.jcis.2012.02.052.

[109] L. Zhang, D. Wu, Y. Chen, X. Wang, G. Zhao, H. Wan, et al., Surface modification of polymethyl methacrylate intraocular lenses by plasma for improvement of antithrombogenicity and transmittance, Appl. Surf. Sci. 255 (2009) 6840—6845. Available from: https://doi.org/10.1016/j.apsusc.2009.03.029.

[110] A. Pimenta, A. Vieira, R. Colaço, B. Saramago, M. Gil, P. Coimbra, et al., Controlled release of moxifloxacin from intraocular lenses modified by Ar plasma-assisted grafting with AMPS or SBMA: an in vitro study, Coll. Surf. B: Biointerf. 156 (2017) 95—103. Available from: https://doi.org/10.1016/ j.colsurfb.2017.04.060.

[111] D. Banoriya, R. Purohit, R. Dwivedi, Advanced application of polymer based biomaterials, Mater. Today: Proc. 4 (2017) 3534—3541.

[112] E. Allen, S. Bayne, T. Donovan, T. Hansson, J. Klooster, J. Kois, Annual review of selected dental literature, J. Prosthet. Dent. 76 (1996) 56—93.

[113] S. Najeeb, M. Zafar, Z. Khurshid, F. Siddiqui, Applications of polyetheretherketone (PEEK) in oral implantology and prosthodontics, J. Prosthod. Res. 60 (2016) 12—19. Available from: https://doi.org/10.1016/j.jpor.2015.10.001.

[114] A. Schwitalla, T. Zimmermann, T. Spintig, I. Kallage, W. Müller, Fatigue limits of different PEEK materials for dental implants, J. Mech. Behav. Biomed. Mater. 69 (2017) 163—168. Available from: https://doi.org/10.1016/j.jmbbm.2016.12.019.

[115] P. Amrollahi, B. Shah, A. Seifi, L. Tayebi, Recent advancements in regenerative dentistry: a review, Mater. Sci. Eng. C 69 (2016) 1383—1390. Available from: https://doi.org/10.1016/j.msec.2016.08.045.

[116] A. Poulsson, D. Eglin, S. Zeiter, K. Camenisch, C. Sprecher, Y. Agarwal, et al., Osseointegration of machined, injection moulded and oxygen plasma modified PEEK implants in a sheep model, Biomaterials. 35 (2014) 3717—3728. Available from: https://doi.org/10.1016/j.biomaterials.2013.12.056.

[117] S. Cha, Y. Park, Plasma in dentistry, Clin. Plasma Med. 2 (2014) 4—10. Available from: https://doi.org/10.1016/j.cpme.2014.04.002.

[118] D. Boonyawan, P. Waruriya, K. Suttiat, Characterization of titanium nitri-de—hydroxyapatite on PEEK for dental implants by co-axis target magnetron sputtering, Surf. Coat. Technol. 306 (2016) 164—170. Available from: https://doi.org/10.1016/j.surfcoat.2016.05.063.

[119] Y. Zhang, Q. Yu, Y. Wang, Non-thermal atmospheric plasmas in dental restoration: improved resin adhesive penetration, J. Dent. 42 (2014) 1033—1042. Available from: https://doi.org/10.1016/j.jdent.2014.05.005.

[120] M. Chen, Y. Zhang, M. Sky Driver, A. Caruso, Q. Yu, Y. Wang, Surface modification of several dental substrates by non-thermal, atmospheric plasma brush, Dent. Mater. 29 (2013) 871—880. Available from: https://doi.org/10.1016/j.dental.2013.05.002.

CHAPTER 15

Plasma Modified Polymeric Materials for Biosensors/ Biodevice Applications

Lavanya Jothi and Gomathi Nageswaran
Department of Chemistry, Indian Institute of Space Science and Technology, Thiruvanathapuram, Kerala, India

15.1 INTRODUCTION

A long chain molecule made up of a repeated pattern of monomers, polymers, have been an inseparable material in biomedical platforms. Typically these materials are advantageous in terms of lightweight, flexibility, corrosion-resistance, high chemical inertness, and ease of processing [1]. Such an advantageous material plays a vital role in the biomedical field, for instance biosensors, artificial organs, disposable clinical apparatus, medical devices, and cardiovascular implants, such as artificial heart valves and stent and sewing rings [2–10].

15.2 NEED FOR SURFACE MODIFICATION OF POLYMERS

Surface modification is the process of modifying the surface of a material by changing the physical or chemical characteristics from the ones originally found on the surface of the material, which can be carried out in a nanoscale or bulk level [11]. In general, the most important need of surface modification has been to convert nonpolar, hydrophobic surfaces to polar, hydrophilic, and water wettable surfaces [12,13]. In biomedical applications, the major hurdle is the long-term thrombogenic property of polymers. Therefore, an attempt will be required to enhance antithrombogenic properties through suitable surface modification techniques by introducing some functional groups onto the outer-most surface which

Non-Thermal Plasma Technology for Polymeric Materials
DOI: https://doi.org/10.1016/B978-0-12-813152-7.00015-9

tailor the surface properties—such as wear resistance, corrosion resistance, biocompatibility, wettability, scalability, printability, and dye uptake—its resistance to glazing, its adhesion to other materials, or its interaction with a biological environment while maintaining the desirable bulk properties of the polymer. In biosensor, surface modified polymer films act as a good interface between the biological components and the transducers. Especially in electrochemical biosensor conducting, polymers are amenable for providing a platform to immobilize biomolecules and for rapid electron transfer [14−17]. An example concerns the immobilization of DNA and antibodies in Zeonor, a cyclic polyolefin slide [18]. Moreover, another distinct purpose of surface modification is found on implant surfaces for both dental and orthopedic applications to exhibit biological, mechanical, and morphological compatibilities to receiving vital hard/soft tissue, resulting in promoting osseointegration and also to improve blood compatibility.

15.3 OVERVIEW OF PLASMA POLYMER SURFACE MODIFICATION

Modification can be done by different methods with an objective to altering a wide range of surface properties, such as surface roughness, hydrophilicity, surface charge, biocompatibility, and reactivity [19−21]. The introduction of such unique properties can be achieved via various chemical methods and physical methods, for instance corona treatment, radiation assisted polymerization, wet chemical, phase separation etc., However, the conventional surface modification methods have certain drawbacks; for instance, the use of solvents, consuming more time, high temperature, huge capital cost, lack of reproducibility, and stability. Therefore, the polymer surface modified by the plasma-assisted polymerization has distinct properties compared with conventional surface modification techniques [22,23]. And the obtained polymeric films from plasma surface modification are pinhole free, conformal, highly crosslinked, having unique surface chemistry, and strongly adhering with the substrate [24−27]. The introduction to polymer, properties and surface modification of polymer has already been reported in various book and reviews [28−33]. In this chapter, we briefly describe the application of plasma surface modified polymer in biosensors and biodevice

15.4 STRATEGIES FOR PLASMA SURFACE MODIFICATION OF POLYMER IN BIOMEDICAL APPLICATION

15.4.1 Influence of Plasma Treatment Parameters

Plasma, the fourth state of matter, is generated when a gas is exposed to an electromagnetic field at low pressure and near ambient temperature. When the plasma consists of ionized gases with equal density of positive and negative particles was discharged with oxygen, nitrogen, and ammonia-containing gas. The excited molecules and radicals generated in the plasma, attack the surface of the material and create defect sites that act as prime sites for functionalization with various chemical functional groups, including carboxyl, amine, and hydroxyl functional groups. Plasmas can be classified as low pressure and atmospheric plasmas [34]. Both plasmas can be used for surface modification of materials. Atmospheric plasma can be generated under atmospheric conditions without the constraint of the vacuum chamber; it can be more easily applied to practical production-line operations. At the end of plasma processes, the resulting materials are known to increase the wettability, usually low, of conventional polymers, and are also utilized to favor adhesion and growth of cells on polymers [35,36].

The parameters that affect the surface of polymers via plasma modification are as follows: the type of plasma discharge, its power, pressure, treatment time, the gas used in the plasma chamber, flow rate of the gas, and the precursor. For instance, the thermal plasma method operated at high temperature, such as chemical vapor deposition, is not suitable for the fabrication of thin films for biomedical applications, while it affects the bulk property of the polymer and also it causes cell death. So nonthermal or cold plasma, that is, plasma-enhanced chemical vapor deposition (PECVD) is preferred for surface modification purposes.

In general, the thickness, film density, and roughness of the polymer surface can be varied by changing the conditions of plasma parameters, for instance sputtered precursor, type of gas in a plasma chamber, and their flow rates. Among the various types of plasma, atmospheric pressure plasma treatment is easier to process polymers, since there is no need for expensive equipment such as vacuum pumps, etc. [37]. Especially dielectric barrier discharge (DBD) and glow discharge (GD) are the two types of discharges at atmospheric pressure, which are easily employed for the surface modification of various polymeric biomaterials [38−41]. In GD, the plasma was generated in the space between two electrodes, which is filled with the process gas. In the case of DBD, AC voltage power supply is used to generate plasma, since it requires capacitive coupling. Hence, less energy is required to form the reactive species in the gas and a substrate can be kept at a low temperature. The uniformity of the coatings is maintained owing to a uniform dispersion of nanosecond discharges occurring on the electrode surface.

For uniform coatings, most of the materialists and researchers use low pressure plasma treatment [42,43]. Radio frequency plasma (RF, generally 13.56 MHz) [44,45] and microwave plasma (MW, generally 2.45 GHz) [46,47] induced by the applied electric field are frequently used low pressure plasma treatment methods.

15.5 APPLICATION AS BIO-INTERFACE IN BIOSENSOR AND BIOMEDICAL DEVICE

15.5.1 Biosensor

The concept of biosensors was born with the pioneering work of Prof. L. C. Clark who invented, in 1956, the first oxygen sensor and later on, in 1962, the first biosensor: an amperometric enzyme electrode for glucose [48]. A biosensor is a device that works on direct coupling between a biomolecule and a signal transducer equipped with an electronic amplifier [49]. It consists of two main integrated components: (1) a biological element (such as enzymes, antibodies, ssDNA, tissues, microorganisms, etc.) or a synthetic receptor that recognizes and selectively binds to the target of interest; and (2) the transducer or detector element that converts the binding reaction to a measurable quantity (Fig. 15.1) which is displayed

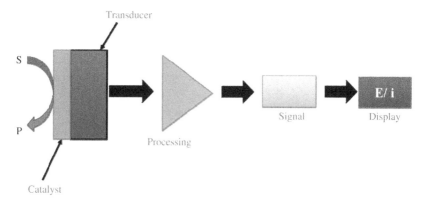

Figure 15.1 Principal stages in the operation of an electrochemical sensor.

by a signal processor or reader device in a user-friendly way. Based on signal transduction, biosensors can be classified into four different basic groups: electrochemical, optical, immunosensor, aptamer, and DNA sensors. This section provides a detailed study on the application of plasma surface modified polymers in various types of biosensors.

15.5.1.1 Electrochemical Biosensor

Electrochemical biosensors are the biosensors that transform biochemical information, such as analyte concentrations into an analytically useful signal: current or voltage. On the other hand, this type of sensor does not require a separate transduction element, it directly detects the concentration of the analyte from the measured electrical response proportional to the analyte concentration obtained from the electrode. Conducting polymers (CPs), in particular, are especially amenable for the development of an electrochemical biosensor in providing a good platform from immobilizing biomolecules and for fast electron transfer.

Firstly, due to the advantageous behavior, the biosensor device made of polyaniline [50] with good chemical stability, high conductivity, and low operation potential, can provide fast electron transfer between electrode and analyte. Hence, Zhang et al. [51] synthesized a nanocomposite of SnO_2@3D-rGO by simultaneous reduction of 3D graphene oxide (3D-GO) and by translating $SnCl_4$ into SnO_2 (Fig. 15.2). Meanwhile, to enhance the adhesive strength between the nanostructured composites and surfaces of electrode, without decreasing the electrochemical properties and biosensing ability of the biosensors. The electrode was plasma polymerized by conducting thin polymer films. The fabricated material was applied to detect glucose electrochemically. Compared with other usual glucose sensing, the fabricated electrochemical sensor based on SnO_2@3D-rGO@pPANI provides three benefits: (1) various nitrogen functional moieties introduced on the surface of the nanocomposite provides good stability of the electrode materials in the aqueous solution; (2) the porous nanostructure fabricated from 3D-rGO offers a large loading capacity of glucose oxidase; and (3) high electrochemical performance. Therefore, the proposed biosensor exhibited a much higher sensitivity than that formed from individual components, namely, SnO_2@3D-rGO and pPANI. This biosensor demonstrated a low detection limit of 0.047 ng/mL(0.26 nM) ($S/N = 3$) within the concentration range of 0.1 ng/mL to 5 μg/mL good selectivity, reproducibility, stability, and practicability.

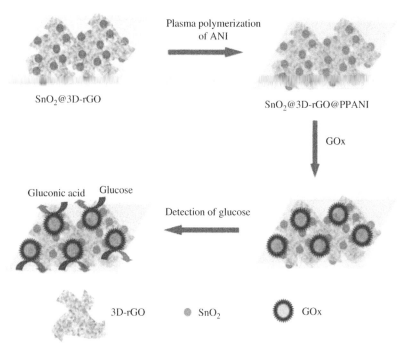

Figure 15.2 Schematic of the preparation of SnO2@3D-rGO@pPANI nanocomposite-based biosensors for glucose detection. *Reproduced with permission from Wu, S., Su, F., Dong, X., Ma, C., Pang, L., Peng, D., ... et al. (2017). Development of glucose biosensors based on plasma polymerization-assisted nanocomposites of polyaniline, tin oxide, and three-dimensional reduced graphene oxide. Appl. Surf. Sci., 401(Supplement C), 262–270. https://doi.org/10.1016/j.apsusc.2017.01.024 copyright 2017, Elsevier.*

Subsequently, there are some strategies reported by Muguruma et al. [52–54] based on the fabrication of plasma surface modified polymer films for electron-transfer-mediated biosensors. One such strategy was the application of acetonitrile combined with vinylferrocene plasma polymerized films containing redox sites. In this strategy, vinylferrocene plasma polymerized films were first deposited onto a sputtered platinum electrode. Later, an acetonitrile plasma polymerized films was superimposed onto the vinylferrocene layer on the modified electrode. With the addition of glucose, an increase in current was observed at $+0.45$ V potential, indicating the successful construction of a biosensor with rapid electron-transfer system. This enhancement in peak current associated with the bio electrochemical reaction is due to the layer of plasma polymerized acetonitrile film sandwiched between the immobilized enzyme and plasma polymerized vinylferrocene. Indeed, without the plasma polymerized acetonitrile film, significant peak current resulting from the enzymatic reaction was not observed.

The other strategy is utilization of plasma modified dimethyl amino methyl ferrocene (DMAMF) deposited onto a sputtered electrode applied to detect glucose. In this biosensor device, various nitrogen functional moieties introduced during plasma modification make the surface of plasma polymerized film hydrophilic, with a water-contact angle of about 80o. Finally, Muguruma et al. immobilize array of GOx on this modified material and the electrochemical measurements showed the enhancement in the peak current with successive addition of glucose, indicating the system with fast electron transfer.

15.5.1.2 Surface Plasmon Resonance Biosensor

Surface plasmon resonance is a technique for detecting changes in the refractive index at the surface of a sensor. The sensor is comprised of a glass substrate and (any metal), for example, thin gold coating. Light passes through the substrate and is reflected off the gold coating. At certain angles of incidence, a portion of the light energy couples through the gold coating and creates a surface plasmon wave at the sample and gold surface interface. The angle of incident light required to sustain the surface plasmon wave is very sensitive to changes in refractive index at the surface (due to mass change), and it is these changes that are used to monitor the association and dissociation of biomolecules.

Parylene-N film was surface modified by Choi et al. [55] via plasma surface modification. The modified polymer was applied for the immobilization of proteins through physical adsorption. Since, immunoassays are used for the detection of target analytes in complex sample mixtures such as human blood by using highly specific antigen–antibody interactions [56,57]. The surface coverage of protein immobilized on the polymer was demonstrated by comparing the immobilization efficiencies of different modified surfaces using model proteins with different surface charges, such as streptavidin (pI = 5, negatively charged at pH 7), horseradish peroxidase (pI = 6.6, nearly neutral at pH 7), and avidin (pI = 10, positively charged at pH 7). The model proteins with both acidic and basic pIs showed higher immobilization efficiencies on the plasma surface modified parylene-N film compared to the untreated parylene-N film and polystyrene. The plasma-surface modified parylene-N film created a 25 percent more hydrophilic surface with long-term stability. The fabricated electrode material was applied as an SPR biosensor (Fig. 15.3) for the detection of the human hepatitis virus surface antigen (HBsAg), in the concentration range from 10 pg/mL to 1 μg/mL, and the limit of detection was estimated to be less than 10 pg/mL. In comparison with the

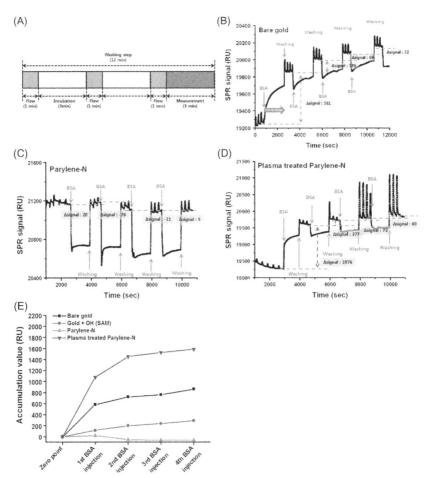

Figure 15.3 Response of SPR biosensor according to the adsorption of bovine serum albumin (BSA) to differently modified sensor surfaces: (A) Flow control during the washing step of SPR measurement. (B) Bare gold. (C) Plasma-treated parylene-N film. (D) Comparison of protein adsorption to differently modified sensor surfaces. (E) Comparison of SPR signal of the adsorbed BSA by repeated treatment to differently prepared surfaces. *Reproduced with permission from Chung, J.W., Kim, S.D., Bernhardt, R., & Pyun, J.C. (2005). Application of SPR biosensor for medical diagnostics of human hepatitis B virus (hHBV). Sensors and Actuators B: Chemical, 111(Supplement C), 416–422. https://doi.org/10.1016/j.snb.2005.03.055 [56]. copyright 2005, Elsevier.*

commercial ELISA kit with the limit of detection of 40 ng/mL, the SPR biosensor with the plasma-treated parylene-N film could achieve more than 1000-fold improved sensitivity in comparison with the conventional ELISA kit, and such an improved sensitivity was considered to be resulted from the enhanced immobilization of receptor antibodies against HBsAg by using the plasma-treated parylene-N film.

15.5.1.3 Immunosensor

Immunosensors are compact analytical devices in which the event of formation of antigen-antibody complexes is detected and converted, by means of a transducer, to an electrical signal, which can be processed, recorded, and displayed. Different transducing mechanisms are employed in immunological biosensors, based on signal generation (such as an electrochemical or optical signal) or properties changes (such as mass changes) following the formation of antigen-antibody complexes. In this chapter, the basics of immunosensors are presented focused on the different transduction techniques used in immunosensing.

In 2013, Ardhaoui et al. [58] immobilized laccase via a simple and rapid method based on the plasma polymer functionalized carbon electrode and applied for biosensor application. In this work, high density amine groups were introduced on rough carbon surface electrodes using a low pressure inductively excited RF tubular plasma reactor under mild plasma conditions (low plasma power [10 W], few minutes, N/C ratio up to 0.18) to deposit allylamine. Laccase from Trametes versicolor was physisorbed and covalently bound to these allylamine modified carbon surfaces. At a period of 30 min the plasma coated electrode showed best efficiency in laccase activities and current outputs measured in the presence of 2,2′-azinobis-(3-ethylbenzothiazole-6- sulfonic acid) (ABTS). The results also showed that for all the tested electrodes the activities and current outputs of the covalently immobilized laccases were twice higher than the physically adsorbed ones. The sensitivity attained in the presence of ABTS and DMP respectively were determined by performing amperometric method (Fig. 15.4) 4.8 and 2.7 μA/mg L. An excellent stability of this laccase biosensor was observed for over 6 months.

15.5.1.4 Aptamer Sensors

Aptamers are an excellent example of functional molecules selected in vitro. In 1990, two groups independently developed in vitro selection and amplification for the isolation of RNA sequences that could

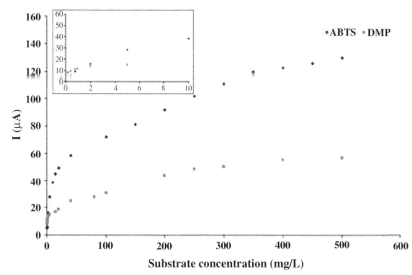

Figure 15.4 Catalytic current responses of covalently immobilized laccase PPAA electrodes (30 min) obtained by chronoamperometry as function of ABTS and DMP concentrations. Inset: detail of linear range of the biosensor for both substrates. Experimental conditions: 50 mM acetate buffer/100 mM NaClO$_4$, pH 4.2, room temperature. *Reproduced with permission from Ardhaoui, M., Bhatt, S., Zheng, M., Dowling, D., Jolivalt, C., & Khonsari, F.A. (2013). Biosensor based on laccase immobilized on plasma polymerized allylamine/carbon electrode. Materials Science and Engineering: C, 33(6), 3197—3205. https://doi.org/10.1016/j.msec.2013.03.052 copyright 2013, Elsevier.*

specifically bind to target molecules [59,60]. These functional RNA oligonucleo-tides were then termed aptamers, derived from the Latin aptus, meaning "to fit" [61]. Later, DNA-based aptamers were also found [62].

In 2016, Fang et al. [63] reported a novel, label-free electrochemical aptasensor for detecting proteins in whole blood based on a three-component nanocomposite, in which ferriferous oxide and three-dimensional graphene nanocomposite were modified with the plasma-polymerized 4-vinyl pyridine (Fe$_3$O$_4$@3D-rGO@PP4VP). In this novel sensing strategy, large amounts of amino groups in PP4VP facilitated the immobilization of aptamer strands via the strong electrostatic interaction between positively charged ammonium groups of the nanocomposites and negatively charged phosphate groups of aptamers. In the presence of thrombin, LYS (LYS), and platelet-derived growth factor-BB (PDGF-BB), the adsorbed aptamer strands on the developed nanocomposite surface caught the targeted proteins at the electrode interface. The aptamer preferred to be a barrier for electrons and inhibited electron transfer, leading to the decreased peak current of cyclic voltammetry measurements and the increased electron transfer resistance of electrochemical impedance spectroscopy. The determination of the thrombin, PDGF-BB, and LYS concentrations with this novel strategy showed low detection limits of 4.5, 29.4, and 14 pg/mL, and the analytical ranges extend from 0.01 to 50, 0.1 to 100, and 0.1 to 200 ng/mL, respectively. The resultant aptasensor exhibited high selectivity, acceptable reproducibility, and stability toward thrombin. The aptasensor could be used to detect thrombin in whole blood samples, thereby suggesting its possible application in clinical settings.

15.5.1.5 DNA Sensor

Recently, the conductive polymers and its nanocomposites are attractive materials for the development of electrochemical DNA sensors. Development of electrochemical sensors for DNA detection is motivated by applications in many fields: DNA diagnostics, gene analysis, fast detection of biological warfare agents, and forensic applications [63]. DNA sensor has been fabricated by immobilization of a DNA probe onto a surface to recognize its complementary DNA target sequence by hybridization. Transduction of DNA hybridization, can be measured optically, electrochemically, or using mass-sensitive devices.

Wang et al. [64] fabricated a biosensor for the detection of highly toxic metal mercury (Hg), based on graphene nanostructures by highly sensitive and selective DNA sensors. The proposed sensor consists of multiple layers of nanostructured plasma-polymerized allylamine (PPAA) and nanosheets of graphene. The PPAA acts as a transducer element for both immobilization of DNA and sensing the target analyte. The Hg^{2+} ions became intercalated into the DNA poly-ion complex membrane based on T-Hg^{2+}-T coordination chemistry.

Experimental results revealed that the frequency variation of the quartz chip of our system increased with increased Hg^{2+} level in the sample and had limits of detection of 0.031 and 0.017 nM determined by QCM and electrochemical measurements (Fig. 15.5), respectively. The strategy afforded high selectivity of Hg^{2+} against other interfering metal ions. In addition, the developed DNA sensor for the determination of Hg^{2+} ions could be reproduced up to 10 cycles at approximately 88 percent recovery.

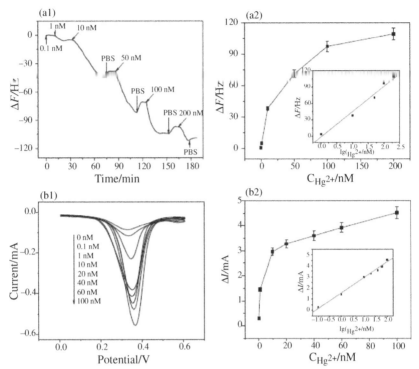

Figure 15.5 (A1) Frequency response profiles for the detection of Hg^{2+} at different concentrations (0.1, 1, 10, 50, 100, and 200 nM), and (A2) linear relationship between ΔF and the logarithm of concentration of Hg^{2+}. (Inset: linear calibration curve for ΔF versus $log(CHg^{2+}/nM)$); (B1) DPV curves of G-PPAA film modified with P1 after hybridization with P3-Au-NPs in the presence of different concentrations of Hg^{2+} at the range from 0 to 100 nM and (B2) plot of the difference in the value of the peak currents against the Hg^{2+} concentration. (Inset: linear calibration curve for ΔI vs log (CHg^{2+}/nM)). *Reproduced with permission from Wang, M., Liu, S., Zhang, Y., Yang, Y., Shi, Y., He, L., ... et al. (2014). Graphene nanostructures with plasma polymerized allylamine biosensor for selective detection of mercury ions. Sensors and Actuators B: Chemical, 203(Supplement C), 497–503. https://doi.org/10.1016/j.snb.2014.07.009 copyright 2014, Elsevier.*

Yang et al. [64] utilized the RF technology to develop a multilayered polymeric DNA sensor with the help of gold and magnetic nanoparticles. Poly (p-xylylene) (Parylene) and polyethylene naphtholate (PEN), the flexible polymeric materials, were used as substrates to replace the conventional rigid substrates such as glass and silicon wafers. Thioglycolic acid (TGA) was used on the surface of the proposed biochip to form a thiolate-modified sensing surface for DNA hybridization.

Gold nanoparticles (AuNPs) and magnetic nanoparticles (MNPs) were used to immobilize on the surface of the biosensor to enhance overall detection sensitivity. In addition to gold nanoparticles, the magnetic nanoparticles has been demonstrated the applicability for RF DNA detection.

The performance of the proposed biosensor was evaluated by the shift of the center frequency of the RF biosensor because the electromagnetic characteristic of the biosensors can be altered by the immobilized multi-layer nanoparticles on the biosensor. The experimental results show that the detection limit of the DNA concentration can reach as low as 10 pM, and the largest shift of the center frequency with triple layer AuNPs and MNPs can approach 0.9 and 0.7 GHz, respectively (Fig. 15.6). Such an achievement implies that the developed biosensor can offer an alternative inexpensive, disposable, and highly sensitive option for application in bio-medicine diagnostic systems because the price and size of each biochip can be effectively reduced by using fully polymeric materials and multilayer-detecting structures.

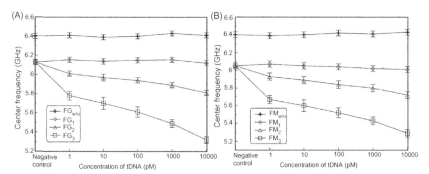

Figure 15.6 Measurements of the center frequency shift using AuNPs and MNPs with various tDNA concentrations: (A) the center frequency of self-assembly AuNPs versus various tDNA concentrations; (B) the center frequency of using self-assembly MNPs versus various tDNA concentrations. (The FG1, FG2 and FG3 denote the S21 response of monolayer, double-layer and triple-layer AuNPs structures, respectively; the FM1, FM2, and FM3 denote S21 response of mono-layer, double-layer and triple-layer MNPs structures, respectively.). *Reproduced with permission from Yang, C.-H., Kuo, L.-S., Chen, P.-H., Yang, C.-R., & Tsai, Z.-M. (2012). Development of a multilayered polymeric DNA biosensor using radio frequency technology with gold and magnetic nanoparticles. Biosens. Bioelectron., 31(1), 349–356. https://doi.org/10.1016/j.bios.2011.10.044 [65] copyright 2012, Elsevier.*

15.5.2 Biodevice

15.5.2.1 Implants

Biomaterials are materials intended to interface with biological systems to evaluate, treat, augment, or replace any tissue, organ, or function of the body [66]. In recent years, polymers with excellent properties have attracted much attention in bone substitution materials, some bone fixing materials, and dental materials [67].

Unfortunately, insufficient osteoblast compatibility and bioactivity of this polymeric material acts as a major hurdle in a wider range of biomedical applications as implant material [68]. These types of limitations can be overcome by surface modification, which is a desirable approach to enhance the biocompatibility and bioactivity of polymer. Since, surface modification can reduce the wear and improve frictional behavior of surface regions while maintaining desirable bulk properties of the underlying substrate. It can improve service life of components and devices significantly [69].

Since early 1060s polymeric materials have been used in joint implants. Common ones are ultrahigh molecular weight poly ethylene, poly tetra fluro ethylene, and urethane for knee, hip, jaw, or heart valves. Like metal and ceramic implants, a similar issue in polymer is adhesion which is one of the major focuses for development.

The goal of this study is to show the use of plasma technology to enhance the adhesion of polymers. Two different plasma techniques (low pressure plasma activation and atmospheric pressure plasma polymerization) are performed on ultra-high molecular weight polyethylene to increase the adhesion between: (1) the polymer and polymethylmethacrylate (PMMA) bone cement, and (2) the polymer and osteoblast cells. Both techniques are performed using a dielectric barrier discharge (DBD). Various previous paper showed that low pressure plasma activation of polymer results in the incorporation of oxygen–containing functional groups, which leads to an increased surface wettability. Atmospheric pressure plasma polymerization of methylmethacrylate (MMA) on polymer results in a PMMA-like coating, which could be deposited with a high degree of control of the chemical composition and layer thickness.

Generally, the ammonia plasma treatment technique was developed from the plasma polymerization method in the early 1990s. Plasma polymerization modifies the surface of polymers by introducing various oxygen- and nitrogen-containing volatile chemical moieties such as methanol, amine, and amide compounds. Therefore, various amine and amide groups moieties introduced during ammonia plasma modification alter the surface chemistry of the polymer. These groups play an important role in endothelial cells adhesion and growth on a modified surface. This section describes various plasma surface modified polymers in the field of bioimplants.

Poly (butylene succinate) (PBSu), is a biodegradable aliphatic polyester with better processability compared to other polymers. It possesses superior mechanical properties which are comparable to those of polyethylene. Such an advantageous material suffers from insufficient biocompatibility and bioactivity after implantation into the human body. In order to enhance the biocompatibility of the polymer, Wang et al. surface modified the polymer by H_2O or NH_3 plasma immersion ion implantation. The differences in surface chemistry observed after plasma treatments improve the hydrophilicity, osteoblast compatibility, and roughness of PBSu significantly. The results (Fig. 15.7) suggest that $C-OH$ groups benefit osteoblast adhesion and proliferation more than $C-NH_2$ groups.

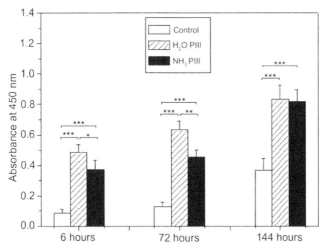

Figure 15.7 The cell viability assay of hFOB 1.19 cells incubated on control PBSu, H_2O plasma immersion ion implantation PBSu and NH_3 plasma immersion ion implantation PBSu for 6, 72 and 144 h. The absorbance of the diluted Cell Counting Kit solution at 450 nm ($A = 0.324$) has been deducted from each data point. Statistical significance indicated by $*P < .05$, $**P < .01$ and $***P < .001$. *Reproduced with permission from Wang, H., Ji, J., Zhang, W., Zhang, Y., Jiang, J., & Wu, Z. (2009). Biocompatibility and bioactivity of plasma-treated biodegradable poly (butylene succinate). Acta Biomater., 5(1), 279−287. http://doi.org/10.1016/j.actbio.2008.07.017 [70]. copyright 2008, Elsevier.*

It is also found that both H_2O and NH_3 plasma immersion ion implantation can transform the bioinert PBSu surface into a bioactive biomaterial.

Polyurethane is a type of shape memory polymer, with advantageous mechanical properties including large recovery strains and low recovery stresses. These types of polymer are promising biomedical implant materials due to their ability to recover to a predetermined shape from a temporary shape induced by thermal activation close to human body temperature. Despite its benefits as a biomaterial, upon contact with living tissues the polymer induces problematic issues including inflammation and the foreign body reaction. In 2017 Cheng et al. [71] surface modified the polymer by plasma immersion ion implantation. This method is a surface modification process using energetic ions that generate radicals in polymer surfaces leading to carbonization and oxidation and the ability to covalently immobilize proteins without the need for wet chemistry.

Due to the generation of radicals and subsequent oxidation after plasma surface treatment, a dramatic change of surface chemistry was observed. The polyurethane-type shape memory polymer after plasma immersion ion implantation treatment it displays a significant enhancement in wettability and direct covalent bonding of protein. This improved protein immobilization is the direct consequence of the formation of highly reactive radicals in the modified surface by the ion bombardment. Although the density of radicals would be expected to increase with increasing plasma immersion ion implantation treatment time, it was found that the protein immobilization saturated after plasma immersion ion implantation treatment of 200 s, suggesting that a full protein monolayer was formed for the incubation protocol used here.

Another strategy is based on plasma polymerization system, this system features a pulsed-DC power supply, which is used to modify the surface of polyurethane, by employing low-cost hexamethyldisiloxane and tetrafluoromethane as precursors (Fig. 15.8).

The process parameters of plasma-polymerized HMDSO/CF_4 (pp-HC) were appropriately adjusted with both hierarchical scale topography and CF and CF_2 functional groups. According to the obtained microstructures, mechanical properties, water contact angles, surface morphology, cytotoxicity, and blood compatibility of the pp-HC films after treatment, the pp-HC films were robust and noncytotoxic and prevent thrombosis when in directly contact with blood.

With an increasing monomer flow rate, the water contact angle and surface roughness of the deposited pp-HC films increased with a decrease

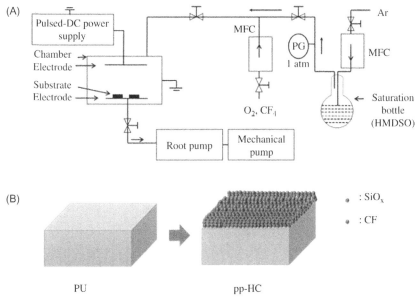

Figure 15.8 Schematic drawings of (A) the plasma polymerization equipment, and (B) the pp-HC film with SiOx nanoparticles and fluorocarbons coated on the PU substrate. *Reproduced with permission from Hsiao, C.R., Lin, C.W., Chou, C.M., Chung, C.J., & He, J.L. (2015). Surface modification of blood-contacting biomaterials by plasma-polymerized superhydrophobic films using hexamethyldisiloxane and tetrafluoromethane as precursors. Appl. Surf. Sci., 346, 50–56. http://doi.org/10.1016/j.apsusc.2015.03.208 copyright 2015, Elsevier.*

in the transmittance. These results indicates the super hydrophobic and antifouling characteristics of the coating were retained even after it was rubbed 20 times with a steel wool tester. The in vitro WST-1 cell proliferation test results revealed that the pp-HC films exhibited favorable myoblast cell proliferation (Fig. 15.9). Moreover, the pp-HC-coated specimens exhibited extremely low platelet adhesion even after a long incubation time, attributable to their hierarchical-scale surface roughness and super hydrophobicity. These quantitative findings imply that the pp-HC coating can potentially prevent the formation of thrombi and provide an alternative means of modifying the surfaces of blood–contact biomaterials.

Poly(ether ether ketone) (PEEK) is a rigid semi-crystalline thermoplastic polymer. Being the best known and an important member of the polyaryletherketone group (PAEK) it has excellent sliding and very good mechanical properties, broad chemical resistance, and bone-like stiffness and is widely used in biomedical fields. However, the bio-inert surface of

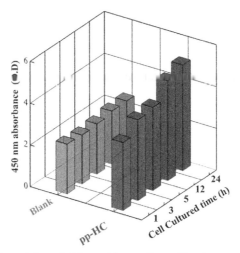

Figure 15.9 Results of the c2c12 myoblast cell proliferation test cultured on the blank PU substrate and pp-HC coated specimens (deposited at fH = 30 sccm) for 1, 3, 5, 12, and 24 h. *Reproduced with permission from Hsiao, C.R., Lin, C.W., Chou, C.M., Chung, C.J., & He, J.L. (2015). Surface modification of blood-contacting biomaterials by plasma-polymerized superhydrophobic films using hexamethyldisiloxane and tetrafluoromethane as precursors. Appl. Surf. Sci., 346, 50−56. http://doi.org/10.1016/j. apsusc.2015.03.208 [72] copyright 2015, Elsevier.*

PEEK tends to hinder its biomedical applications when direct osteointegration between the implants and the host tissue is desired.

In this section, we describe the modification of the PEEK polymer carried out with the aim of improving its performance for biomedical applications. Here, polymer samples were exposed to plasmas of oxygen, argon, hydrogen, and deuterium by placing them on a substrate holder to which 15 μs pulses of 10 and 15 kV were applied. It was observed that the treatment with oxygen decreased the contact angle with water drops to nearly zero while argon modification decreased the contact angle to half that of the untreated material. The deeper the treatment the longer the time required to recover the original hydrophobic character. We also found that the hydrogen treatment produced the deepest modification and that the electrical resistivity of the treated material was the same for hydrogen and argon, but was higher for oxygen.

In 2015 Zheng et al. [73] demonstrated a dual modification method, which combines the laser and plasma surface treatment to combine advantages of both chemical states and microstructures for osteoblasts responses. In this method, the plasma treatment introduces surface carboxyl groups

(−COOH) onto the PEEK surface, the laser treatment constructs micro-structures over the PEEK surface. These results indicated that −COOH as well as microgrooves containing micropores or microcraters structure are constructed on PEEK surface and plasma treatment has no apparent effect on the morphology of microstructures produced by laser micromachining. Unexpectedly, the superior mechanical properties of PEEK were maintained irrespective of the treatment used. Compared to native PEEK and single treated PEEK, dual modified PEEK is more favorable for preosteo-blasts (MC3T3-E1) adhesion, spreading and proliferation.

Moreover, it is widely accepted that improved initial adhesion of osteoblasts or osteoblast precursor cells to orthopedic implant surfaces may lead to improved bone-implant integration and longer-term stability. The osteoblasts (see Fig. 15.10) on PEEK-L-COOH not only exhibit good adhesion and spread-out morphology, but also appear to tightly cling to the PEEK surface, and more importantly, cell pseudopodia protrude into the micropores or microcraters, indicating dual treated PEEK implants may further enhance bone-implant integration compared to the bare or the single-treated PEEK after implant in vivo. Therefore, PEEK implant with dual modification surface has potential use in orthopedic or dental implants.

In 2016 Wiacek et al. [74] prepared a novel chitosan-coated PEEK bio-materials by surface modifying PEEK via air plasma modification. Owing to the appreciable antimicrobial activity and low toxicity, chitosan has been utilized either alone or together with other polymers (or more frequently biopolymers) in biomedical, chemical, and food industries. During the plasma treatment, the surface of PEEK interacts with air plasma cause hydrogen separation from polymeric chains and free radical creation. Created free radicals can interact with oxygen or nitrogen from air and, thus, the new functional moieties can be incorporated into the surface of PEEK. Furthermore, the C−C and C−H bonds can be disrupted and crosslinking by activated species of gas can be formed. Also, the air plasma containing oxygen can be used for incorporation of hydroxyl groups on the polymer surface, on the other hand, nitrogen-containing plasma improves wettability, biocompatibility, and printability of polymeric surfaces and new material with different combinations can be produced.

After plasma modification the surface is oxidized and the strong hydrophilic character of PEEK surface (water contact angle is equal to 14.6 degrees) is changed to become more hydrophobic (54.3 degrees) close the value for untreated PEEK (73.2 degrees). As the amount of

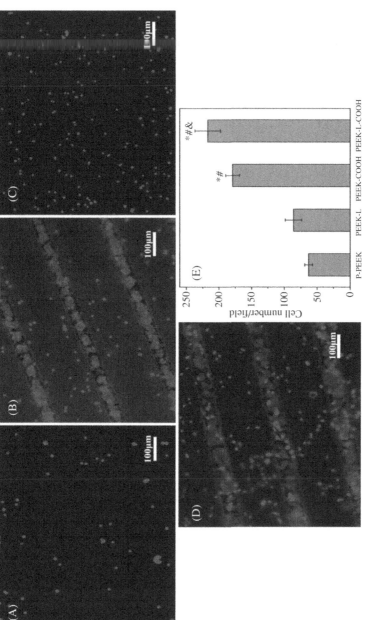

Figure 15.10 MC3T3-E1 preosteoblasts adhesion measured on P-PEEK (A), PEEK-L (B), PEEK-COOH (C) and PEEK-L-COOH (D) after incubation for 4 h. (E) Preosteoblasts measured by counting cell nuclei in at least 3 microscope fields under a fluorescence microscope. Statistical significance is indicated by *$P < .05$ compared to P-PEEK, #$P < .05$ compared to PEEK-L and $P < .05$ compared to PEEK-COOH. *Reproduced with permission from Zheng, Y., Xiong, C., Wang, Z., Li, X., & Zhang, L. (2015). A combination of CO 2 laser and plasma surface modification of poly(etheretherketone) to enhance osteoblast response. Appl. Surf. Sci., 344, 79–88. http://doi.org/10.1016/j.apsusc.2015.03.113 copyright 2015, Elsevier.*

polar groups on the polymer surface increases after air plasma modification noticeably, the interactions between the PEEK surface and chitosan molecules are stronger too. Moreover, all roughness parameters of plasma treated PEEK surface also increase, which could promote the chitosan molecules adhesion to plasma modified surface. The surface apparent free energy determined by two approaches (LWAB and CAH) exhibit the similar values and same tendency of the energy changes. Moreover, from the calculations, the increased polar component of the surface free energy is responsible for the wettability changes of the studied polymer surfaces. Significant increase in the surface roughness of PEEK after plasma modification extends PEEK adhesive features and as an effect stronger bonds with the chitosan molecules and formation of dense film were possible, but excellent mechanical properties were retained. The hydrophobic film of chitosan closely packed on the biomaterial surface solely modifies the surface and, as a result, the polymer again becomes more hydrophobic. Chitosan contributes antibacterial properties to the polymer surface and broadens the spectrum of PEEK applications. We hope that such materials could be the generation of bio-interactive polymers leading future oriented therapeutic substances.

Polydimethylsiloxane (PDMS) is one of the high-performance polymers, with unique physical and chemical properties like flexible, thermotolerant, resistant to oxidation, ease of fabrication, tunable hardness, and other desirable properties. This material has attracted increasing interest from materials scientists and engineers in the field of biomedical application. However, the inherently hydrophobic property of PDMS, that is, the basic $(-OSi(CH_3)_2-)_n$ structure in PDMS with many methyl groups $(-CH_3)$, and quite small surface-free energy of PDMS impede its application particularly as contact lenses and implants. The inherent hydrophobicity and inadequacy biocompatibility have hitherto restricted the application of PDMS in biomedical engineering.

In 2016, Tong et al. [75] enhanced the hydrophilicity and biocompatibility of PDMS through the plasma immersion ion implantation technique. Under the optimal plasma immersion ion implantation conditions, active functional groups like abundant O-containing functional groups and "herringbone" topographical features are formed on the PDMS surface, tailor the surface hydrophilicity and consequently biological properties. The PDMS samples after plasma immersion ion implantation have better cytocompatibity and lower geno toxicity than both the untreated and plasma-exposed sample. Also the performance is even comparable to that observed from the culture plate. Plasma immersion ion implantation is demonstrated to be a viable and effective way to improve the surface properties of PDMS in biomedical applications.

The property of biodegradable polymers along with inorganic bioactive particles has emerged as a promising strategy to lead materials with applications ranging from structural implants to tissue engineering scaffolds, particularly in the field of bone tissue engineering.

In 2013, Larrañaga et al. [76] modified the surface of bioactive glass particles through plasma polymerization of acrylic acid, for the enhancement of thermal stability of bioglass filled composite systems. The developed poly (acrylic acid) layer a medical (co)polyesters which is conventionally manufactured by thermoplastic processing techniques, such as injection molding or extrusion. The surface-modified bioglass particles, hinders the degradation reaction between the $Si-O^-$ groups present in the surface of the particles and the $C = O$ groups of the polymer's backbone and, thus, improved the thermal stability of the studied composites.

The optimization of this process may result in the production of completely surface modified particles and would permit the manufacturing of bioglass particles filled polyester composites by the conventional thermoplastic techniques, avoiding the risk of thermal degradation during processing.

15.6 CONCLUSION

We have described the application of surface modified polymers using plasma technology in the field of biosensor and biodevice and summarized as follows.

1. We have described biosensor concepts and discussed variations in biosensor designs. The typical structure of biosensors is designed to accommodate closely integrated biological components and transducers. Plasma surface modified polymers are a good interface between the biological components and the transducers. This plasma possessed several advantages compared to the well-established other conventional methods. In spite of the usefulness of plasma surface modified polymer in biosensors, there are only a few reports concerning this.

   In the case of biomaterials:

2. Most of the plasma surface modified polymers could be made so that they are chemically inert and show no acute toxicity in biological systems.

3. Plasma surface modified polymers can be used to effectively modify the surface properties of materials, thus, it is possible to control the interaction between a biomaterial and a biological system.
4. This process can provide a means of controlling transport of substances in and out of biomaterials.
5. Plasma surface modified polymers can be used as protective coating of biomedical devices.
6. It may be used to improve adhesion of different materials used in composite biomedical devices.

15.7 FUTURE TRENDS

Nonthermal plasma technology can greatly enhance the interaction of polymeric material, however, a better understanding of these interactions is of crucial importance. The in-depth knowledge in the fundamentals of plasma technology will further develop research to achieve better control over incorporation of various functional moieties, cell adhesion, proliferation, and differentiation. A better control of the modified surface (thickness, chemical composition) should also be achieved and the development of new nonthermal plasma sources, such as microplasmas will certainly create new possibilities and push the limits of current surface-control. As new plasma sources are continuously being developed, it will probably become possible to modify polymers at atmospheric pressure in an industrial setting. Most likely, even controlled patterns of functional groups will be robotically drawn on polymers in the near future. From the given literature overview, it should become clear to the reader that plasma-assisted surface modification is mostly performed on two-dimensional polymer substrates. In the future, this research field will most certainly evolve and the acquired expertise will be useful when applying plasma-assisted surface modification techniques to 3D porous scaffolds. Some preliminary studies have already demonstrated the applicability of plasma technology for the surface modification of the interior of porous scaffolds, however, it is expected that this research topic will become more and more important in the near future. At the same time, new scaffold production techniques will be developed and the authors, therefore, strongly believe that the combination of highly porous, 100 percent interconnected scaffolds with state-of-the-art plasma surface modification technology will greatly improve the success of biosensors and biodevices.

ACKNOWLEDGMENT

The authors would like to thank Indian Institute of Space science and Technology for its generous support and Dr. Linsha V and Mr. Sujith Vijayan for their help with the editing of this chapter.

REFERENCES

[1] V. Saxena, B.D. Malhotra, Prospects of conducting polymers in molecular electronics, Curr. Appl. Phys. 3 (2) (2003) 293–305. Available from: http://doi.org/https://doi.org/10.1016/S1567-1739(02)00217-1.

[2] C. Hassler, T. Boretius, T. Stieglitz, Polymers for neural implants, J. Polym. Sci. Part B: Polym. Phys. 49 (1) (2011) 18–33. Available from: http://doi.org/10.1002/polb.22169.

[3] D.B. Hazer, E. Kılıçay, B. Hazer, Poly(3-hydroxyalkanoate)s: diversification and biomedical applications: a state of the art review, Mater. Sci. Eng. C 32 (4) (2012) 637–647. Available from: http://doi.org/https://doi.org/10.1016/j.msec.2012.01.021.

[4] C. Luo, G.Z. Cao, I.Y. Shen, Development of a lead-zirconate-titanate (PZT) thin-film microactuator probe for intracochlear applications, Sensors Actuators A: Phys. 201 (Suppl. C) (2013) 1–9. Available from: http://doi.org/https://doi.org/10.1016/j.sna.2013.06.027.

[5] Y. Ma, R. Wang, X. Cheng, Z. Liu, Y. Zhang, The behavior of new hydrophilic composite bone cements for immediate loading of dental implant, J. Wuhan Univ. Technol. Mater. Sci. Ed. 28 (3) (2013) 627–633. Available from: http://doi.org/10.1007/s11595-013-0742-1.

[6] K.S. Min, S.H. Oh, M.-H. Park, J. Jeong, S.J. Kim, A polymer-based multichannel cochlear electrode array, Otol. Neurotol. 35 (7) (2014). Retrieved from http://journals.lww.com/otology-neurotology/Fulltext/2014/08000/A_Polymer_Based_Multichannel_Cochlear_Electrode.14.aspx.

[7] P. Moalli, B. Brown, M.T.F. Reitman, C.W. Nager, Polypropylene mesh: evidence for lack of carcinogenicity, Int. Urogynecol. J. 25 (5) (2014) 573–576. Available from: http://doi.org/10.1007/s00192-014-2343-8.

[8] P. Parida, A. Behera, S.C. Mishra, Classification of biomaterials used in medicine, Int. J. Adv. Appl. Sci. 1 (3) (2012) 125–129. Available from: http://doi.org/10.11591/ijaas.v1i3.882.

[9] A. Rahimi, A. Mashak, Review on rubbers in medicine: natural, silicone and polyurethane rubbers, Plast. Rubber Compos. 42 (6) (2013) 223–230. Available from: http://doi.org/10.1179/1743289811Y.0000000063.

[10] J. Zhou, X. Huang, D. Zheng, H. Li, T. Herrler, Q. Li, Oriental nose elongation using an l-shaped polyethylene sheet implant for combined septal spreading and extension, Aesthetic Plast. Surg. 38 (2) (2014) 295–302. Available from: http://doi.org/10.1007/s00266-014-0299-1.

[11] A.A. John, A.P. Subramanian, M.V. Vellayappan, A. Balaji, S.K. Jaganathan, H. Mohandas, et al., Review: physico-chemical modification as a versatile strategy for the biocompatibility enhancement of biomaterials, RSC Adv. 5 (49) (2015) 39232–39244. Available from: http://doi.org/10.1039/C5RA03018H.

[12] A.S. Hoffman, Modification of material surfaces to affect how they interact with blooda, Ann. N. Y. Acad. Sci. 516 (1) (1987) 96–101. Available from: http://doi.org/10.1111/j.1749-6632.1987.tb33033.x.

[13] J.G.A. Terlingen, G.A.J. Takens, F.J. Van Der Gaag, A.S. Hoffman, J. Feijen, On the effect of treating poly(acrylic acid) with argon and tetrafluoromethane plasmas: kinetics and degradation mechanism, J. Appl. Polym. Sci. 52 (1) (1994) 39−53. Available from: http://doi.org/10.1002/app.1994.070520105.

[14] A. Kotwal, C.E. Schmidt, Electrical stimulation alters protein adsorption and nerve cell interactions with electrically conducting biomaterials, Biomaterials. 22 (10) (2001) 1055−1064. Available from: http://doi.org/https://doi.org/10.1016/S0142-9612(00)00344-6.

[15] B. Lakard, L. Ploux, K. Anselme, F. Lallemand, S. Lakard, M. Nardin, et al., Effect of ultrasounds on the electrochemical synthesis of polypyrrole, application to the adhesion and growth of biological cells, Bioelectrochemistry 75 (2) (2009) 148−157. Available from: http://doi.org/https://doi.org/10.1016/j.bioelechem.2009.03.010.

[16] J.Y. Lee, C.A. Bashur, A.S. Goldstein, C.E. Schmidt, Polypyrrole-coated electrospun PLGA nanofibers for neural tissue applications, Biomaterials. 30 (26) (2009) 4325−4335. Available from: http://doi.org/https://doi.org/10.1016/j.biomaterials.2009.04.042.

[17] G.G. Wallace, M. Smyth, H. Zhao, Conducting electroactive polymer-based biosensors, TrAC Trends Analyt Chem 18 (4) (1999) 245−251. Available from: http://doi.org/https://doi.org/10.1016/S0165-9936(98)00113-7.

[18] S. Laib, B.D. MacCraith, Immobilization of biomolecules on cycloolefin polymer supports, Anal. Chem. 79 (16) (2007) 6264−6270. Available from: http://doi.org/10.1021/ac062420y.

[19] A.P. Alekhin, G.M. Boleiko, S.A. Gudkova, A.M. Markeev, A.A. Sigarev, V.F. Toknova, et al., Synthesis of biocompatible surfaces by nanotechnology methods, Nanotechnol. Russia 5 (9) (2010) 696−708. Available from: http://doi.org/10.1134/S1995078010090144.

[20] G. London, K.-Y. Chen, G.T. Carroll, B.L. Feringa, Towards dynamic control of wettability by using functionalized altitudinal molecular motors on solid surfaces, Chem. A Europ. J. 19 (32) (2013) 10690−10697. Available from: http://doi.org/10.1002/chem.201300500.

[21] S. Bertazzo, W.F. Zambuzzi, H.A. Da Silva, C.V. Ferreira, C.A. Bertran, Bioactivation of alumina by surface modification: a possibility for improving the applicability of alumina in bone and oral repair, Clin. Oral. Implants. Res. 20 (3) (2009) 288−293. Available from: http://doi.org/10.1111/j.1600-0501.2008.01642.x.

[22] L.S. Barbarash, E.N. Bolbasov, L.V. Antonova, V.G. Matveeva, E.A. Velikanova, E.V. Shesterikov, et al., Surface modification of poly-ε-caprolactone electrospun fibrous scaffolds using plasma discharge with sputter deposition of a titanium target, Mater. Letters 171 (Suppl. C) (2016) 87−90. Available from: http://doi.org/https://doi.org/10.1016/j.matlet.2016.02.062.

[23] S.I. Tverdokhlebov, E.N. Bolbasov, E.V. Shesterikov, L.V. Antonova, A.S. Golovkin, V.G. Matveeva, et al., Modification of polylactic acid surface using RF plasma discharge with sputter deposition of a hydroxyapatite target for increased biocompatibility, Appl. Surf. Sci. 329 (Suppl. C) (2015) 32−39. Available from: http://doi.org/https://doi.org/10.1016/j.apsusc.2014.12.127.

[24] J. Garcia-Torres, D. Sylla, L. Molina, E. Crespo, J. Mota, L. Bautista, Surface modification of cellulosic substrates via atmospheric pressure plasma polymerization of acrylic acid: structure and properties, Appl. Surf. Sci. 305 (Suppl. C) (2014) 292−300. Available from: http://doi.org/https://doi.org/10.1016/j.apsusc.2014.03.065.

[25] D. Sakthi Kumar, Y. Yoshida, Dielectric properties of plasma polymerized pyrrole thin film capacitors, Surf. Coat. Technol. 169 (Suppl. C) (2003) 600−603. Available from: http://doi.org/https://doi.org/10.1016/S0257-8972(03)00118-X.

[26] K.S. Siow, L. Britcher, S. Kumar, H.J. Griesser, Plasma methods for the generation of chemically reactive surfaces for biomolecule immobilization and cell colonization - a review, Plasma Process. Polym. 3 (6−7) (2006) 392−418. Available from: http://doi.org/10.1002/ppap.200600021.

[27] M. Bashir, J.M. Rees, W.B. Zimmerman, Plasma polymerization in a microcapillary using an atmospheric pressure dielectric barrier discharge, Surf. Coat. Technol. 234 (Suppl. C) (2013) 82−91. Available from: http://doi.org/https://doi.org/10.1016/j.surfcoat.2013.01.041.

[28] N. De Geyter, R. Morent, T. Desmet, M. Trentesaux, L. Gengembre, P. Dubruel, et al., Plasma modification of polylactic acid in a medium pressure DBD, Surf. Coat. Technol. 204 (20) (2010) 3272−3279. Available from: http://doi.org/https://doi.org/10.1016/j.surfcoat.2010.03.037.

[29] N. De Geyter, R. Morent, C. Leys, L. Gengembre, E. Payen, Treatment of polymer films with a dielectric barrier discharge in air, helium and argon at medium pressure, Surf. Coat. Technol. 201 (16) (2007) 7066−7075. Available from: http://doi.org/https://doi.org/10.1016/j.surfcoat.2007.01.008.

[30] D. Hegemann, H. Brunner, C. Oehr, Plasma treatment of polymers for surface and adhesion improvement, Nucl. Inst. Methods Phys. Res. Sect. B: Beam Interact. Mater. Atoms 208 (Suppl. C) (2003) 281−286. Available from: http://doi.org/https://doi.org/10.1016/S0168-583X(03)00644-X.

[31] T. Jacobs, N. De Geyter, R. Morent, T. Desmet, P. Dubruel, C. Leys, Plasma treatment of polycaprolactone at medium pressure, Surf. Coat. Technol. 205 (Suppl. 2) (2011) S543−S547. Available from: http://doi.org/https://doi.org/10.1016/j.surfcoat.2011.02.012.

[32] J. Lai, B. Sunderland, J. Xue, S. Yan, W. Zhao, M. Folkard, et al., Study on hydrophilicity of polymer surfaces improved by plasma treatment, Appl. Surf. Sci. 252 (10) (2006) 3375−3379. Available from: http://doi.org/https://doi.org/10.1016/j.apsusc.2005.05.038.

[33] R. Morent, N. De Geyter, T. Desmet, P. Dubruel, C. Leys, Plasma surface modification of biodegradable polymers: a review, Plasma Process. Polym. 8 (3) (2011) 171−190. Available from: http://doi.org/10.1002/ppap.201000153.

[34] H.A. Karahan, E. Özdoğan, Improvements of surface functionality of cotton fibers by atmospheric plasma treatment, Fib. Polym. 9 (1) (2008) 21−26. Available from: http://doi.org/10.1007/s12221-008-0004-6.

[35] S.I. Ertel, A. Chilkoti, T.A. Horbetti, B.D. Ratner, Endothelial cell growth on oxygen-containing films deposited by radio-frequency plasmas: the role of surface carbonyl groups, J. Biomater. Sci. Polym. Ed. 3 (2) (1992) 163−183. Available from: http://doi.org/10.1163/156856291X00269.

[36] T.R. Gengenbach, X. Xie, R.C. Chatelier, H.J. Griesser, Evolution of the surface composition and topography of perfluorinated polymers following ammonia-plasma treatment, J. Adhesion Sci. Technol. 8 (4) (1994) 305−328. Available from: http://doi.org/10.1163/156856194X00267.

[37] D.P. Dowling, S. Maher, V.J. Law, M. Ardhaoui, C. Stallard, A. Keenan, Modified drug release using atmospheric pressure plasma deposited siloxane coatings, J. Phys. D. Appl. Phys. 49 (36) (2016) 364005. Retrieved from http://stacks.iop.org/0022-3727/49/i=36/a=364005.

[38] H. Chim, J.L. Ong, J.-T. Schantz, D.W. Hutmacher, C.M. Agrawal, Efficacy of glow discharge gas plasma treatment as a surface modification process for three-dimensional poly (D,L-lactide) scaffolds, J. Biomed. Mater. Res. A. 65A (3) (2003) 327−335. Available from: http://doi.org/10.1002/jbm.a.10478.

[39] A. Chirokov, A. Gutsol, A. Fridman, Atmospheric pressure plasma of dielectric barrier discharges, Pure Appl. Chem. 77 (2) (2005) 487−495. Available from: http://doi.org/10.1351/pac200577020487.

[40] C. Amorosi, V. Ball, J. Bour, P. Bertani, V. Toniazzo, D. Ruch, et al., One step preparation of plasma based polymer films for drug release, Mater. Sci. Eng. C 32 (7) (2012) 2103−2108. Available from: http://doi.org/https://doi.org/10.1016/j.msec.2012.05.045.

[41] B.-O. Aronsson, J. Lausmaa, B. Kasemo, Glow discharge plasma treatment for surface cleaning and modification of metallic biomaterials, J. Biomed. Mater. Res. 35 (1) (1997) 49−73. Available from: http://doi.org/10.1002/(SICI)1097-4636(199704)35:1 < 49:AID-JBM6 > 3.0.CO;2-M.

[42] S. Osaki, M. Chen, P.O. Zamora, Controlled drug release through a plasma polymerized tetramethylcyclo-tetrasiloxane coating barrier, J. Biomater. Sci. Polym. Ed. 23 (1−4) (2012) 483−496. Available from: http://doi.org/10.1163/092050610X552753.

[43] Y. Yuan, C. Liu, Y. Zhang, M. Yin, C. Shi, Radio frequency plasma polymers of n-butyl methacrylate and their controlled drug release characteristics: I. Effect of the oxygen gas, Surf. Coat. Technol. 201 (15) (2007) 6861−6864. Available from: http://doi.org/https://doi.org/10.1016/j.surfcoat.2006.09.020.

[44] A. Costoya, F.M. Ballarin, J. Llovo, A. Concheiro, G.A. Abraham, C. Alvarez-Lorenzo, HMDSO-plasma coated electrospun fibers of poly(cyclodextrin)s for antifungal dressings, Int. J. Pharm. 513 (1) (2016) 518−527. Available from: http://doi.org/https://doi.org/10.1016/j.ijpharm.2016.09.064.

[45] J. Jiang, J. Xie, B. Ma, D.E. Bartlett, A. Xu, C.-H. Wang, Mussel-inspired protein-mediated surface functionalization of electrospun nanofibers for pH-responsive drug delivery, Acta Biomater. 10 (3) (2014) 1324−1332. Available from: http://doi.org/https://doi.org/10.1016/j.actbio.2013.11.012.

[46] C. Labay, J.M. Canal, C. Canal, Relevance of surface modification of polyamide 6.6 fibers by air plasma treatment on the release of caffeine, Plasma Process. Polym. 9 (2) (2012) 165−173. Available from: http://doi.org/10.1002/ppap.201100077.

[47] J.C. Park, B.J. Park, D.W. Han, D.H. Lee, I.S. Lee, S.O. Hyun, et al., Fungal sterilization using microwave-induced argon plasma at atmospheric pressure, J. Microbiol. Biotechnol. 14 (1) (2004) 188−192. Available from: http://doi.org/10.1063/1.1613655.

[48] G.S. Wilson, R. Gifford, Biosensors for real-time in vivo measurements, Biosens. Bioelectron. 20 (12) (2005) 2388−2403. Available from: http://doi.org/http://dx.doi.org/10.1016/j.bios.2004.12.003.

[49] J. Wang, Amperometric biosensors for clinical and therapeutic drug monitoring: a review, J. Pharm. Biomed. Anal. 19 (1) (1999) 47−53. Available from: http://doi.org/http://dx.doi.org/10.1016/S0731-7085(98)00056-9.

[50] V. Mazeiko, A. Kausaite-Minkstimiene, A. Ramanaviciene, Z. Balevicius, A. Ramanavicius, Gold nanoparticle and conducting polymer-polyaniline-based nanocomposites for glucose biosensor design, Sensors Actuators B: Chem. 189 (Suppl. C) (2013) 187−193. Available from: http://doi.org/https://doi.org/10.1016/j.snb.2013.03.140.

[51] S. Wu, F. Su, X. Dong, C. Ma, L. Pang, D. Peng, et al., Development of glucose biosensors based on plasma polymerization-assisted nanocomposites of polyaniline, tin oxide, and three-dimensional reduced graphene oxide, Appl. Surf. Sci. 401 (Suppl. C) (2017) 262−270. Available from: http://doi.org/https://doi.org/10.1016/j.apsusc.2017.01.024.

[52] A. Hiratsuka, K. Kojima, H. Muguruma, K.-H. Lee, H. Suzuki, I. Karube, Electron transfer mediator micro-biosensor fabrication by organic plasma process, Biosens.

Bioelectron. 21 (6) (2005) 957−964. Available from: http://doi.org/https://doi.org/10.1016/j.bios.2005.03.002.

[53] Ase, Y.K., Uguruma, H.M., Amperometric Glucose Biosensor Based on Mediated Electron Transfer between Immobilized Glucose Oxidase and Plasma- polymerized Thin Film of Dimethylaminomethylferrocene on Sputtered Gold Electrode, 20 (August), 2004, 1143−1146.

[54] H. Muguruma, Y. Kase, H. Uehara, Nanothin ferrocene film plasma polymerized over physisorbed glucose oxidase: high-throughput fabrication of bioelectronic devices without chemical modifications, Anal. Chem. 77 (20) (2005) 6557−6562. Available from: http://doi.org/10.1021/ac0501691.

[55] Y.-H. Choi, G.-Y. Lee, H. Ko, Y.W. Chang, M.-J. Kang, J.-C. Pyun, Development of SPR biosensor for the detection of human hepatitis B virus using plasma-treated parylene-N film, Biosens. Bioelectron. 56 (Suppl. C) (2014) 286−294. Available from: http://doi.org/https://doi.org/10.1016/j.bios.2014.01.035.

[56] J.W. Chung, S.D. Kim, R. Bernhardt, J.C. Pyun, Application of SPR biosensor for medical diagnostics of human hepatitis B virus (hHBV), Sensors and Actuators B: Chemical 111 (Suppl. C) (2005) 416−422. Available from: http://doi.org/https://doi.org/10.1016/j.snb.2005.03.055.

[57] P.B. Luppa, L.J. Sokoll, D.W. Chan, Immunosensors—principles and applications to clinical chemistry, Clinica Chimica Acta 314 (1) (2001) 1−26. Available from: http://doi.org/https://doi.org/10.1016/S0009-8981(01)00629-5.

[58] M. Ardhaoui, S. Bhatt, M. Zheng, D. Dowling, C. Jolivalt, F.A. Khonsari, Biosensor based on laccase immobilized on plasma polymerized allylamine/carbon electrode, Mater. Sci. Eng. C 33 (6) (2013) 3197−3205. Available from: http://doi.org/https://doi.org/10.1016/j.msec.2013.03.052.

[59] A.D. Ellington, J.W. Szostak, In vitro selection of RNA molecules that bind specific ligands, Nature 346 (6287) (1990) 818−822. Retrieved from https://doi.org/10.1038/346818a0.

[60] C. Tuerk, L. Gold, Systematic evolution of ligands by exponential enrichment: RNA ligands to bacteriophage T4 DNA polymerase, Science 249 (4968) (1990). 505 LP-510. Retrieved from http://science.sciencemag.org/content/249/4968/505.abstract.

[61] S.D. Jayasena, Aptamers: an emerging class of molecules that rival antibodies in diagnostics, Clin. Chem. 45 (9) (1999). 1628 LP-1650. Retrieved from http://clinchem.aaccjnls.org/content/45/9/1628.abstract.

[62] A.D. Ellington, J.W. Szostak, Selection in vitro of single-stranded DNA molecules that fold into specific ligand-binding structures, Nature 355 (6363) (1992) 850−852. Retrieved from https://doi.org/10.1038/355850a0.

[63] S. Fang, X. Dong, S. Liu, D. Penng, L. He, M. Wang, et al., A label-free multi-functionalized electrochemical aptasensor based on a Fe3O4@3D-rGO@plasma-polymerized (4-vinyl pyridine) nanocomposite for the sensitive detection of proteins in whole blood, Electrochim. Acta 212 (Suppl. C) (2016) 1−9. Available from: http://doi.org/https://doi.org/10.1016/j.electacta.2016.06.128.

[64] M. Wang, S. Liu, Y. Zhang, Y. Yang, Y. Shi, L. He, et al., Graphene nanostructures with plasma polymerized allylamine biosensor for selective detection of mercury ions, Sensors Actuators B: Chem. 203 (Suppl. C) (2014) 497−503. Available from: http://doi.org/https://doi.org/10.1016/j.snb.2014.07.009.

[65] C.-H. Yang, L.-S. Kuo, P.-H. Chen, C.-R. Yang, Z.-M. Tsai, Development of a multilayered polymeric DNA biosensor using radio frequency technology with gold and magnetic nanoparticles, Biosens. Bioelectron. 31 (1) (2012) 349−356. Available from: http://doi.org/https://doi.org/10.1016/j.bios.2011.10.044.

[66] R. Plonsey, The biomedical engineering handbook, The Biomedical Engineering Handbook, Second Edition. 2 Volume Set, CRC Press, 1999. Available from: http://doi.org/doi:10.1201/9781420049510.sec1.

[67] A.J.T. Teo, A. Mishra, I. Park, Y.-J. Kim, W.-T. Park, Y.-J. Yoon, Polymeric biomaterials for medical implants and devices, ACS Biomater. Sci. Eng. 2 (4) (2016) 454—472. Available from: http://doi.org/10.1021/acsbiomaterials.5b00429.

[68] M. Boutonnet Kizling, S.G. Järås, A review of the use of plasma techniques in catalyst preparation and catalytic reactions, Appl. Catal. A: General 147 (1) (1996) 1—21. Available from: http://doi.org/https://doi.org/10.1016/S0926-860X(96)00215-3.

[69] P. Favia, R. d'Agostino, Plasma treatments and plasma deposition of polymers for biomedical applications, Surf. Coat. Technol. 98 (1) (1998) 1102—1106. Available from: http://doi.org/https://doi.org/10.1016/S0257-8972(97)00285-5.

[70] H. Wang, J. Ji, W. Zhang, Y. Zhang, J. Jiang, Z. Wu,). Biocompatibility and bioactivity of plasma-treated biodegradable poly (butylene succinate), Acta Biomater. 5 (1) (2009) 279—287. Available from: http://doi.org/10.1016/j.actbio.2008.07.017.

[71] X. Cheng, A. Kondyurin, S. Bao, M.M.M. Bilek, L. Ye, Applied Surface Science Plasma immersion ion implantation of polyurethane shape memory polymer: surface properties and protein immobilization, Appl. Surf. Sci. 416 (2017) 686—695. Available from: http://doi.org/10.1016/j.apsusc.2017.04.179.

[72] C.R. Hsiao, C.W. Lin, C.M. Chou, C.J. Chung, J.L. He, Surface modification of blood-contacting biomaterials by plasma-polymerized superhydrophobic films using hexamethyldisiloxane and tetrafluoromethane as precursors, Appl. Surf. Sci. 346 (2015) 50—56. Available from: http://doi.org/10.1016/j.apsusc.2015.03.208.

[73] Y. Zheng, C. Xiong, Z. Wang, X. Li, L. Zhang, A combination of CO_2 laser and plasma surface modification of poly(etheretherketone) to enhance osteoblast response, Appl. Surf. Sci. 344 (2015) 79—88. Available from: http://doi.org/10.1016/j.apsusc.2015.03.113.

[74] Wia, A.E., Worzakowska, M., 2016. Effect of low-temperature plasma on chitosan-coated PEEK polymer characteristics Agnieszka Ewa Wia, 78, 1—13. https://doi.org/10.1016/j.eurpolymj.2016.02.024.

[75] L. Tong, W. Zhou, Y. Zhao, X. Yu, H. Wang, P.K. Chu, Enhanced cytocompatibility and reduced genotoxicity of polydimethylsiloxane modified by plasma immersion ion implantation, Coll. Surf. B: Biointerf. 148 (2016) 139—146. Available from: http://doi.org/10.1016/j.colsurfb.2016.08.057.

[76] A.A. Ensafi, Z. Ahmadi, M. Jafari-Asl, B. Rezaei, Graphene nanosheets functionalized with Nile blue as a stable support for the oxidation of glucose and reduction of oxygen based on redox replacement of Pd-nanoparticles via nickel oxide, J. Electroanal. Chem. 173 (2015) 619—629. Available from: http://doi.org/10.1016/j.electacta.2015.05.109.

CHAPTER 16

Plasma Modified Polymeric Materials for Scaffolding of Bone Tissue Engineering

K.S. Joshy[1], S. Snigdha[1] and Sabu Thomas[1,2]
[1]International and Inter University Centre for Nanoscience and Nanotechnology (IIUCNN),
Mahatma Gandhi University, Kottayam, Kerala, India
[2]School of Chemical Sciences, Mahatma Gandhi University, Kottayam, Kerala, India

16.1 INTRODUCTION

Tissue engineering has been developing at a very fast pace and this technology has been applied to bring improvements to various clinical conditions, such as spinal bone injuries, fractures, and age-related degeneration of bones, etc. [1]. Artificial constructs made using the principles of material engineering and biological sciences are used for regenerating new tissue [1]. The field of tissue engineering has forged ahead in leaps and bounds over the past decade; this has opened up the potential to regenerating almost every tissue and organ of the human body. Tissue engineering and regenerative medicine is a dynamic area of research and development and has great potential for coming up with new treatments for various maladies affecting the population. This growing field incorporates many disciplines, such as cell biology, biomaterials science, imaging, characterization of surfaces, and cell material interactions. Tissue engineered scaffolds are designed in order to restore, maintain, or improve tissue or organs that are faulty or lost due to trauma or congenital disorders. The general approach employed for developing tissue engineered scaffolds involves: (1) introduction of cells/cellular analogues into the target site of the organism; (2) implantation of tissue along with various trophic factor at the target site; and (3) introducing cells in conjugation with the matrices. The third strategy is characteristic of the concept of tissue engineering, involving the incorporation of live cells on a natural or synthetic extracellular substrate in order to fabricate an implantable conduit for the organism [2].

Non-Thermal Plasma Technology for Polymeric Materials
DOI: https://doi.org/10.1016/B978-0-12-813152-7.00016-0
439

Bone tissue engineering has been successfully carried out in various parts of the world and it typically involves the use of a scaffold, osteoblasts or osteoblast precursors, and tropic factors that will aid in bone development and mineralization at the site of implantation. Highly porous 3D scaffolds have gained a lot of popularity in bone tissue engineering in recent years. This could be attributed to their ability to aid in cell seeding, proliferation, and tissue development that is characteristic of three dimensional (3D) structures found in higher animals. An ideal scaffold must be biocompatible, biodegradable, mechanically robust, and it should aid in new tissue formation; scaffolds should also be suitable enough to promote cell−biomaterials interaction, cell adhesion, permit gas/nutrient/tropic factor transportation, and possess controllable degradation. It is also desirable that the scaffolds be nontoxic, release nontoxic by-products, and cause minimal amount of inflammation in the host organism. Design and fabrication of scaffold materials is of major significance in biomaterials research [3]. Bioengineered scaffolding has been observed to play a major role in revival and repair of tissue and organs. The past few decades have seen the development of practically applicable scaffolds for tissue engineering and regenerative medicine. More and more scaffolds are being developed that have tunable optimal characteristics, such as mechanical strength, degradation rate, porosity, controllable shape and size, and easy reproducibility [4]. Some verities of scaffolds display bioactive properties which effects regeneration in tissues and organs that do not regenerate spontaneously, such scaffolds are referred to as "regeneration templates." Biological scaffolds are derived from human and animal tissues that have been cleaned and sterilized, while synthetic scaffolds are prepared from various polymers.

The first biologically active scaffold was reported in 1974 and was observed to have good degradation and very low antigenicity in vivo, and thrombo resistant behavior in vitro [5], The first patent for describing biologically active scaffolds was granted in 1977 [6]. A detailed description for synthesizing biologically active scaffolds was provided in 1980 [7]. Reports describing induction of tissue regeneration using scaffolds for skin regeneration in animals [8,9] and humans [10], peripheral nerve regeneration [11], and conjunctiva regeneration [12] highlighted the importance of such scaffolds. Biomaterials have been extensively used for creating synthetic cellular frameworks, often referred to as scaffolds, matrices, or constructs. The fast-paced development in the design of biomaterial scaffolds has led to the advancement of artificial scaffolds that closely mimic the extracellular matrix. These material scaffolds can, thus, be used to restore the function of damaged tissues and organs in humans and, thereby, help in improving the quality of life of the patient as well as the caregiver. Biomaterials are extensively used in the form of bone plates, dental implants, ligaments, sutures, joint replacements, vascular grafts, heart valves, and intraocular lenses; these materials are also employed for making various medical devices like pacemakers, and biosensors [13,14].

16.2 POLYMERIC MATERIALS FOR SCAFFOLD FABRICATION

Various materials such as metals, polymers, and ceramics, and their combinations have been used to replace and regenerate damaged bone tissue. Metals and ceramics are disadvantaged due to their lack of degradability in a biological environment, and difficulty in processing however, polymers can be easily tailored for specific needs. Their design flexibility makes them attractive candidates for bone tissue engineering. Biodegradability is an attractive feature in polymers and can be introduced through molecular design. Polymers degrade by different pathways in the body, such as by hydrolysis in the body's aqueous environment, and also by various cellular or enzymatic pathways. These reasons have made polymeric materials highly sort-after and are being widely researched for bone tissue engineering and regeneration [15,16]. Polymers for biomedical applications are selected on the basis of their various physical and chemical properties, such as molecular weight, degradation rate, solubility, and degree of polymerization, shape and structure, affinity to water, lubricity, surface properties, water uptake, and erosion mechanisms. Polymeric scaffolds possess many desirable properties, such as high surface-to-volume ratio, variable pore size, high porosity, biodegradation, and mechanical properties. Their biocompatibility, chemical versatility, and biological properties play a significant role in their application for tissue engineering and organ substitution.

Scaffolds can be of synthetic or biological origin, and degradable or nondegradable, depending on the intended use. Three main types of polymers are used as materials for preparing scaffolds: (1) naturally occurring polymers; (2) biodegradable synthetic polymers; and (3) nonbiodegradable synthetic polymers. Naturally occurring polymers were among the first biodegradable biomaterials used clinically [17]. Natural materials show better interaction with tissue/cellular components owing to the bioactive properties. Natural polymers can be classified as proteins, polysaccharides, or polynucleotides (DNA, RNA). The commonly used protein biomaterials include silk, collagen, gelatin, fibrinogen, elastin, keratin, actin, and myosin. Cellulose, amylose, dextran, chitin, and glycosaminoglycans make up the commonly used polysaccharide materials.

Synthetic polymers are very popular in the biomedical sector due to the ease of tailoring them for specific applications. Synthetic polymers are usually more economical than their biological counterparts. Synthetic polymers can be synthesized in bulk quantities under controlled conditions, and have uniform properties and a long shelf life. This enables them to possess predictable and reproducible physical and chemical properties, which could be stable over long periods of time. Many commercially available polymers closely mimic the physicochemical and mechanical properties exhibited by biological tissues [18]. Copolymers are some of the commonly utilized synthetic polymers in biomedical engineering. It is gaining popularity for the synthesis of scaffolds and belongs to a class of microbial polyesters [19].

16.3 SCAFFOLD DESIGN AND FABRICATION

The design of 3D scaffolds for bone tissue engineering requires the fine tuning of various parameters. A bone scaffold can be considered ideal if it possesses adequate porosity to house osteoblasts or osteoprogenitor cells, sufficient mechanical strength to bear load, support cell proliferation, promote differentiation, and bone tissue formation. High porosity ($\sim 90\%$) is highly desirable for all tissue engineering applications, including bone tissue engineering. Connectivity between pores can assist uniform cell seeding and distribution, and the diffusion of nutrients and metabolites through the scaffold matrix. The scaffold has to possess sufficient mechanical stability in order to support new bone tissue formation. The degradation rate of the scaffold has to be engineered to correspond to the rate at which new bone formation occurs. This will aid maintaining the structural integrity for tissue formation. In addition, the scaffold's surface chemistry should aid in bone cell adhesion and proper functioning [20].

16.4 PLASMA-SURFACE MODIFICATION OF BIOMATERIALS

Bio-integration is the ultimate and most desirable outcome expected from an implant. Bio-integration occurs only when the implant and host tissue interface are in harmony without inducing any inflammatory response or formation of unusual tissues. Thus, it is crucial to engineer scaffolds with the most suitable surface properties. However, these scaffolds should also possess bulk material properties, such as being mechanically robust, to function efficiently in the physiological environment. As it is quite difficult to design biomaterials fulfilling both needs, a common approach is to fabricate biomaterials with adequate bulk properties followed by a special treatment to enhance the surface properties. In this way, ideal biomaterials with surface attributes that are decoupled from the bulk properties can be made. Furthermore, the surface properties can be selectively modified to enhance the performance of the biomaterials. For instance, by altering the surface functionality using thin film deposition, the optimal surface, chemical, and physical properties can be attained [21]. Hence, surface modification of biomaterials is becoming an increasingly popular method to improve device multifunctionality, tri-biological and mechanical properties, as well as biocompatibility of artificial devices while obviating the needs for large expenses and a long period of time to develop new materials. Altering the surface functionality using thin film deposition has become one of the key methods in biomaterials engineering. Usually more than one approach can satisfy the biomaterials requirements, and the ultimate selection must take into account the process reliability, reproducibility, and products yield [22]. A plasma which can be regarded as the fourth state of matter is composed of highly excited atomic, molecular, ionic, and radical species. It is typically obtained when gases are excited into energetic states by radio frequency (rf), microwave, or electrons from a hot filament discharge. Plasma is a highly unusual and reactive chemical environment in which many plasma—surface reactions occur. The high-density of ionized and excited species in the plasma can change the surface properties of normally inert materials, such as ceramics. Plasma-based techniques combining the advantages of conventional plasma and ion beam technologies are effective methods for medical implants with complex shapes. In particular, modification of the surface energy of the materials can improve the adhesion strength, surface and coating properties, and biocompatibility, just to name a few properties [23]. Succinctly speaking, plasma-based techniques offer the following advantages with regard to biomaterials engineering:

1. The benefits of plasma processing arise from a good understanding of plasma physics and chemistry learnt from other fields such as microelectronics; for example, plasma homogeneity and effects of nonuniform plasma on the substrate surface.

2. Plasma engineering is usually reliable, reproducible, nonline-of-sight, relatively inexpensive, and applicable to different sample geometries as well as different materials, such as metals, polymers, ceramics, and composites. Plasma processes can be monitored quite accurately using in situ plasma diagnostic devices.

3. Plasma treatment can result in changes of a variety of surface characteristics, for example, chemical, tri-biological, electrical, optical, biological, and mechanical. Proper applications yield dense and pinhole-free coatings with excellent interfacial bonds due to the graded nature of the interface [24].

4. Plasma processing can provide sterile surfaces and can be scaled up to industrial production relatively easily. On the contrary, the flexibility of nonplasma techniques for different substrate materials is smaller.

5. Plasma techniques are compatible with masking techniques to enable surface patterning, a process that is commonly used in the microelectronics industry.

As a result, plasma-surface modification (PSM), as an economical and effective materials processing technique, is gaining popularity in the biomedical field. It is possible to change in continuum the chemical composition and properties such as wettability, metal adhesion, dyeability, refractive index, hardness, chemical inertness, lubricity, and biocompatibility of materials surfaces. In the biomedical context, "good biocompatibility" refers to that a prosthesis or biomaterial device is nontoxic, does not induce deleterious reactions from the bio-medium, performs properly all the functions they have been designed for, and has a reasonable lifetime. The application of plasma-based techniques is quite diverse, and examples of applications include cleaning/sterilization, coating or depositing, and implantation modification of surface chemistry of a substrate. Orthopedic prostheses are made harder and more wear-resistant by ion implantation into the articulating surfaces. Orthodontic appliances, surgical instruments, and venous catheters are treated to improve friction, fretting resistance, and biocompatibility [25].

16.5 WET-SPUN SCAFFOLDS

Wet-spinning (WS) was a suitable technique for fabricating fibers made of polymers derived from natural sources, like chitosan and collagen, and found wide application in the biomedical field for those polymers which are susceptible to thermal degradation when processed by other fiber spinning techniques. The main advantage of this technique involves it providing an opportunity for lading variety of therapeutics (e.g., antibiotics and chemotherapeutics), by means of drug incorporation methods. In addition, various approaches have been investigated for assembling wet-spun polymeric fibers for the development of biodegradable scaffolds with a 3D network of macropores suitable for tissue engineering applications. A number of studies of in vitro bone regeneration have shown that the resulting interconnected porous architectures, characterized by high surface-to-volume ratio, as well as support bone-forming cell colonization. Because of the versatility of WS in material selection, drug-loading method, and porosity tailoring at different dimensional scales, an increasing amount of literature has reported on the employment of this technique for processing a great variety of biomedical polymers, obtained either by synthetic routes or from natural sources. Besides wet-spun hollow fibers that have been clinically employed for decades in extracorporeal blood treatment applications, promising results have been obtained for the development of biodegradable fibers and fibrous assemblies investigated as drug release systems and tissue engineering scaffolds. Recent advances in automation of WS scaffold fabrication have led to the development of novel processes enabling one to obtain 3D wet-spun porous structures with predefined external shape and internal network of macropores [26].

16.6 SCAFFOLD FUNCTIONALIZATION TO SUPPORT A TISSUE BIOCOMPATIBILITY

Protein adsorption on surfaces of biomaterials and medical implants is an essential aspect of the cascade of biological reactions taking place at the interface between a synthetic material and the biological environment. Type, amount, and conformation of adsorbed proteins mediate subsequent adhesion, proliferation, and differentiation of cells; and they are believed to steer foreign body response and inflammatory processes. Protein signaling regulates cell phenotype and, thus, tissue structure and function. Biomaterial vehicles are being designed to incorporate and locally deliver various molecules involved in this signaling, including both growth factors and peptides that mimic whole proteins. Controlling the concentration, local duration, and spatial distribution of these factors is the key to their utility and efficacy. A promising strategy being extensively investigated is the encapsulation of growth factors or collagen in biodegradable polymer matrices that release, over time, the incorporated biomolecules. The covalent and noncovalent functionalization of scaffold surface with biomimetic molecules, including bioactive proteins, peptides, and polysaccharides are used to improve scaffold biocompatibility [27].

16.7 SCAFFOLDS FOR BONE REPAIR

The need for implants to repair large bone defects is driving the development of porous synthetic scaffolds with the requisite mechanical strength and toughness in vivo. Recent developments in the use of design principles and novel fabrication technologies are paving the way to create synthetic scaffolds with promising potential for reconstituting bone in load-bearing sites. This section reviews the state-of-the-art in the design and fabrication of bioactive glass and ceramic scaffolds that have improved mechanical properties for structural bone repair. An implant for bone repair should have a combination of properties that satisfy a stringent set of requirements. Ideally, the implant should be biocompatible, bioactive, or biodegradable—with a degradation rate comparable to the rate of new bone formation. The implant should have a porous 3D architecture capable of supporting cell and tissue infiltration, transport of nutrients, and development of capillaries. It must have the requisite mechanical properties for supporting loads experienced by the bone to be replaced. Considerable improvements in scaffold design, fabrication, and evaluation are necessary to meet these stringent requirements, particularly for applications in the repair of load-bearing bone [28].

16.8 IMPACT OF NONTHERMAL PLASMA SURFACE MODIFICATION ON POROUS CALCIUM HYDROXYAPATITE CERAMICS FOR BONE REGENERATION

Plasma can be classified as thermal or nonthermal on the basis of its occurrence. Thermal plasma is extensively used for surface modification in which metals or ceramics are sprayed onto the surface of another material. The thermal plasma technique improves the biocompatibility of the material so that it can be successfully employed in dental or orthopedic implants. The potential use of nonthermal plasma involves medical applications in that it can be directly applied in living tissues during sterilization, blood coagulation, wound healing, and tissue regeneration. Recent studies showed that nonthermal plasma-treated biomaterials can be applied for cartilage regeneration and bone repair. A subsequent study showed the use of nonthermal plasma-treated hydroxyl apatite can be effectively used as a bone substitute. Hydroxyapatite (HA) ceramics is a proven material for bone tissue regeneration. The low-pressure plasma treatment on interconnected porous calcium hydroxyapatite (IP-CHA) can enhance the hydrophilicity and the osteoconductivity of the IP-CHA disc. In vivo implantation of plasma-treated IP-CHA demonstrated superior bone in growth than untreated IP-CHA. They did not change osteoblast cell propagation; it improved osteogenic segregation of seeded marrow mesenchymal stem cells. In vitro X-ray photoelectron spectroscopy (XPS) exposed that this plasma treatment raises levels of oxygen, rather than nitrogen, on the plasma-treated IP-CHA surface. These findings propose that plasma treatment is a simple and effortless processing, and can considerably progress the osteoconductive potential of frequently-used artificial bones, such as IP-CHA [29].

16.9 RECENT ADVANCES IN BIOMEDICAL APPLICATIONS OF SCAFFOLDS IN WOUND HEALING AND DERMAL TISSUE ENGINEERING

Bilayered collagen gels seeded with human fibroblasts can be used as an upper part of artificial skin and has been commercialized as "ApligrafVR," containing the lower part as human keratinocytes. Another collagen derivative, "BiomendVR," is commonly used in the periodontal tissue regeneration. Fibrinogen, another naturally occurring compound containing fibrin, can be used as a natural wound-healing scaffold. So, fibrin can be considered as a biological-based polymer known as an outstanding latent in tissue regeneration and wound healing because it, alone or in combination with other factors, has been regarded as a biological scaffold for stem or primary cells to stimulate bone, ligament, tendons, liver, cardiac tissue, cartilage, nervous tissue, ocular tissue, and skin. Fibronectin, a glycoprotein-derived from bovine or human plasma is another scaffold forming material for wound healing. Gelatin is another natural polymer widely used as a scaffold forming material in tissue engineering. Elastin, an insoluble protein is also an important material in tissue engineering. Electrospun composite of collagen and tropoelastin can be widely used in dermal tissue engineering. Chitosan, keratin, Hyaluronic acid, and alginate are further important materials widely used as biomaterials for wound healing and skin-substituted scaffolds. Another class of biomaterials widely used as skin substitutes and wound healing is hybrid scaffolds of synthetic polymers and biological polymers. Examples include PLLA−collagen, Poly(ethylene oxide)−chitosan Carboxyethyl chitosan/PVA, Chitosan/collagen/PEO, AgNPs/chitosan/ PEO, PCL−collagen, Pullulan−collagen, and Collagen−elastin. The major mechanism behind this substituting scaffold is that they stimulate and secrete the wound growth factor by which epithelialization is achieved. They are able to release the growth factors and cytokines incorporated in dressings because bioengineered materials are capable of adapting to their environment [30].

16.10 3D PRINTED SCAFFOLDS

Currently, there is a considerable requirement for synthetic bone replacement materials employed in bone tissue engineering (BTE). 3D printing is an emerging technique used in conventional implants based on medical datasets. Hydroxyapatite with high resolution and complex internal structures was produced through this technique. The main advantage of 3D printing is that a one-step method can be used for the manufacture of an implant from the 3D dataset without the use of an additional mold. The porous, ceramic structure which is necessary for the growth of bone cells for synthetic scaffolds that promote good vascularization and attachment of bone cells can be provided by hydroxyapatite granules through the use of 3D printing. The stoichiometric similarity of hydroxyapatite that mimics the inorganic part of natural bone makes this material important in BTE. The matrices fabricated by 3D printing can be used for BTE using the patient's cells seeded onto the scaffolds. The scaffolds act as 3D templates for primary cell attachment and the following tissue formation. Recent investigations showed that 3D printed scaffolds show improved cell proliferation, and studies show that in-depth growth of cells between hydroxyapatite granules. So 3D printing-based HA scaffolds are very suitable for bone replacement and demonstrate the application of 3D printed scaffolds for BTE [31].

16.11 PLASMA TREATMENT OF POLYMERS

Plasma treatment is a convenient and cost-effective method for altering the surface properties of polymeric materials by introducing the desired functionalities onto the surface and improving the hydrophilic properties and permeability of polymer surfaces [32]. This approach has been applied to a variety of polymers in order to improve adhesion (for layers coated on treated substrates or for bonding processes where adhesives are used to form mechanical joints between treated parts), improve wetting and inking in coating and printing processes, and in the development of biomaterials. The extent of cell attachment and growth on polymer surfaces are important factors that directly influence the capacity of cells to proliferate and differentiate on these polymeric scaffolds. Adhesion has received much attention, and a discussion of metal—polymer bonding, polymer—polymer bonding, and polymer-matrix composites can be found in a review of plasma surface modification by Liston et al. [33].

One of the important areas of application for plasma surface modification is metallized plastics. Examples of the industrial use of plasmas to promote adhesion of thin metal films to treated polymer surfaces are found in abundance in patent literature. Specific applications disclosed include silver—polyester adhesion for ion selective electrode structures [34], copper—polyimide adhesion for microelectronics, aluminum—polycarbonate adhesion [35], adhesion of electrodeposited copper, nickel or chromium to acrylonitrile—butadiene—styrene (ABS), polyphenylene oxide (PPO) automotive parts, and aluminum—polymer adhesion for thin film capacitors. Polymer surface modification by flame and corona discharge treatment, as well as low-pressure plasma, and have also been disclosed for treatment of polyolefins prior to metalization for making aluminized barrier materials for the food packaging industry. These metallized plastics applications lend themselves readily to low-pressure plasma treatments, which are implemented immediately prior to vacuum metalization processes. For some roll-to-roll processes, however, effective treatments are done at atmospheric pressure [36] before loading the rolls into a vacuum coating machine. In addition to metal—polymer adhesion, polymer—polymer adhesion has also been the object of polymer surface modification technology. In the photographic film industry, reduced pressure plasma treatments have been disclosed for promoting adhesion of aqueous polymer coatings (gelatin-containing layers, in particular) to polymer supports in general and, more recently, to polyethylene (PE) [37] and polyester supports.

Polymer surface modification by plasma polymerization of thin layers on to polymer supports has been disclosed for forming selectively permeable membranes for gas separation. Similar processes have also been disclosed for improving the biocompatibility of implant materials and for controlling biofouling [38,39].

16.11.1 Plasma Implantation

Implantation can introduce elements into the surface of the materials without thermodynamic constraints. Plasma implantation into polymers is described in this section. The performance of a biomaterial can be improved by modifying the surface properties which, in turn, enhance the performance of a biomaterial. A variety of methods have been employed to alter the surface properties, like changing the surface functionality using thin film deposition. By using thin film deposition, the surface properties as well as physical and chemical properties of a biomaterial can be altered. Thus, surface modification can be considered as an efficient method to improve the performance of a device by enhancing certain properties, like the mechanical, biological, and functional properties. Hence surface modification is a key process for the engineering of biomaterials for medical devices. More than one modification technique has been adopted depending on the requirement of the biomaterial, such as consistency, reproducibility, and yield of the products [40].

Hence some sort of surface modification is essential to attain properties suitable for specific applications. Different surface modification techniques have been employed, such as plasma and laser treatment, UV lamp modification, etching, grafting, metallization, and ion sputtering etc., and each method has its own merits and drawbacks. Appropriate methods induce specific properties to the polymers which make them suitable for a variety of biomedical applications. Plasma treatment is a very popular method used for polymer surface modification. Plasma comprises of highly excited radical species. These are obtained by the excitation of gases to high energy state by means of electrons produced by hot filament discharge, radio frequency, or microwave radiations. The highly activated and energetic species in plasma can change even surface properties inert materials like ceramics.

Plasma-surface modification techniques combining the advantage of both conventional plasma irradiation and ion beam techniques is an effective method for applying even complex shapes of medical implants [41]. The surface modification enhances materials adhesion strength, coating and surface properties, and biocompatibility, etc. Plasma treatment can be regarded as an effective material processing technique is attaining popularity in the biomedical field. Plasma modification results in changes in properties and composition of materials like adhesion of metals, wettability, refractive index, hardness, chemical inertness, and lubricity, etc. It is a rather diverse technique, and finds applications including cleaning/sterilization, coating or depositing, and implantation modification of surface chemistry of a substrate.

16.12 PLASMA-TREATED POLYMERIC SCAFFOLDS

Plasma treatment is a suitable and cost-effective method for altering the surface properties of a polymeric material. Plasma-induced hydrophilicity to the polymers by introducing suitable functionalities. For polymeric scaffolds, the cell attachment and growth are two important processes that directly influence the capacity of cells to proliferate and differentiate on these scaffolds.

A study by Paula M. Lopez-Perez et al. [42] proposed plasma-induced polymerization as an effective method for the functionalization of 3D porous scaffolds. By using this method they are effectively grafted sulfonic or phosphonic groups onto complex shaped materials without altering the surface topography or morphology. They also found this method could be effectively applied to different materials and monomers. They observed that the introduction of negatively charged groups, like sulfonic and phosphonic groups, results in a remarkable difference in cell adhesion and proliferation. They further found that grafting of negatively charged units, surface grafting by means of functional group grafting by plasma-induced polymerization is an important tool for BTE.

Scaffolds of electrospun polycaprolactone (PCL) prepared by Nina Recek et al. in another study using various plasma treatments. They have created plasma from various sources like O_2, NH_3, or SO_2 gas under identical conditions. They have analyzed the plasma treatment on polymers using X-ray photoelectron spectroscopy.

Polycaprolactone (PCL) or polylactic acid (PLA) are broadly used polymers for tissue engineering scaffolds. These polymers are biocompatible and biodegradable, so they are not harmful to the body and it is not necessary to remove them after implantation. In in vivo conditions they slowly degrade to produce nontoxic products. But in order to enhance the cell adhesion and cell proliferation properties for scaffolding applications, these polymers should be surface modified. So, in order to induce functional groups on the surface of these polymers, plasma treatment as an important technique used for surface modification has been employed. Surface wettability and surface energy, which drastically alter the protein and cell adhesion properties of these polymers, can be improved by using a suitable gas in plasma and was generated by an RF generator coupled to the coil via a matching network. In order to change the surface properties, the polymer is treated with NH_3, O_2, and SO_2 plasma. Treatment with plasma results in changes of properties like chemical composition, morphology, and hydrophobicity, etc. The cell adhesion and cell viability of the polymer is significantly changed by treatment with O_2 and NH_3 plasmas compared with untreated polymers. Low cell adhesion resulted after treatment with SO_2 plasma, which changes the fibrous morphology of the polymer. Plasma-treated PCL surfaces show better cell adhesion and proliferation confirmed by high signals in the cell viability assay [43].

Due to the morphological similarity to the extracellular matrix (ECM), electrospun nanofibrous scaffolds have been broadly used in numerous biomedical applications for tissue engineering. There is an urgent need for these electrospun matrices in cardiovascular implants because the native endothelium of a nanostructured surface that imitates the native endothelium, in turn, encourages endothelialization and diminishes the complications of thrombosis and implant failure. In a study by Anna Maria Pappa et al. [44], electrospun nanofibrous scaffolds of poly-ε-caprolactone (PCL) were fabricated, which serves as a coating for cardiovascular implants. For improved cell growth, the surface of these scaffolds were modified with oxygen plasma treatment. The conditions of surface treatment were optimized to get a surface which is favorable for cell growth. The cell adhesion studies and cytocompatibility studies demonstrate the efficiency of these plasma-treated biomimetic tissue engineering scaffolds as efficient coatings for cardiovascular implants.

Wei Zhu et al. [45] synthesized cold atmospheric plasma modified electrospun scaffolds with embedded microspheres for improved cartilage regeneration. Articular cartilage is a thin layer of tissue which acts as the connective tissue of diarthrodial joints and provides sufficient lubrication in bones and can bear the repetitive strain during a person's lifetime. There are three main types of injuries that occur in articular cartilage. Injuries to cartilaginous tissue can be classified as superficial matrix disruption, partial-, and full-thickness defects; among which self-repairing is not applicable in the case of partial thickness defects and full-thickness defects. 3D tissue engineered scaffolds provide an alternative for repairing defects in cartilage tissue. So, in this study, a novel surface modification technique, cold atmospheric plasma, has been used for the modification of electrospun polycaprolactone polymers for scaffold preparation. After CAP treatment, in order to achieve a more biomimetic and bioactive surface for enhanced cell growth, nanocrystalline hydroxyapatite/chitosan scaffold was induced. Dielectric barrier discharge of helium gas was used as the plasma source in the CAP device. A sequence of new CAP-modified cartilage scaffolds were fabricated in this study and the morphology revealed an interconnected porous surface topography with consistently distributed microspheres. These microsphere-embedded surfaces provided enhanced MSC growth, 3D infiltration, and chondrogenesis. Improved adhesion-mediating protein adsorption is another significant advantage of CAP treatment. Thus, the combination of CAP, a new plasma surface modification technique, along with bioactive factor embedded microspheres and the electrospinning technique holds great promise for a variety of tissue regeneration applications. David E. Robinson et al. [46] fabricated a synthetic skin composite of polycaprolactone (PCL) scaffolds made from electrospun PCL using the plasma polymerization technique. Plasma polymerization provides a thin uniform conformal layer over the polymer surface. The plasma polymer coating produced enhanced biomolecule binding on the polymer scaffolds. The source of plasma polymerization was carried out by using monomer, allylamine, and subsequent immobilization of biomolecules like heparin and fibroblast growth factor-2 [46] (Figs. 16.1−16.2).

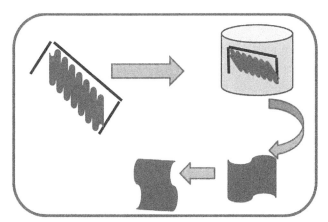

Figure 16.1 (A) Schematic diagram of the scaffolds mounted for coating; (B) placement of scaffolds within the reactor for coating; and (C) scaffold preparation.

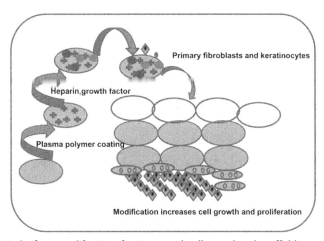

Figure 16.2 Surface modification for improved cell-populated scaffolds.

In this study, a simple and cost-effective protocol was developed to synthesize cell-seeded scaffolds with enhanced cell growth and proliferation. This protocol aims to fabricate a skin composite embedded with human fibroblasts and keratinocytes in a polymeric 3D scaffold. A recent study by Pieter Cools et al. [47] fabricated plasma-coated 3D scaffolds of acrylic acid for cartilage tissue engineering applications. In this study, poly (ethylene oxide terephthalate/poly (butylene terephthalate) surfaces were modified with two plasma modification techniques. The source for the first plasma activation is from a helium discharge which induced nonspecific polar functionalities. A plasma polymer coating using a carboxylic acid precursor, acrylic acid was used in the second approach. There is a significant change in properties, like wettability, due to the presence of oxygen-containing functional groups and also the generation of the glucoaminoglycans (GAG) matrix which stimulated the movement of cells during the surface of the plasma-coated scaffolds. Plasma activation by He stimulated the formation of GAGs, but it did not result in the migration of chondroblasts. Both the treatments resulted in the improvement of many important properties of the polymer scaffold. Thus, the acrylic acid-based plasma coating technique is considered as a potential surface modification technique which demonstrates potential as a surface modification technique for cartilage tissue engineering applications.

Calcium phosphate cement (CPC) is a highly promising material for orthopedic   applications   due   its   injectability   and   excellent osteoconductivity. A study by Michael D. Weir et al. encapsulated human bones marrow mesenchymal stem cell (hBMSC) in CPC for bone tissue engineering. In this study, they show the layer by layer encapsulation of micro/nano beads in scaffolds for tissue engineering applications. In this study, hBMSCs and alginate hydrogel beads were encapsulated inside the CPC, CPC−chitosan, and CPC−chitosan−fiber scaffolds. The degradable fibers of chitosan and chitosan itself can be used to mechanically reinforce the scaffolds. Their study showed that the percentage of survival of cells and cell density of hBMSCs inside CPC containing constructs and those without CPC give similar results, which indicates that incorporation of CPC setting did not harm the growth of cells inside the hBMSCs [48].

16.13 FUTURE DIRECTIONS

Medical research continues to explore new scientific frontiers for diagnosing, treating, curing, and preventing diseases at the molecular/genetic level. Important advances have been made in the clinical use of medical implants and other devices. Presently, emphasis is placed on the design of polymeric scaffolds, that is, materials that obtain specific, desired, and timely responses from surrounding cells and tissues. The need for alternative solutions to meet the demand for replacement organs and tissue parts will continue to drive advances in tissue engineering. Polymer scaffolds have all the prospectives to provide a new means to control the physical and chemical environment of the biological system. There are several advantages to use biological polymers over widely utilized synthetic polymer as tissue engineering scaffolds. Despite these recent improvements to the mechanical properties, porosity, and bioactivity of scaffolds, future research is needed to overcome many remaining limitations in the fabricating process. We believe no one material will satisfy all design parameters in all applications, but a wide range of materials will find uses in various tissue engineering applications. The overall challenges in scaffold design and fabrication is an opportunity for new exciting application-oriented research in scaffold design, which includes polymer assembly, surface topography or chemical cues, nano-/macrostructure, biocompatibility, biodegradability, mechanical properties, directing cell function, and induced formation of natural tissue.

REFERENCES

[1] J.P. Vacanti, R. Langer, Tissue engineering: the design and fabrication of living replacement devices for surgical reconstruction and transplantation, Lancet 354 (1999) S32−S34.

[2] D.M. López Pérez, et al., Plasma-induced polymerization as a tool for surface functionalization of polymer scaffolds for bone tissue engineering: an in vitro study, Acta Biomater. 6 (9) (2010) 3704−3712.

[3] D. Ronca, et al., Bone tissue engineering: 3D PCL-based nanocomposite scaffolds with tailored properties, Proc. CIRP 49 (2016) 51−54.

[4] O. Nedela, P. Slepička, V. Švorčík, Surface modification of polymer substrates for biomedical applications, Materials 10 (10) (2017) 1115.

[5] C. Oehr, Plasma surface modification of polymers for biomedical use, Nucl. Instrum. Methods Phys. Res Sect. B: Beam Interact. Mater. Atoms 208 (2003) 40−47.

[6] K. Rezwan, et al., Biodegradable and bioactive porous polymer/inorganic composite scaffolds for bone tissue engineering, Biomaterials 27 (18) (2006) 3413−3431.

[7] X. Liu, X.M. Peter, Polymeric scaffolds for bone tissue engineering, Ann. Biomed. Eng. 32 (3) (2004) 477−486.

[8] B. Dhandayuthapani, et al., Polymeric scaffolds in tissue engineering application: a review, Intern. J. Polym. Sci. 2011 (2011).

[9] P.K. Chu, et al., Plasma-surface modification of biomaterials, Mater. Sci. Eng. R: Rep. 36 (5−6) (2002) 143−206.

[10] J.M. Grace, J.G. Louis, Plasma treatment of polymers, J. Dispersion Sci. Technol. 24 (3−4) (2003) 305−341.

[11] E.M. Liston, Plasma treatment for improved bonding: a review, J. Adhesion 30 (1−4) (1989) 199−218.

[12] E.M. Liston, L. Martinu, M.R. Wertheimer, Plasma surface modification of polymers for improved adhesion: a critical review, J. Adhesion Sci. Technol. 7 (10) (1993) 1091−1127.

[13] R. Augustine, N. Kalarikkal, S. Thomas, Advancement of wound care from grafts to bioengineered smart skin substitutes, Progress Biomater. 3 (2−4) (2014) 103−113.

[14] R. Augustine, et al., Electrospun poly (ε-caprolactone)-based skin substitutes: In vivo evaluation of wound healing and the mechanism of cell proliferation, J. Biomed. Mater. Res. Part B: Appl. Biomater. 103 (7) (2015) 1445−1454.

[15] M. Modic, et al., Aging of plasma treated surfaces and their effects on platelet adhesion and activation, Surface Coatings Technol. 213 (2012) 98−104.

[16] M.P. Prabhakaran, et al., Surface modified electrospun nanofibrous scaffolds for nerve tissue engineering, Nanotechnology 19 (45) (2008) 455102.

[17] J.M. Goddard, J.H. Hotchkiss, Polymer surface modification for the attachment of bioactive compounds, Progress Polym. Sci. 32 (7) (2007) 698−725.

[18] H. Cao, N. Kuboyama, A biodegradable porous composite scaffold of PGA/β-TCP for bone tissue engineering, Bone 46 (2) (2010) 386−395.

[19] R. Vasita, S.K. Dhirendra, Nanofibers and their applications in tissue engineering, Intern. J. Nanomed. 1 (1) (2006) 15.

[20] M. Singh, et al., Microsphere-based seamless scaffolds containing macroscopic gradients of encapsulated factors for tissue engineering, Tissue Eng. Part C: Methods 14 (4) (2008) 299−309.

[21] A. Jaklenec, et al., Novel scaffolds fabricated from protein-loaded microspheres for tissue engineering, Biomaterials 29 (2) (2008) 185−192.

[22] M. Sokolsky-Papkov, et al., Polymer carriers for drug delivery in tissue engineering, Adv. Drug Delivery Rev. 59 (4-5) (2007) 187−206.

[23] T.G. Vargo, et al., Synthesis and characterization of fluoropolymeric substrata with immobilized minimal peptide sequences for cell adhesion studies. I, J. Biomed. Mater. Res. Part A 29 (6) (1995) 767–778.

[24] E.M. Liston, L. Martinu, M.R. Wertheimer, Plasma surface modification of polymers for improved adhesion: a critical review, Plasma Surface Modif. Polym. Relevance Adhesion (1994) 3–39.

[25] M.I. Sabir, X. Xu, L. Li, A review on biodegradable polymeric materials for bone tissue engineering applications, J. Mater. Sci. 44 (21) (2009) 5713–5724.

[26] D. Puppi, E. Chiellini, Wet-spinning of biomedical polymers: from single fibers production to additive manufacturing of 3D scaffolds, Polym. Intern. (2017).

[27] Q.Z. Chen, et al., The surface functionalization of 45S5 Bioglass®-based glass-ceramic scaffolds and its impact on bioactivity, J. Mater. Sci. Mater. Med. 17 (11) (2006) 979–987.

[28] Q. Fu, et al., Toward strong and tough glass and ceramic scaffolds for bone repair, Adv. Funct. Mater. 23 (44) (2013) 5461–5476.

[29] Y. Moriguchi, et al., Impact of non-thermal plasma surface modification on porous calcium hydroxyapatite ceramics for bone regeneration, PLoS One 13 (3) (2018) e0194303.

[30] A. Rahmani Del Bakhshayesh, et al., Recent advances on biomedical applications of scaffolds in wound healing and dermal tissue engineering, Artific. Cells Nanomed. Biotechnol. 46 (4) (2018) 691–705.

[31] B. Leukers, et al., Hydroxyapatite scaffolds for bone tissue engineering made by 3D printing, J. Mater. Sci. Mater. Med. 16 (12) (2005) 1121–1124.

[32] L.S. Nair, T.L. Cato, Polymers as biomaterials for tissue engineering and controlled drug delivery, Tissue engineering I, Springer, Berlin, Heidelberg, 2005, pp. 47–90.

[33] Z. Kai, D. Ying, C. Guo-Qiang, Effects of surface morphology on the biocompatibility of polyhydroxyalkanoates, Biochem. Eng. J. 16 (2) (2003) 115–123.

[34] G.-Q. Chen, Q. Wu, The application of polyhydroxyalkanoates as tissue engineering materials, Biomaterials 26 (33) (2005) 6565–6578.

[35] K. Zhao, et al., Polyhydroxyalkanoate (PHA) scaffolds with good mechanical properties and biocompatibility, Biomaterials 24 (6) (2003) 1041–1045.

[36] A. Salerno, et al., Effect of micro-and macroporosity of bone tissue three-dimensional-poly (ε-caprolactone) scaffold on human mesenchymal stem cells invasion, proliferation, and differentiation in vitro, Tissue Eng. Part A 16 (8) (2010) 2661–2673.

[37] A. Salerno, et al., Processing/structure/property relationship of multi-scaled PCL and PCL–HA composite scaffolds prepared via gas foaming and NaCl reverse templating, Biotechnol. Bioeng. 108 (4) (2011) 963–976.

[38] A.E.D.M. Salerno, et al., Tailoring the pore structure of PCL scaffolds for tissue engineering prepared via gas foaming of multi-phase blends, J. Porous Mater. 19 (2) (2012) 181–188.

[39] Shokrolahi, F, et al., (2011). Fabrication of poly (urethane urea)-based scaffolds for bone tissue engineering by a combined strategy of using compression moulding and particulate leaching methods, 645–658.

[40] V.J. Mkhabela, S.S. Ray, Poly (ε-caprolactone) nanocomposite scaffolds for tissue engineering: a brief overview, J. Nanosci. Nanotechnol. 14 (1) (2014) 535–545.

[41] J.L. Lowery, Characterization and modification of porosity in electrospun polymeric materials for tissue engineering applications, Massachusetts Institute of Technology, Diss, 2009.

[42] C. Mingyu, et al., Surface modification and characterization of chitosan film blended with poly-L-lysine, J. Biomater. Applicat. 19 (1) (2004) 59–75.

[43] N. Recek, et al., Cell adhesion on polycaprolactone modified by plasma treatment, Intern. J. Polym. Sci. 2016 (2016).

[44] H. Cui, J.S. Patrick, The role of crystallinity on differential attachment/proliferation of osteoblasts and fibroblasts on poly (caprolactone-co-glycolide) polymeric surfaces, Front. Mater. Sci. 6 (1) (2012) 47–59.

[45] W. Zhu, et al., Cold atmospheric plasma modified electrospun scaffolds with embedded microspheres for improved cartilage regeneration, PLoS One 10 (7) (2015) e0134729.

[46] D.E. Robinson, et al., Plasma polymer and biomolecule modification of 3D scaffolds for tissue engineering, Plasma Processes Polym. 13 (7) (2016) 678–689.

[47] P. Cools, et al., Acrylic acid plasma coated 3D scaffolds for cartilage tissue engineering applications, Sci. Rep. 8 (1) (2018) 3830.

[48] M.D. Weir, H.K.X. Hockin, Human bone marrow stem cell-encapsulating calcium phosphate scaffolds for bone repair, Acta Biomater. 6 (10) (2010) 4118–4126.

INDEX

Note: Page numbers followed by "*f*" and "*t*" refer to figures and tables, respectively.

Made in the USA
Monee, IL
31 March 2021